Back from the Brink:

Refining the Threatened Species Recovery Process

Back from the Brink:

Refining the Threatened Species Recovery Process

Edited by

Sally Stephens and Stephanie Maxwell

Published by

Surrey Beatty & Sons Pty Limited

in association with

Australian Nature Conservation Agency

The National Library of Australia
Cataloguing-in-Publication entry:

Back from the Brink: Refining the Threatened Species Recovery Process

Includes index.

ISBN 0 949324 69 8

1. Endangered species — Australia — Congresses. 2. Endangered species — New Zealand — Congresses. 3. Plant conservation — Australia — Congresses. 4. Wildlife conservation — Australia — Congresses. 5. Plant conservation — New Zealand — Congresses. 6. Wildlife conservation — New Zealand — Congresses. I. Australian Nature Conservation Agency. II. Title.

333.9542160994.

Released December 1996

PRINTED IN AUSTRALIA BY
SURREY BEATTY & SONS PTY LIMITED
43 Rickard Road, Chipping Norton, NSW 2170

CONTENTS

Appraising threatened species recovery efforts: practical recommendations

Tim. W. Clark[1,2]

Conservation of threatened plants, animals, and ecological communities is a vital task for any country concerned with its sustainability, and given real world contexts, formulating and implementing management policy for sustainability are complex tasks. Three general aspects of current conservation and recovery operations need explicit, systematic attention and improvement. First, *problem orientation* is an essential task to help managers clarify the conservation goal, define problems with respect to goal attainment, and examine alternative ways of achieving the goal. A brief problem-orientation exercise is illustrated for sustaining biodiversity, and three principal strategies for meeting goals, science-based, practice-based, and prototyping, are identified. Second, it is necessary to establish a thorough, ongoing *appraisal mechanism* in order to learn systematically from past conservation experience and to make practical recommendations. Summarized here are about 50 appraisals of endangered species and biodiversity policies and programmes, case studies, and recovery planning and teams. Third, it is essential to recast the challenge of threatened species recovery in terms of *learning,* at individual, organizational, and policy levels, as a basis for accelerating improvements. *Recommendations* are offered for developing a strategy to ensure the sustainability of biological diversity based on problem orientation, appraisal, and learning.

INTRODUCTION

AMONG the most vital tasks for any country concerned with its long-term sustainability are the recovery and conservation of threatened species and ecological communities or ecosystems. Recovery is one key part of a commitment to conserving biodiversity; the other is proactive management of common species to prevent their becoming threatened. A balanced species and ecosystem management approach is needed to accomplish these dual goals. Currently, rates of species and ecosystem endangerment and extinction are on the rise (Lawton and May 1995), as are demands to improve the effectiveness, efficiency, and equitability of our efforts to create a sustainable world. New strategies and tactics must be successfully implemented along with the new conservation biology and social science tools that are presently being brought to the problem. Important background on the endangered species/biodiversity loss problem, especially as it pertains to Australia, is given by Nilsson (1986), Kennedy (1991), Clark *et al.* (1990), World Conservation Monitoring Centre (1992), Davis *et al.* (1993), Australian Nature Conservation Agency (1993, 1994), Toyne (1994), and Flannery (1994).

To address the biodiversity loss problems succesfully, we must be clear about our goal and the alternatives to achieve it, we must learn from experience, and we must continue to employ state-of-the-art principles and practices. A good starting point is to study the recent literature appraising past approaches to conservation to find lessons and streamline ongoing efforts. For many years, and especially since passage of the *Endangered Species Act 1973* (ESA) in the United States, the *Flora and Fauna Guarantee Act 1988* (FFGA) in Victoria, and the *Endangered Species Protection Act 1992* (ESPA) at the national level in Australia, considerable implementation experience has accrued, and numerous authors have described these policies, their programmatic elements, and their application in actual cases. The hard-earned experience of thousands of people regarding the recovery process should be brought together with the literature into a "configurative picture" of the whole. This would facilitate learning and improvement. An opportunity now exists to organize the experience and literature and learn from them.

The widely scattered literature on ways to improve recovery processes is brought together here as part of my appraisal and recommendations. I also review some general literature on policy implementation, organization and management, and critical problem solving. Collectively, these sources offer many

[1]School of Forestry and Environmental Studies, Yale University, New Haven, Connecticut 06511, USA.
[2]Northern Rockies Conservation Cooperative, Jackson, Wyoming 83001, USA.
Pages 1–22 *in* BACK FROM THE BRINK: REFINING THE THREATENED SPECIES RECOVERY PROCESS ed by S. Stephens and S. Maxwell. Surrey Beatty & Sons, 1996.

Table 1. A problem-oriented exercise on the threatened species recovery process[1]. The table is intentionally left incomplete so that practitioners can fill in details of their particular cases.

Problem-solving tasks	Application to threatened species conservation
1. **Goals.** What outcomes do we prefer?	Sustainability of biodiversity
• What are the **problems** with respect to these goals? Problems are discrepancies between goals and actual or anticipated states of affairs.	Problems: Principles and practices Are they insufficiently clear? Are they insufficiently consensual? Other probems?
2. **Alternatives.** What alternatives are available to the participants and others to solve the problems?	A. Science-based (e.g., research) B. Practice-based (e.g., review practice) C. Prototyping (innovation)
3. **Evaluation** of alternatives: Would each contribute toward solution of the problem or not?	
• **Trends**: Did it work or not work when tried in the past on relevant occasions?	
• **Conditions**: Why, or under what conditions, does it work or not work?	
• **Projections**: Would it work satisfactorily under these conditions?	
4. **Repeat** the procedure to refine and update goals, alternatives, and evaluation.	

[1]**Note**: This simple problem-orientation procedure is adpated from Harold Lasswell's "A Pre-View of Policy Sciences" (1971b). There are many variations.

perspectives and suggestions to reform and improve the recovery and conservation processes.

My terms of reference for this paper require that I step outside my role as a biologist to take a broad look at the "big picture". I will look at our collective response to the biodiversity loss problem, that is, the whole system of professional, technical, organizational, and policy efforts to restore species and communities and prevent common species from becoming threatened.

Specifically, this paper first introduces a problem-orientation exercise that clarifies what outcomes we prefer — biodiversity sustainability — and alternatives to achieve them. Second, it discusses the need for a thorough appraisal mechanism to determine how well the recovery process is working. As part of this approach, the paper summarizes about 50 policy and program reviews, case studies, and recovery planning and team assessments. Third, the paper recasts the challenge of threatened species recovery in terms of professional, organizational, and policy learning. Finally, it offers eight practical recommendations to improve on-the-ground conservation and management. Sixteen appendices summarize diverse recommendations from other authors.

My view, in brief, is that we face a formidable task. Given the complexity of the task and real world contexts, it is not surprising that policies and programs fail to meet our high ideals and lofty aims to sustain biodiversity in all situations. The conservation and management community needs a systematic learning mechanism to help us clarify our goals, explore alternatives, evaluate past and ongoing efforts, identify lessons from experience, and incorporate them in successive field efforts. The overall process recommended here is an evolutionary one that requires a permanent commitment to learning and a strategy of practice-based improvements. My view is tentative; I hope this paper is a prelude to further systematic collection, analysis, and discussion of relevant data, learning, and recovery process improvements.

PROBLEM ORIENTATION AND CONCEPTIONS OF THE RECOVERY PROCESS

Problem orientation

To solve any problem, problem solvers must have an adequate strategy (Clark 1992). Lasswell (1971b:39) identified five tasks that constitute a comprehensive strategy for understanding and addressing any problem, including the conservation of biodiversity and restoration of threatened species. Problem solvers — individuals or groups — can properly orient to a problem by undertaking these tasks:

1. *Goal clarification:* What outcomes do we prefer? What future states are to be realized as far as possible in the social process? What are the problems with respect to these goals?
2. *Trend description:* To what extent have past and recent events approximated the preferred outcomes? What discrepancies are there? How great are they?
3. *Analysis of conditions:* What factors have affected (or conditioned) or caused the direction and magnitude of the trends?
4. *Projection of developments:* If current policies are continued, what is the probability that we will realize our goals?
5. *Invention, evaluation, and selection of alternatives:* What intermediate objectives and strategies will help realize the preferred goals?

Lasswell (1971b:39) noted that "the most productive procedure is to examine the whole problem by returning again and again to the [five] separate tasks". There are several different ways to use problem orientation to address threatened species problems. Table 1 suggests one approach.

In orienting to the problem of the loss of biodiversity, several considerations emerge. First, I think we agree that the common goal (or value we wish to realize) is the sustainability of biodiversity (see Mace and Lande 1991). The various policies and programs, such as the ESPA and the Commonwealth Endangered Species Advisory Committee and the Australian Nature Conservation Agency's Endangered Species Program, are instrumental means to achieve that goal. But problems exist with respect to the goal. (In the sense used here, a "problem" is the discrepancy between goals and actual or anticipated states of affairs.) The problem is that the instruments need to be more effective because the loss of biodiversity is accelerating faster than our capabilities to stop or reverse it. This problem exists in part because the conservation principles and practices needed to achieve the goal are unclear or lack consensus.

Principles represent the theory behind the practices of any discipline, and practices are the activities to which practitioners devote their time and energy (Senge 1990). Principles help both beginners and experienced professionals to understand the rationale behind their discipline and to make sense of the discipline's practices. For example, "structure affects behaviour" is a central principle underlying systems thinking. Clearly, the structure of bureaucracy affects organizational behaviour, including that of agencies charged with recovering threatened species, and we must understand those effects. This principle can guide an organization in manipulating its structure to improve its behaviour and, ultimately, its conservation success. Practices are the most obvious part of any discipline and the primary focus of practitioners of a discipline. Beginners must work at learning practices, but with experience, practices become automatic as do the underlying assumptions. Learning requires developing new principles and carrying out new practices. Learning always involves both — thinking and doing — and this is the rationale for distinguishing principles from practices (Senge 1990).

Second, there are at least three complementary alternatives to solve the problem of biodiversity loss (Table 1). *Science-based approaches* look to science to derive and upgrade principles and practices. *Practice-based approaches* look to current professionals and actual recovery programs, review state-of-the-art principles and practices, and look for ways to improve them. *Prototyping approaches* look to innovation to improve principles and practices. Alternatives must be evaluated based on what has worked in the past or is likely to work in the future and under what conditions. Evaluation must be carefully carried out before irretrievably committing scarce resources.

Finally, the problem orientation process, which offers an explicit, systematic way to address the threatened species process, should be repeated as long as time and resources permit to clarify goals, alternatives, and evaluations further.

Recommendation No. 1: Carry out a rigorous, systematic, explicit problem orientation of each threatened species recovery effort, no matter how large or small. This approach will help participants to set goals, define and analyse problems with respect to achieving the goals, and invent and evaluate alternative solutions.

Conceptualizing the recovery process: two views

Another major consideration in current conservation efforts is the existence of two major conceptualizations of the recovery process. Participants and observers may use one or the other without being fully aware of the fact. In turn, their expectations, demands, and actions also differ, which may lead to miscommunication, conflict, and possibly failure if these viewpoints are not clarified.

The first conceptualization sees the recovery process largely as a technical task requiring that biologists be given a relatively free hand to formulate the challenge and overcome it. The major constraints are perceived to be lack of scientific information about the species or ecosystem, lack of money, or various political obstacles (e.g., the public, politicians, developers), all external to the program. Examples of this standpoint abound (e.g., McFarlane 1992; Butler and Merton 1992). Within this view, the way to achieve recovery is to carry out needed studies, gain the missing knowledge, decide on its management implications, implement them, and otherwise maximize money flowing in and minimize political interference. Practitioners often emphasize a "science-based" approach. They believe they are agents of objective, value-neutral science, and are often unaware of the social matrix within which they work. Despite their skills in scientific methods, they are often unskilled in social processes. This is the view of "technical rationality" (Schön 1983; also see Alvarez 1993; Clark 1993; Mattson and Craighead 1994; Groves 1994; Yaffee 1994b; National Research Council 1995).

The second conceptualization sees recovery as a multifaceted task with both technical and social process dimensions. In this view, the major constraint is perceived to be lack of effective social processes that would integrate science and values. Examples of this view are less evident in the literature (e.g., Kellert 1985, 1994; Clark 1989; Backhouse and Clark 1995; Miller *et al.* 1996). Within this view, the way to achieve recovery is to address, simultaneously and explicitly, socioeconomic, organizational, and political as well as biological dimensions of the task. This requires that practitioners develop awareness of their own roles in the social process and carry out tailored learning processes as described below. This "practice-based" approach encourages people to observe what actually works, both technically and socially, and apply experience and lessons successfully. This conceptualization requires a broader, more inter-disciplinary set of skills than the technical rationalist uses. This view is called "reflective practice" by Schön (1983) or "policy-oriented practice" by Lasswell (1971a,b).

These two views are clear in the literature on the recovery process, endangered species and biodiversity policy, descriptions of programs, and accounts of specific cases. The "science-based" (or "technical rationalist") view dominates much of past and current thinking and action on species recovery and biodiversity conservation. The minority "practice-based" (or "reflective practitioner") conceptualization is emphasized in this paper. Of course, other views of species recovery, such as religious views, also exist.

Recommendation No. 2: Be clear about the assumptions and approaches that people bring to the recovery process, science-based/technical rational or practice-based/reflective practice.

APPRAISING PAST AND ONGOING RECOVERY PROCESSES

From a practical standpoint, recovery alternatives must be invented, evaluated, selected, and enacted. Each proposed alternative must first undergo trend description, analysis of conditions, and projection of developments to see how it might contribute toward solution of the problem. In short, an appraising function is needed. Appraisal must be organized so that that analysis can be carried out easily, practically, and interactively.

The appraisal function

The appraisal process works best when it is both independent and continuous (Lasswell 1971b). Independence minimizes the likelihood that appraisal will be self-promotional by special interests, public or private. Independence is improved by involving multiple, even competing, teams of appraisers, some of which should be outside the influence of the agency with a direct stake in the results. Continuity should take the form of semi-annual or annual appraisal and publications. Continuity helps to clarify, establish, and spread standards of good practice and their practical, field-tested applications. This encourages practitioners to employ the standards through current models (i.e., principles and practices) or innovative models (e.g., prototyping), or to explain why not.

Appraisal is a matter of great policy importance in any conservation effort. The core issue is to determine how well past and ongoing efforts have been carried out and to assign responsibility and accountability for success or failure. This is a matter of judgement of past events. The major challenge of appraisal is to set up a vehicle capable of examining the complex issues at hand while remaining loyal to common goals and minimizing the temptation to distort findings. Getting totally independent appraisals is difficult, in part, because the specific definition of goals may be controversial and the appraisal standards uncertain, especially over short periods of time. The realities of bureaucratic life and contentious policy arenas, which certainly characterize biodiversity conservation, may present difficulties for organizing an appraisal mechanism as a basis for clarifying standards, learning, and improvement. Ideally, government agencies should conduct ongoing appraisals of their own activities along with their other functions (Lasswell and McDougal 1992).

Recommendation 3: Organize an appraisal function for all policies, programs, and field-level conservation efforts. It should be as independent and continuous as possible. Results should clarify standards of good practice, or best current practice. All participants should be encouraged to meet these standards or to invent more effective ones.

Organizing the literature

Organizing the large and growing literature on recovery processes for the purpose of apprasial is essential. This section reviews some literature on recovery, examines recovery planning and teams, and offers ways to overcome common obstacles to improving recovery processes. The apprasials reviewed here were undertaken by diverse authors with different motives, perspectives, and standards; many were independent of government agencies responsible for implementing threatened species conservation policy. Few of them used the problem-oriented approach described above. As a result, the following review is eclectic.

Recognizing that all policies, programs, and cases follow more or less six predictable stages provides a structure to organize the literature (Brewer and deLeon 1983). The steps are: (1) initiation or start up, (2) estimation or problem definition, (3) selection or choosing a solution, (4) implementation or carrying out actions, (5) evaluation or appraising results, and (6) termination or stopping the effort or changing it to solve a new problem. All literature on the recovery process addresses one or more of these six steps in one way or another. For example, initiation is addressed by Backhouse and Clark (1995), estimation and selection by J. Clark (1994), implementation by Wallace (1994), evaluation by Reading and Miller (1994), Bean *et al.* (1991), and termination by Backhouse (1992). In turn, suggestions for improving recovery are keyed to these six steps.

Appraisal questions can be asked about policies and programs at each of these six steps. For example, when were these acts or programs initiated, by whom, and for what reasons? What kinds of information were used to estimate the problem, how reliable was it, and how was it used? Was the problem definition complete or did it obviously leave out important considerations? How were the technical and political issues brought together so that a preferred course of action was selected? What kind of program was set up with what rules and operational procedures to implement the selected policy? What was evaluated as implementation proceeded, by whom, for what reasons, how often, and with what effect? If the effort was successful, how was the policy and program terminated, or was the effort transformed into a different program to meet the same or a different goal? These are only a few of the possible questions that can be posed.

The following section reviews some literature concerning policy development and implementation to see what is recommended to improve the threatened species recovery process. Most appraisals are from the American ESA experience. Key points from other diverse appraisals are listed in Appendices 1–16.

Policy reviews

Several authors have reviewed state or national policy initiatives, that is, broad or strategic statements of intent to accomplish specified aims (Brewer and deLeon 1983). Prominent among these are reviews of the ESA by the U.S. General Accounting Office (1981, 1988, 1992, 1993a,b), Yaffee (1982), Rohlf (1989), Tobin (1990), Kohm (1991), Comer (1993), Clark *et al.* (1994, Appendix 3), the National Research Council (1995, Appendix 2), Beedy (1995, Appendix 4), Anonymous (1995, Appendix 5), Administrative Conference of the United States (1995, Appendix 6), Clark *et al.*, unpubl. (Appendix 7), and Tear *et al.* (1995, Appendix 8).

Houck (1993), exploring effects of the Act on the Departments of Interior and Commerce, concluded that the ESA reads as an inflexible, prohibitive policy, but operates in fact as a discretionary system for both federal and private activity. This discrepancy has permitted the overwhelming majority of human development proposals to proceed without difficulty. In Houck's view, federal agencies responsible for implementing the ESA have altered each major provision, listing, critical habitat designation, consultation, exemption, takings, and recovery, from dictates and prohibitions into accommodations and permissions. The practical result is an act that is considerably more flexible than its reputation or its language. Houck concluded that "the risk is, of course, that the compromises arrived at through agency interpretation of the Act will only prolong the process of species decline" (p. 358). He offered no direct suggestions to improve the Act.

Second, J. Clark (1994) reviewed the history of U.S. federal endangered species protection and major provisions of the ESA, for example, listing process, protection for listed species, limits of federal actions, and co-operation with the states. He also reviewed criticisms of the ESA, including its imprecise standards for delineating species status, inappropriate units of protection, undue attention to high-profile species, insufficient protection of habitat, inadequate recovery planning process, inadequate agency consultation, tendency to ignore uncertainty (i.e., wishful thinking), and insufficient consideration of economic factors. He suggested amending the Act to give greater protection to ecosystems, invertebrates, and plants, for example. He suggested that agency discretion could be limited, that standards for judicial review and the basis for standing to sue could be opened up, and that a multi-directional approach could also be promoted.

Third, the Victorian FFGA was reviewed five years after its passage by Wilson and Clark (1995) and Mansergh *et al.* (1995). The FFGA was the first broadly-based biodiversity conservation Act in Australia with aims to protect species, genetic diversity, and communities, as well as taking a preventive approach to potential threats to flora and fauna. To mid-1994, 188 taxa, 14 communities, and 12 potentially threatening processes had been listed. In all, 53 Action Statements covering 66 items had been produced and another 25 were in advanced form. No critical habitats had been designated and no interim conservation orders made. These authors suggested that implementation could be improved by critical habitat determinations; widespread use of techniques like population viability assessment (PVA), conservation genetics, and decision analysis to set goals and make decisions; assessing Action Statement effectiveness; provision of adequate resources; and periodic evaluation of FFGA implementation.

No overall appraisals of the ESPA have been completed to date. Its implementation is likely to follow patterns seen in the ESA and FFGA.

Collectively, these appraisals show that formal conservation policy is frequently altered through implementation into "real" policy (Bardach 1977; Pressman and Wildavsky 1978; Nakamura and Smallwood 1980). Numerous suggestions have been made to strengthen the Acts and improve their implementation (see Ascher and Healy 1990; Wondolleck *et al.* 1994; Appendices 1–16). In addition, ways to improve endangered species policy are evident from reviews of programs and species-specific cases as described below.

Program reviews

The ESA, FFGA, and ESPA are implemented through programs applied to individual species, communities, and threatening processes. A program is a tactical means to achieve policy objectives, according to Brewer and deLeon (1983). At the heart of many programs are recovery planning and teams, which are discussed separately below. Policy and guidelines are described for the ESA in the U.S. Fish and Wildlife Service (1990) and for the ESPA by the Endangered Species Unit (1994). A variety of program reviews have been published (e.g., Culbert and Blair 1989; Dixon and Cook 1989; Tear *et al.* 1993; Miller *et al.* 1994, 1996; Yaffee 1994a,b; Clark *et al.* 1994); three of these are examined here.

First, 314 recovery plans under the ESA were examined by Tear *et al.* (1993:976, 1995, Appendix 8), who noted that the ultimate goal is to "restore the listed species to a point where they are viable, self-sustaining components of their ecosystem". The authors noted that few species have been recovered because of the ESA. Recovery goals for population sizes and numbers of populations have too often been set at or below what existed in the wild at the time the recovery plan was written. Thus, 28% to 37% of threatened and endangered species are being "managed for extinction" (p. 976). Current plans do not try to save too much; to the contrary, they attempt to save too little. These authors saw this as a fundamental flaw in recovery plans, and they recommended ensuring "that a minimum of three viable representatives of each vegetation cover type [be] preserved in each ecoregion in which they occur" (1993:977, see also Scott *et al.* 1993).

Second, Yaffee (1994a,b) reviewed the Northern Spotted Owl *Strix occidentalis* program since 1945. This watershed event in American natural resource management and environmental policy is more about values, economics, and decision making than biology. Yaffee concluded that the program was fragmented and often impotent, short-term oriented and reactive, conservative and status-quo seeking, and not particularly creative or good at seeking and dealing honestly with information needed to make wise choices. Elected officials, the media, and interest groups tended to exacerbate matters. Individuals both inside and outside the program showed cognitive biases and value preferences that reinforced officials, the media, and special interests. In short, the owl recovery program set up and operated by the agencies tended to look backward and used time-honoured means to deal with new problems, an approach that failed. Although the agencies could potentially have risen above all this, they did not in the owl case.

Yaffee noted that his assessment was not very encouraging for the future. But he offered several useful suggestions for improving programs: (1) shift focus to ecosystem-level issues and clarify objectives to meet the new aims; (2) change professional, organizational, and policy incentives, thus providing opportunities for enhanced co-operation; and (3) enhance the capacity of individuals and institutions, such as agency professionals and bureaucracies, to maximize chances they will be successful. He concluded hopefully by saying that "new policy approaches, altered models of collective choice decision-making, new patterns of political coalition-formation, new methods of information collection and analysis, new ways of managing forest systems, and new ways of administering and leading resource agencies are all there for the taking" (p. 375).

Third, individual contributors to Clark *et al.* (1994, Appendix 3) reviewed programs for 12 species in the United States and Australia. All authors described implementation difficulties, including starting programs too late, poor planning and plans, weak recovery teams, interference by administrators and politicians, weak implementation, and lack of evaluation. The contributors listed scores of lessons or suggestions to improve recovery programs, ranging across technical, informational, contextual, team leadership, valuational, social, economic, and policy improvements. The editors boiled them down to eight "meta-lessons" (Appendix 3).

Collectively, program reviews show that programs are difficult to carry out successfully, but that many options exist for improvement. Another view of the recovery process is possible by looking at species-specific applications.

Case study reviews

Many case studies focusing on individual species have been published (e.g., Butler and Merton 1992; Walters 1992; Alvarez 1993; Reading 1993; Yaffee 1994a; Warshall 1994; Robertson *et al.* 1994; Clark, in press). The cases focus on plants (e.g., Falk and Holsinger 1992; Bowles and Whelan 1994) and invertebrates (e.g., New 1984; Yen *et al.* 1990; Murphy 1991; New 1991; Yen and New 1995; New and Yen 1995) as well as vertebrates (e.g., Clark *et al.* 1994; Backhouse and Clark 1995). Summarizing this literature in terms of its significance for policy and program improvements is difficult however, because most accounts are written from the "science-based" or "technical rationality" viewpoint and thus omit descriptions of professional and organizational behaviour, decision making, and policy processes. Offering general improvements based on these cases is problematic, but several issues emerge from case study reviews.

First, most case studies utilize what Caughley (1994:215) calls the "small population paradigm" for recovery, which deals with the effects of smallness on the persistence of a population, but he also identified the "declining population paradigm," which deals with the causes of smallness and its cure. Examining the role of research in conservation biology, Caughley argues that employing both paradigms would be most fruitful. One Australian case, that of the Eastern Barred Bandicoot *Perameles gunnii*, illustrates the two-paradigm approach, and Caughley (1994:432) called the application of PVA in the bandicoot case an "excellent account" and a "cheering exception." The bandicoot case has been used in Victoria to illustrate "model" programs, but it should not be misconstrued as an optimum set of structures and procedures to be emulated. Instead, this program is an effort explicitly set up for learning about itself as well as about bandicoot recovery. Overviews of bandicoot recovery are given by Clark and Seebeck (1990), Backhouse (1992), Backhouse *et al.* (1994b), Reading *et al.* (1995, 1996), Clark *et al.* (1995a), and Seebeck and Patrick (in press). A "practice-based" and "prototyping" approach was taken as described below.

The concerted bandicoot effort began in 1988, even though other work had preceded it. By 1988, the species had suffered a 99% loss of range and abundance in Victoria; about 200 animals remained in the wild. An "extinction map" was used to diagnose the recovery challenge (Fig. 1; Clark *et al.* 1990). An effort was made over the next few years to gain reliable scientific knowledge on the status, history, and dynamics of the wild population and its habitat. Rigorous quantitative studies of demographics and genetics were carried out (Robinson *et al.* 1993; Robinson 1995; Clark *et al.* 1995b). PVA and habitat models were developed (Lacy and Clark 1990; Reading *et al.* 1996). Decision analysis was used to explore the efficacy of various management options (Maguire *et al.* 1990). Management interventions were tried in the field (e.g., habitat enhancement, predator removal, predator-excluding fencing, public education). Captive breeding and translocations were carried out (Backhouse *et al.* 1994a; Humphries and Seebeck 1995; Myroniuk 1995). Community groups were enlisted, social surveys conducted, and personal briefings to city, shire, and state officials were made (Reading *et al.* 1995). Field monitoring and recovery teams were organized and reorganized, and plans, implementation, and evaluations carried out.

After 1991, a "core decision group" was set up to streamline decision making and accountability. A new, streamlined recovery plan was written with short-term, one-year, and 10-year goals (Backhouse 1992, 1993). Work groups were, in part, freed from bureaucratic constraints, empowered, and charged with monthly and annual goals to meet. Today, the mainland bandicoot population is about 750 individuals in an upgraded, intensively managed captive breeding population, two increasing, confined semi-wild colonies, two newly reintroduced wild colonies, and a few surviving wild animals at the original site. Two new reintroductions are planned. Although the species is not yet recovered, the bandicoot case illustrates most of the interactive biological, social, economic, organizational, and policy elements inherent in most recovery programs. The programme is outstanding in that it explicitly recognized all these factors and treated them constructively and simultaneously (see Reading *et al.* 1991; Clark *et al.* 1995a; Humphries and Seebeck 1995). Other species cases need to be similarly developed and described for the largest possible audience as a basis for lessons, learning, and improvement.

Second, despite apparent differences between plant and animal cases, many facets of restoration apply equally or similarly (Bowles and Whelan 1994). Threatened plants and animals share many problems and solutions. Restoration ecology aims to re-establish or rehabilitate damaged or lost populations or species assemblages native to the area of interest (Jordan *et al.* 1987). Conceptual issues range from the "sociopolitical,

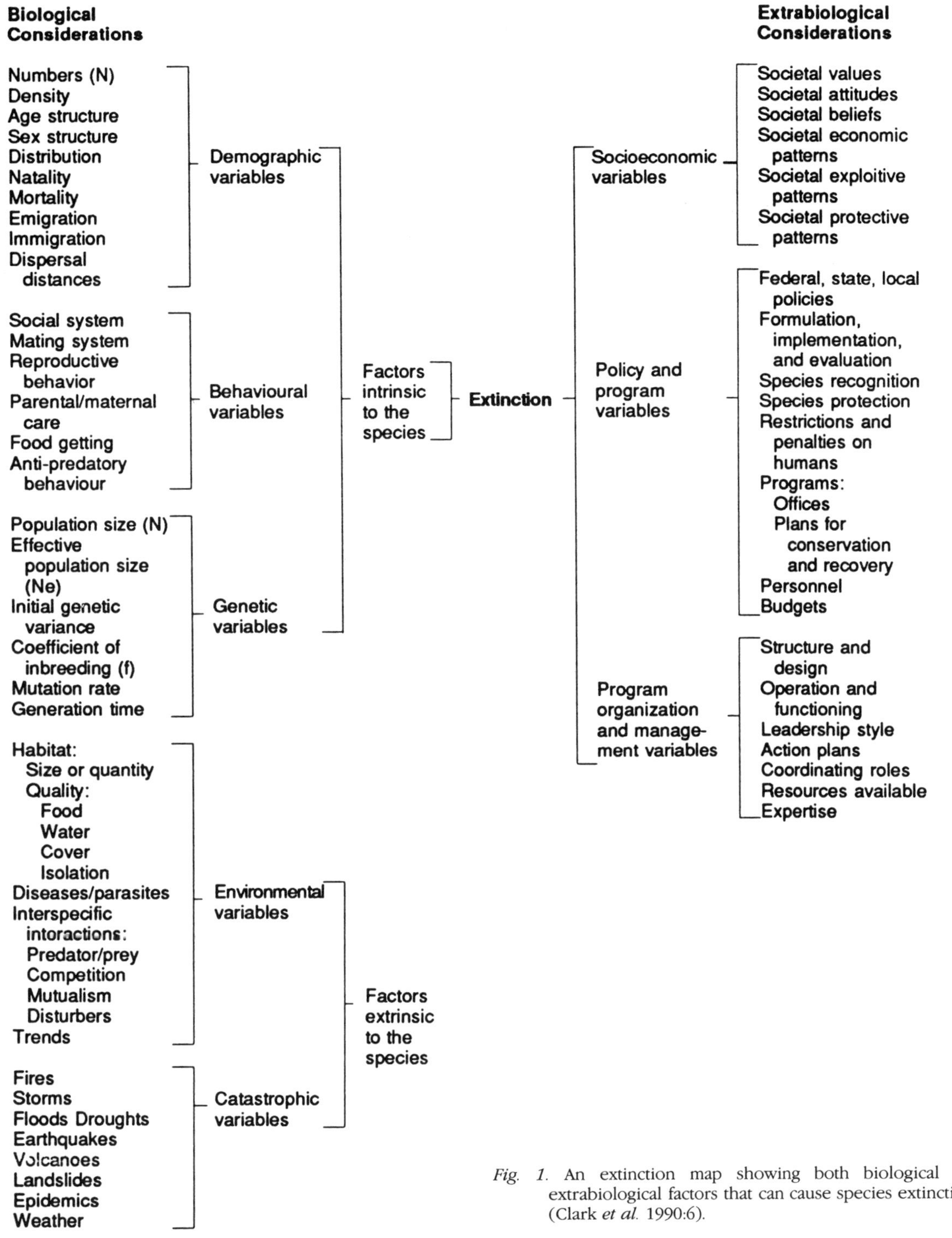

Fig. 1. An extinction map showing both biological and extrabiological factors that can cause species extinctions (Clark *et al.* 1990:6).

including organization and management of restoration efforts, to the biological, such as PVA, conservation genetics, metapopulation biology, landscape ecology, etc." (Bowles and Whelan 1994:1). Restoration goals can range from "species-specific autecological requirements, to organization of species assemblages, to landscape level processes" (p. 139). Successful implementation depends on addressing factors like habitat loss, persecution (e.g., collecting, trapping, hunting), small population size, inbreeding, susceptibility to disease, translocations, captive propagation, PVA, metapopulation theory, dedication and long-term commitment. Experimental approaches that "maximize efficiency of effort, money, and most important, biological material" should be used more for both plants and animals (Bowles and Whelan 1994:245). Metapopulation management requires multiple site management, commitment to long-term monitoring of population viability, and planned movement of organisms among sites.

Case studies tell us what actually happens in the field, which always includes a host of non-biological issues in addition to the currently recognized biological variables. Knowledge of professional problem-solving options, organization and management theory, and policy processes are all essential to recovery. On the basis of case studies, we can see that recovery in the field is challenging and much professional discretion is involved. At the heart of field work is usually a recovery plan and a team.

Recovery planning and teams

Recovery planning and teams set goals, guide recovery, and often carry out the restoration work. What should the plan include? How should goals be set? What are realistic goals? What are relevant actions needed to recover the species or ecosystem? How should land use or other conflicts be resolved? How should flexibility be ensured? What is the relationship of planning and the team to local, state, and national publics? How will recovery be integrated with other programs (e.g., state forest management plans, care groups)? How will financing occur? And how will actions be monitored and evaluated? Several authors have examined recovery planning exercises and the teams that carry them out, and some authors have offered guidelines to improve their operation. Four are briefly examined below.

First, species must be listed and ranked for recovery. The U.S. Fish and Wildlife Service (FWS) developed guidelines for listing and assigning priority for recovery actions (Federal Register 1983). The guidelines consider the degree of threat (high, medium, low; imminent, potential), availability of information (no field work or field work required), taxonomy (monotypic family or genus, species, subspecies; classification), and potential for conflict (high, medium, low).

However, some authors have raised questions about the objectivity and consistency of application of these principles in the listing of species as endangered or threatened. Easter-Pilscher (1993) investigated priority criteria for selecting candidate species for listing and for selecting listed species for recovery efforts. She used the Federal Register's final rules for four classes of animals (mammals, n = 119; birds, n = 39; fish, n = 60; reptiles, n = 54) listed between 1 January 1975 and 21 June 1991. All pertinent biological information in the Federal Register was compiled and analysed. Her analysis showed an abundance of missing information throughout the final listings. Biological criteria were not defined and were inconsistently applied by the FWS both across and within classes. Descriptive variable categories were not defined relative to each other and did not correlate with the values of related quantitative variables. There were no clearly defined biological thresholds distinguishing endangered species from threatened species. A jackknifed discriminate analysis of all variables misclassified 96 of the 272 listed species examined. She concluded that "systematic consideration of biological, objective, and defined criteria would significantly improve the scientific validity and the effectiveness of the listing process" (p. vxii). Clearly, listing and ranking/priority systems embody a range of values for threatened species (Ehrenfeld 1976; Thibodeau 1983). Lovejoy (1976:52) discussed triage in species conservation and noted that "we must decide which species will go forever", and that "with the planet's genetic bank account diminishing rapidly, conservationists may have to lose some battles deliberately".

Second, a number of authors besides Tear *et al.* (1993, 1995) have reviewed recovery plans, identified deficiencies, and offered recommendations. Among them, Culbert and Blair's (1989, Appendix 10) appraisal described the origin and current processes of recovery planning under ESA as well as recovery team selection and composition, recovery plan objectives, and implementation issues. They noted that composition of the team is critical to the success of a recovery plan. In the U.S., teams are chosen by the FWS Regional Director. Political factors may weigh heavily; objectivity may be compromised and the quality of the recovery plan, not to mention protection of the species, diminished. They recommended standardizing a procedure for selecting team members: "The appointment of experts who are not agency employees should be obligatory" (p. 3).

Culbert and Blair also noted that many recovery plans are of limited value because they either go too far in accommodating political interests or ignore them altogether. Recovery is jeopardized in both instances. They recommend a two-stage planning process. Stage one, carried out by recovery team biologists, would focus solely on biological issues. Their plan becomes input into stage two, which would include a broad range of advisers to prepare the final plan.

Culbert and Blair also looked at recovery planning from other viewpoints. Concerning species ecology, they noted that plans too often emphasize rebuilding populations within remaining habitat, rather than restoring former habitat. Moreover, they suggested that tasks to learn about the species and threatening factors need to be ranked in order of importance. Competition, predation, disease, and environmental variability need to be addressed. Many plans do not permit adequate movement between populations along corridors. Corridors relevant to the species under consideration must be described in terms of location, geometry (shape, width, length), habitat quality, and management. The authors remarked, too, that more plans should be written using a multiple species approach. Concerning PVAs, which may be qualitative or quantitative depending on available data, they recommended that all recovery plans contain estimates of what constitutes a self-sustaining or recovered population (see Fig. 1 as a guide and numerous examples; e.g., Lacy and Clark 1990; Mace and Lande 1991; Lindenmayer *et al.* 1993).

Culbert and Blair also emphasized the utility of maps in conveying size and location of former and current ranges in addition to other valuable information. They suggested three kinds of maps: (1) a habitat map showing the species' historic and current range; (2) past, present, and future land use and zoning, showing reasons for the species' decline in terms of habitat modification and land use changes; and (3) a recovery plan map synthesizing the analysis and actions needed for species recovery.

They noted that implementation is the most challenging part of the recovery process. "Well designed plans are useless unless they can be put into action" (p. 6). Interagency coordination is required among government and non-government; too often it is precluded or confounded by vague language in plans, disincentive systems, or lack of monitoring. Good implementation relies directly on adequate budgets. "Perhaps the greatest weakness," according to these authors "is the greatest absence of a political feasibility analysis" (p. 7). Public involvement and education can generate support for the plan. Having an educated public pays off in many ways, even if community support is not widespread. Other authors have also addressed these points (e.g., Reading *et al.*, in press; Appendices).

Third, all U.S. recovery plans were evaluated by Tear *et al.* (1995), who found an overall lack of key biological data in plans. They found that achievement of specified recovery goals (see Mace and Lande's 1991 criteria) would not improve the level of endangerment in 60–73% of vertebrate species (also see Wilcove *et al.* 1993). It took an average of five years between steps, listing, recovery plan approval, and subsequent revision. Little emphasis has been placed on multiple-species listings and recovery. Public education was mentioned in 92% of plans. Finally, they offered twelve recommendations (Appendix 8; see Scott *et al.* 1995). The first was that goals should consider both short- and long-term aims, and the second that recovery teams need to agree on the accepted probability of persistence and evaluate and compare recovery options. These concerns are already being addressed in Australia (e.g., Backhouse 1992; Lindenmayer *et al.* 1993).

Falk (1987, 1990) offered an integrated conservation strategy for endangered plants. He first described the historic context of endangerment and conservation, noting that the "multidimensional phenomenon of endangerment poses a constant challenge to devise new and effective ways to protect both species and habitats" (1990:118). Concepts of conservation goals and practices have evolved significantly in this century. He summarized various co-operative onsite and offsite plant protection efforts by examining priority or ranking systems, permitting work, field work, conservation collections, offsite rescue, propagation, revegetation, reintroduction (Maunder 1992), introduction, restoration, basic research, management research, and monitoring wild sites. Falk and Holsinger (1992) noted that there is a movement toward a more diverse plant protection strategy.

Fourth, the organization and management of recovery programs, planning, and teams are of great concern. "While the use of appropriate biological tools is essential to successful recovery, other factors are also indispensable, [including] problem analysis and problem-solving strategies, organization design, work group effectiveness, effectiveness of interpersonal relationships, and clarity and specificity of goals and objectives" (Clark and Cragun 1994:9; Appendix 13). Simply put, inadequacy in any of these factors can result in inefficiency and ineffectiveness, even failure, of recovery.

Clark *et al.* (1989) took a "practice-based" view of the recovery process as an "information processing" system. They recommended that a strong, task-oriented, creative approach be taken, which encourages workers to observe, criticize, and generate new ideas and which rewards achievement of the primary goal of species recovery. A "parallel" organization, a flexible, participatory, problem-solving structure set up alongside traditional bureaucracies, could be used. Task forces and project teams could be set up to staff the parallel organizations. Managerial functions could be improved by including co-ordinating roles to facilitate the flow and use of information, descision making, and active management of the inevitable conflict. These authors concluded that the inability of most organizations (and most recovery efforts) to solve dilemmas, to examine their own structures and management, and to change themselves for improved performance is the major obstacle to improving recovery efforts. In other words, organizations are not self-reflective and lack the knowledge and skills to become reflective. They made recommendations for effecting change for the better in bureaucracies.

Finally, Clark and Westrum (1989, Appendix 14) focused on the use of high-performance teams in recovery. They noted that ecological skills alone are often not enough; also needed are effectively designed and managed programs. Programs and teams should be structured appropriately for their functions and should be properly staffed, led, and buffered from their political environments. High-performance teams should process information well, learn rapidly from their own mistakes, and be creative. Such teams should devote most of their energies to solving external rather than internal problems.

There are clearly many pertinent organization and management considerations in recovery. Given all the considerations, reviews, and discussion to this point, Bowles and Whelan (1994:352) summed it up well: "when restoring endangered species, however, biology is often the least of problems. Restorations occur within a societal context, and therefore, economic and political issues often equal if not surpass the biological issues in determining the course of a restoration . . . Societal demands may make biological issues moot."

Recommendation No. 4: Review all recommendations in existing published appraisals to see if they are applicable to current policies, programs, and specific field-level efforts. Improve efforts as recommended.

PRACTICE-BASED AND PROTOTYPING STRATEGIES: ACTIVE LEARNING FOR IMPROVED RECOVERY

Our responses to the loss of biodiversity have changed over time through practical experience and learning (Gilpin and Soulé 1986). There have been many improvements, but there remains a potential for learning more and upgrading our performance in nearly all settings (see Clark 1996). Practice-based learning and prototyping are two strategies that are likely to be effective in this regard.

Learning in general follows three distinctive stages, according to Senge (1990). Stage one is the development of new cognitive capacities in which people come to see new things and speak a new language. This, in turn, allows them to see more clearly their own and others' assumptions and actions as well as the outcomes of both. Initially, people may find it difficult to translate their new cognitive and linguistic skills into fundamentally new action; they may begin to behave differently, but still operate under the same basic rules, assumptions, and values as before.

Stage two brings about new action rules; people begin to shed old assumptions. They may experiment with new behaviours based on new assumptions so they can see what happens. They begin to use new language to produce new actions and outcomes. Doing this under stress is difficult because the tendency is to revert to the old assumptions, language, and actions.

Stage three results in the establishment of new values and operating assumptions, developing out of people's combining new action rules. In this stage, people fall back on these new ways of understanding the world even under stress and ambiguity. They continue to aid their own and others' learning abilities consciously and actively. People well situated in this stage have internalized the rules into their own practice situations and speak in their own voices.

These three progressive stages of learning should be kept clearly in mind when thinking about improving the threatened species recovery process. Active learning is discussed below at three highly interactive levels: individual, organization, and policy levels. Individual practitioners carry out the practical work of species conservation in the field. These practitioners are employed by and work through organizations, usually government bureaucracies. Individuals and organizations are guided by authoritative policies developed by the public. Therefore, it is essential to examine learning and what can be done to improve learning at all three levels.

Individual learning

Individual professionals are at the heart of the recovery process, and individual learning is the key to all future conservation improvements. This means enhancing the capacity to share cognitive and problem-solving models and making learning easier for people (e.g., through workshops, changing rewards, learning how to learn (Michael 1995).

Many applied or practice-based professionals have evolved a mode of practice and learning that Schön (1983) called reflective practice. Professionals reflect-in-action when they respond to particularly difficult problems not only by following standard rules of inquiry, but also by inventing new rules and questioning their own assumptions. Reflection-in-action is a way of continuously analysing, evaluating and restructuring responses to new situations, creating new solutions, and not being bound by traditional technical or organizational ways of doing things. At its heart is a "generative metaphor" used to think more creatively and effectively about problems that are not accessible to technical, incremental approaches. This kind of improved professional practice can be appreciated, made explicit, and taught (Schön 1987; Clark and Reading 1994; Schön and Rein 1994; Westrum 1994).

Westrum (1994) suggested a special school to aid species recovery based on his discussions with professionals on American recovery teams. He noted a lack of consensus on what a recovery team ought to be and a need to develop an institutional memory about what has worked in the field and what has not. He concluded that professionals are learning from recovery experiences, but that lessons are not being coded in a cogent form or transferred from one team to another. The system as a whole is not learning. He then drew on the U.S. Air Force's "Top Gun" model (Naval Fighter Weapons Schools, Miramar, California). This special school does not focus on turning up new solutions, but maximizes the system's use of those it already has. Similarly, in the species recovery arena, it is likely that many professionals already possess knowledge and skills that can and should be rooted in the system as a whole.

A "top branch" school could concentrate practice-based expertise and training that can be helpful to recovery teams. Attendees could compare notes, exchange practical experiences on tactics, and assist one another in solving actual problems. The very best science could be brought in and shared (National Research Council 1995). Schools could be organized within and between states as well as at the national level so that experiences can be widely shared. Instructors should include actual workers as well as theoreticians. Participants should be drawn from government and NGOs, researchers, managers, activists, and others. The "top branch" school is one way to organize an effective appraisal function, clarify existing principles and practices, and find ways to upgrade them, all for improved recovery.

Recommendation No. 5: Implement a "top branch" training school as suggested by Westrum (1994) to share experiences and illustrate and upgrade principles and practices.

Organizational learning

For recovery efforts to go well, the organizations carrying them out must work well, too. But the organizational dimension of recovery work is often invisible to many scientifically/technically trained professionals (Clark and Cragun 1994). So the organization and management needs of recovery efforts often go unrecognized, unappreciated, and unattended. Recovery efforts can be sub-optimal as a consequence. These issues are briefly examined below by looking at the concept of organizational learning.

Learning is the process by which organizations (or individuals) detect and correct errors, that is, mismatches between expectations and outcomes (Argyris and Schön 1978). Organizational improvements come from learning about past performance, analysing the causes, and then seeking appropriate change (Senge 1990; Leeuw *et al.* 1994). There are two basic kinds of learning. "Single-loop" learning is what organizations do when they develop skills to scan their environment, set goals, gather better information, use it in planning, and monitor their own performance in relation to their goals (Argyris and Schön 1978; Argyris 1982). This kind of error detection and correction takes place within the organization's central cultural norms and values. "Double-loop" learning, on the other hand, recognizes that some "errors" or conflicts challenge the organization's operating assumptions and norms themselves. Some mismatches between expectations and outcomes precipitate debates about core organizational issues; people are forced to ask not only, in the words of Leeuw *et al.* (1994:9), "how well are we doing," but also "does it make sense to do it, even if it is being done well?" Improvements would accelerate if double-loop learning were part of all conservation efforts.

There is abundant evidence that individuals and organizations are differentially good at single-loop learning and very poor at double-loop learning. In species recovery, poor learning can be disastrous. There are many reasons why individuals and organizations are often poor learners and many barriers to good learning (Etheredge and Short 1983; Etheredge 1985; Morgan 1986; Clark and Westrum 1987; Clark and Harvey 1991; Argyris 1993; Michael 1995).

Westrum (1994) offered guidelines for designing recovery teams, noting that success depends on the kind of "human envelop" used to protect and recover species. In his view, wildlife managers should have more than a passing knowledge of organization and management theory to be most effective. Westrum reviewed four reasons why recovery fails: intention, incompetence, ignorance, and ill-fortune. He offered five keys to success: understand the recovery task, build a competent team, get political control of recovery efforts, find "maestros" to lead recovery efforts, and set up "top branch" schools for training high performance teams.

Recommendation No. 6: Design organizations and the programs and teams they administer to learn well, based on the organizational learning literature.

In support of this recommendation, Clark and Cragun (1994) provided some background on organization designs and management processes that are pertinent to species recovery and examined the endangered species "task environment" in organizational terms. They identified the task force/project team model as the most appropriate for recovery work and described the four functions of management (organizing, planning, leading, and controlling) in these teams. They emphasized that successful teams must be task-oriented rather than power-, role-, or people-oriented. They also introduced a procedure to analyse organizational problems and develop action plans (Appendix 13).

Senge's (1990) workbook detailed how to apply organizational learning theory to real organizations. Applying his exercise, which requires that organizations design learning into all structures and operations, would be extremely beneficial to recovery processes. Five simultaneous activities are needed: (1) to think in terms of systems, (2) to master the disciplines, (3) to possess adequate mental models, (4) to build a shared vision, and (5) to learn as a team.

The true value of good professional and organizational learning is that they can be turned into improved policy or social learning.

Policy learning

Any problem dealing with the aims and processes of recovery policy should become the focus of successive adjustments at societal levels. This is policy learning (Rose 1993; Eisner *et al.* 1995). Any group, including governmental bodies at various levels, that wishes to improve its conservation policies should amend not just the content of the formal policy itself (single-loop learning), but also the process by which constructive policy making takes place (double-loop learning, see Lasswell 1971a). Ideally, learning at one level of government is shared with others (Davis *et al.* 1993). Brunner (1996) compared existing policy processes with an alternative (Table 2) that facilitates policy learning.

Policy learning is possible if participants understand that policy is not the same thing as politics (the means by which policies are decided) or laws (a kind of formal policy or rule). Policy includes these but also includes many other aspects of social process. We are most interested here with what many authors have called *real policy,* the actions carried out in recovery implementation and their effects and outcomes.

The policy process is inherently political, to be sure. It centers on values, which differ because special interests promote their own policies which often conflict with other special interests. Humans are unable to combine values, interests, and resources in ways that are not political (Davis *et al.* 1993); politics is inevitable (Lasswell and McDougal 1992). Models of policy learning for biodiversity conservation should recognize this fact. Both formal and real policy involve politics, so our expectations of the recovery process should recognize and be tempered by this fact of life. Significant policy learning takes time, perhaps a decade or more.

Recommendation No. 7: Institute a means to learn systematically and explicitly about the policy process. We must be clear about how the current policy system operates and where and how we want to change things.

Another means of learning at all levels is prototyping, an approach that can directly assist practice-based improvements.

Prototyping as a learning tool

Prototyping is a proven means to enhance performance in situations that are complex and challenging, recovery clearly qualifies as such a task. Lasswell (1971b) uses this term to specify a small-scale, flexible intervention in a social system with identifiable, transferable features. A prototype is a type model, exemplar, or archetype that encourages ongoing, systematic learning and creativity. It permits participants to find out what works and what does not and thus to build on the successes and weed out the failures. The "aim of any prototypic study is to devise a better strategic program" (Lasswell 1971b:190), and thus a key element of any prototype is to describe the program in sufficient detail to be useful to other programs. This has not been done adequately to date in recovery, even though Clark *et al.* (1995a) describe prototyping in threatened species recovery.

Prototypes are not equivalent to pilot studies that are planned in detail. Prototypes are designed to be much more innovative. Although their objectives are usually explicit, numerous ambiguities remain. Prototypes lie halfway between experiments managed by scientists under tightly controlled conditions and interventions made by administrators or elected politicians in the absence of controls. Recovery efforts include many examples where scientists have treated situations as scientific problems while officials have treated them as political problems. In fact, the recovery task lies somewhere in the middle: its conditions cannot be totally controlled by scientists, nor should it be managed by politicians.

This recovery approach is most beneficial in situations of "complexity, uncertainty, and potential for conflict among parties" (Brewer and deLeon 1983:153). Prototyping does not have to consider or measure all

Table 2. A comparison of traditional and pragmatic problem-solving (Brunner 1996).

Traditional	Pragmatism
Science	
The purpose of science is to predict, with precision and accuracy, the consequences of a comprehensive range of possible changes in a system. Science can discover and use objective objective laws of nature for this purpose.	The purpose of science in human affairs is freedom through insights that bring more factors more reliably into conscious awareness for purposes of decision. Such freedom limits forecast accuracy.
Rationality	
Scientific predictions reduce uncertainty. Hence, they are prerequisities to major policy decisions that are rational, comprehensive, and cost-effective with respect to fixed or given national goals.	Given uncertainty and ambiguity, it is more rational to field test multiple, discrete alternatives; to select, diffuse, and replicate the successes; and to allow for the adaptation of goals as well as alternatives.
Politics	
Political consensus depends upon objective scientific knowledge. Scientists deserve a privileged position above politics because their input to policy decisions are objective and value-free.	Political consensus depends on trust and credibility derived from experience. The values and beliefs of scientists, like other participants in politics, should be transparent to the public and their representatives.

conceivable permutations of a proposed effort. It is merely a positive response to the need for innovation and change, a means for improving professional and organizational practices and knowledge in general (Lasswell 1963). "True prototype exercise is controlled by people interested in enlightenment and not power" (Lasswell 1971b:70). Prototyping could be controlled and carried out by recovery teams.

Any prototyping effort for biodiversity conservation is predicated on several conditions. There must be tentative agreement among most participants to try it though not everyone need believe in or even know about the innovative effort in advance. The context must be conducive to the effort even though some degree of conscious and unconscious opposition may exist. Leaders must be committed to trying the innovation. Support may need to be built. The process must be open and creative. The very best people should be involved and respected. The effort should not be turned into a "political football". People who fundamentally want to advance human enlightenment and skill and want a respected place among their colleagues and the wider community are key (Lasswell 1963).

The goal of prototyping in recovery is to reach a level of operation that is clearly a new model. In turn, it can be repeated in pilot programs and eventually on a large scale. Lasswell (1971b:70) noted that successful prototypes encourage others to adopt or establish their fundamental elements, so the main justification for prototyping is to stimulate "the discovery of an improved program and [lay] the foundation for orderly replication of the received prototypical model." Whole efforts or only parts of specific techniques may be transferable depending on the context.

Clark *et al.* (1995a) listed and discussed some tranferable prototypical elements from the Eastern Barred Bandicoot recovery efforts. They noted that advances were possible because of the continuous model building, application, and learning effort they used, which permitted further model modification on an ongoing basis over several years. Participants in the bandicoot recovery effort identified eight elements that are useful to other recovery efforts. First, develop an understanding among participants on key aspects of the recovery effort and seek agreement on goals of a model program. Second, work to build a reliable, comprehensive overview early on of the species, status, threatening processes, and context. Both social and biological scientists and other people are required to do this. Third, centralize all data and make it easily accessible to all participants. Use integrative analytic approaches. Fourth, set up clear, workable, and responsive decision making arrangements. Fifth, work to clarify decision rules, recovery targets, and outline sequential and contingency plans to be flexible and as orderly as possible. Sixth, use multiple teams to build prototypical programs. Seventh, design for surprise, flexibility, and change. And eighth, review and evaluation should be open and ongoing. Thorough appraisals are essential to the entire prototyping effort.

Recommendation No. 8: Use prototyping as a "semi-experimental" or "adaptive management" mechanism to improve practice-based as well as science-based performance. Replicate benefits in other conservation efforts.

CONCLUSIONS

To all societies committed to a sustainable natural resource base, an adaptive human culture, and a commonwealth of human dignity for their citizens, the task of restoring species and communities threatened with extinction is a central challenge. Because of its paramount importance, the conservation and management of living natural resources should be effective, efficient, and equitable. Without an adequate and sustainably managed resource base, the quality of human existence and the status of societies would be signficantly compromised.

Eight recommendations to improve "practice-based" sustainable conservation of biological diversity have been made. Collectively or individually, these recommendations promise to accelerate learning and practical success in the vital task of conserving the nation's essential biological heritage.

All people involved in biodiversity conservation can effect change in the recovery system, at technical, biological, professional, organizational, and policy levels. We must set ourselves up for quick information exchanges (e.g., e-mail). We should develop informal networks for mutual support. We should act as mentors to newcomers and young professionals. Some of these suggestions might be best accomplished by focusing on local opportunities within states. Initial efforts in each case should include mapping the social process as well as the technical, biological details. Finally, it is essential for each individual and each group to make a firm commitment to learning, communication, teamwork, and improvement.

ACKNOWLEDGEMENTS

I am grateful for the generous support and discussions with Gary Backhouse, Robert Begg, Denise Casey, Peyton Curlee, David Lindenmayer, Kim Lowe, Peter Myroniuk, Cathy Patrick, Steve Primm, Richard Reading, John Seebeck, Sally Stephens, Peter Stroud, John Weaver, and many other colleagues in both Australia and the United States. Preparation of this paper was supported by the Northern Rockies Conservation Cooperative and Yale University. Denise Casey (NRCC), Jim Crennan and Sally Stephens (ANCA), and Kim Lowe (Victorian Department of Natural Resources and Environment) critically reviewed the manuscript.

APPENDIX 1

N.B. These appendices comprise lessons, observations, and recommendations from the cited publications. I do not necessarily agree with or endorse all of these recommendations.

Backhouse, G. N. and Clark, T. W., 1995. Case studies and policy initiatives in endangered species recovery in Australia: recommendations. Pp. 110–116 *in* People and nature conservation: perspectives on private land use and endangered species recovery ed by A. Bennett, G. Backhouse, and T. Clark, eds. Transactions of Royal Zoological Society of New South Wales, Australia. Surrey Beatty & Sons: Chipping Norton.

1. Recovery programs aimed at preventing extinction and restoring threatened species need to be commenced much sooner than is frequently done.
2. Professionals involved in endangered species recovery need to recognize that the process is a complex, multifaceted, interdisciplinary task, not simply a biological or technical challenge.
3. Recovery programs need to operate with the best and most reliable knowledge available, both biological and sociological.
4. Recovery programs need to have clearly defined, measurable goals.
5. Give explicit, detailed attention to the implementation process.
6. Recovery programs should undergo regular evaluation.
7. Recovery programs need to have a clearly defined ending or termination.

APPENDIX 2

National Research Council, 1995. *Science and the Endangered Species Act* National Academy Press: Washington 162 pp. + appendices.

The NRC Committee was asked to review the following issues and to evaluate how they relate to the overall purposes of the Endangered Species Act (ESA): definition of species, conservation conflicts between species, role of habitat conservation, recovery planning, risk, and issues of timing. "The overall conclusion is that the ESA is based on sound scientific principles" (p. 3). However, the ESA by itself cannot prevent the loss of many species and their habitats. The major cause of current extinctions is human activity. Some major conclusions and recommendations follow.

1. The species concept

Conclusion: The ESA is clear that species and subspecies of "fish or wildlife or plants", defined in the act to include all members of the plant and animal kingdom, are eligible for protection. The ESA's emphasis on distinct population segments, that is, taxa below the rank of subspecies, is soundly based on science.

Recommendation: To help provide objectivity in identifying population segments, the concept of evolutionary unit (EU) should be adopted. The EU is a segment of biological diversity that contains a potential for a unique evolutionary future.

Recommendation: The EUs of nonvertebrate plants and animals should be covered under the ESA.

2. Habitat

Recommendation: Some core amount of essential habitat should be designated for protection at the time of listing a species as endangered as an emergency, stop-gap measure. Hence, we suggest designation of "survival habitat." This designation should not be used to delay formal designation of critical habitat.

Recommendation: The NRC committee endorses regionally-based, negotiated approaches to the development of habitat plans. Biological data should guide the plan.

3. Recovery

Recommendation: The ultimate goal of the ESA is to recover threatened and endangered species. Plans are developed to guide recovery. Plans must be implemented expeditiously. All plans should include a new section "recovery plan guidance". Guidance should be given for habitat-based recovery approaches; a logical, hierarchical approach to analysis of ecological and genetic data; guidance for demographic modelling, stressing the inherent uncertainty of such modelling; outlining future research needs and how the research will contribute to species and habitat management; and an effective monitoring scheme.

Conclusion: Both plants and animals should receive maximum protection.

Conclusion: Different mechanisms may be needed for avoiding endangerment and achieving recovery on public and private lands.

Conclusion: To permit a rational evaluation of survival and recovery goals, estimates should be provided of probabilities of achieving various goals over various periods. It may be difficult to make estimates in some cases, but even the attempt to make them will have value by requiring an objective analysis and by requiring assumptions to be specified.

4. Conservation conflicts between species

Conclusion: There are few well-documented cases where management practices focusing on particular species protected under the ESA result in direct conflict with the needs of another EU.

Conclusion: The most effective way to avoid conflicts resulting from management plans for individual species is to maintain large enough protected areas to allow the existence of mosaics of habitats and dynamic process of change within these areas.

5. Estimating risk

Conclusion: The concept of risk is central to the implementation of the ESA. With only a few exceptions, biologically explicit, quantitative models for risk assessment have played only a minor role in decisions associated with the ESA. They should play a more central role, especially as guides to research and as tools for comparing the probable effects of various environmental and management scenarios.

Conclusion: Results from population-genetic theory provide the basis for one fairly rigorous conclusion. Populations with long-term mean sizes greater than approximately 1 000 breeding adults can be viewed as genetically secure.

6. Making ESA decisions in the face of uncertainty

Conclusion: In the past, many ESA decisions have failed to meet the guidelines suggested by current scientific thinking, listing species as endangered only when populations had dropped to the point where extinction was imminent and proposing recovery goals that left the species still at high risk of extinction. All groups of plants and animals should be managed by the same risk standards. However, different guidelines may be needed because public and private entities may behave differently and therefore different management may be needed.

Recommendation: To the degree that they can be quantified, the levels of risk associated with endangered status should be higher than those for threatened status. Levels of risk to trigger ESA decisions should be framed as a probability of extinction during a specified period (i.e., x% probability of extinction over the next y years).

Recommendation: ESA decisions are inevitably based on limited information, and so agencies are obliged to act in the face of uncertainty about species status and the impacts of proposed activities.

Recommendation: Because the structure of hypothesis testing related to listing and jeopardy decisions can make it more likely for an endangered species to be denied needed protection than for a non-endangered species to be protected unnecessarily, decisions under the Act should be structured to take explicit account of all types of error that could be made and their consequences, both biological and non-biological. The phrasing of the null hypotheses and setting of error rates should reflect societal as well as scientific judgments about what level of risk is acceptable for which types of errors.

7. Timing

Conclusion: Timing is central to all aspects of recovery. Avoid delays.

8. Beyond the Endangered Species Act

Conclusion: How well has the ESA prevented species from becoming extinct? The committee cannot quantify the answer, but fewer species have become extinct than would have without the ESA. Some changes would probably make the Act more effective and predictable and provide a more objective basis for its implementation.

Conclusion: The ESA cannot by itself prevent all species extinctions, even if modified. Therefore, additional approaches are needed. These should include ecosystem management, reconstruction or rehabilitation of ecosystems, mixed management plans, co-operative management (e.g., habitat conservation plans), and revised economic accounting.

9. Science, policy, and the ESA

The committee has not uncovered any major scientific issue that seriously hinders the implementation of the Act, although its review has suggested several scientific improvements.

A more successful program will require more enlightened commitments on the part of all major parties to achieve success.

Sound science alone will not lead to successful prevention of many species extinctions, but sound science is an essential starting point.

APPENDIX 3

Clark, T. W., Reading, R. P. and Clarke, A. L. (eds), 1994. Endangered species recovery: finding the lessons, improving the process. Island Press: Washington. 450 pp.

1. Endangered species conservation is a multifaceted task of interacting biological, professional, sociological, organizational, economic, political, and policy dimensions.

2. There is no substitute for sound professional training well grounded in state-of-the-art theories and techniques.

3. High performance teams should be employed to carry out the work of species restoration.

4. Government agencies must be made more effective in dealing with endangered species recovery; in some cases, this means an overhaul of the entire government system that is involved.

5. Clearly written prescriptions and guidelines in the ESA [formal policy], founded on modern conservation biology and social sciences, will themselves facilitate implementation by professionals and organizations.

6. Organizations and professionals must initiate conservation actions earlier, set recovery goals high enough to ensure long-term species conservation, and continuously monitor overall progress.

7. Learning should be enhanced at both professional and organizational levels, and mechanisms to promote learning should be institutionalized.

8. Conservation leaders must be encouraged, trained, and empowered.

APPENDIX 4

Beedy, E. C., 1995. Ten ways to fix the *Endangered Species Act*. *Endangered Species Update* **12**(6): 12–14.

The ESA requires careful revision to ensure that it protects the biological heritage that Americans still strongly value, while treating all citizens in a fair, reasonable, and respectful manner. Ten suggestions follow.

1. Facilitate landowner participation in recovery efforts.
2. Eliminate critical habitat designations.
3. Refine the definition of potential habitat.
4. Establish specific criteria for listing and de-listing.
5. Develop recovery plans within one year of listing.
6. Assess take at a population level.
7. Develop an administrative appeals process with the U.S. Fish and Wildlife Service.
8. Streamline the process for obtaining scientific permits.
9. Streamline the Section 10(a) process to parallel the Section 7 process available to federal agencies.
10. Encourage multispecies planning efforts.

APPENDIX 5

Anonymous, 1995. Making the ESA work better. *Endangered Species Bulletin* **XX**(3): 4–7.

Ten principles to improve implementation of the ESA were announced March 6, 1995, by Interior Secretary Bruce Babbitt and Dr D. James Baker, Under Secretary of Commerce. Changes are designed to improve the species recovery rate while minimizing impacts of the ESA on landowners, grant more authority to state and local governments, require greater scientific scrutiny of endangered species decisions, and make implementation more efficient.

1. Treat landowners fairly and with consideration.
2. Minimize social and economic impacts.
3. Create incentives for landowners to conserve species.
4. Provide quick, responsive answers and certainty to landowners.
5. Base ESA decisions on sound and objective scientific information.
6. Prevent species from becoming endangered or threatened.
7. Promptly recover and de-list threatened or endangered species.
8. Provide state, tribal, and local governments with opportunities to play a greater role in carrying out the ESA.
9. Make effective use of limited public and private resources by focusing on groups of species dependent on the same habitat.
10. Promote efficiency and consistency in the Departments of Interior and Commerce.

APPENDIX 6

Administrative Conference of the United States, 1995. Proposed recommendations: improving decision making under the *Endangered Species Act*. Administrative Conference of the US: Washington.

The ESA creates a regulatory framework for managing conflicts between the survival concerns of endangered species and other human interests. Recent examinations of this system of decision making suggest that it appears to work fairly well in most cases, providing appropriate incentives to negotiate and resolve conflicts before they escalate. However, a few highly visible cases have been marked by delay, controversy, and are politically charged. Improvements to prevent these situations might include:

1. Increased use of alternative means of dispute resolution (ADR) and related consensus-building processes in appropriate cases;
2. Increased use of habitat conservation planning, which provides a mechanism by which private landowners and other non-federal parties can collaboratively develop a strategy that provides for both endangered species and development concerns;
3. Enhanced use of an ecosystem-based approach to resource management in order to avoid future endangered species — development conflicts through multi-party, co-operative, regional-scale planning and management.

APPENDIX 7

Clark, T. W., Wallace, R. L. and Kellert, S. R., unpublished. The *Endangered Species Act:* lessons from Balancing on the Brink of Extinction and a framework for analysing and improving performance. 39 pp.

1. Implementing the ESA is typically difficult; its enactment alone does not guarantee its success.
2. Very limited criteria have been used in evaluating the ESA's implementation.
3. Politics dominates all aspects of species restoration.
4. Explanations of programmatic weaknesses seldom include the policy factors that are important for understanding how the program actually worked and how it might be improved.
5. The critical importance of organizations in the ESA's implementation is insufficiently appreciated and understood.
6. Quality leadership is indispensable.
7. Biases are evident in implementation efforts; it is important to recognize these biases and attempt to avoid them.
8. Sometimes there is clear agency non-compliance with the Act.
9. Complexity is inherent in ESA implementation.
10. Uncertainty is highly characteristic of ESA implementation.
11. The courts play a major role in support of ESA policy.
12. The public plays a pivotal role in the Act's implementation and its continued existence.
13. Many endangered species programs reflect tension between roles of federal and state agencies.
14. Most, if not all, ESA programs are underfunded.

APPENDIX 8

Tear, T. H., Scott, J. M., Hayward, P. H. and Griffith, B., 1995. Recovery plans and the *Endangered Species Act:* are criticisms supported by data? *Conserv. Biol.* 9: 182–95.

1. In addition to identifying the threats to a species, establish biological guidelines for long-term recovery in terms of population size, number of populations, and probabilities of persistence over specific periods to distinguish objectively between threatened and endangered species, establish appropriate goals, and conduct viability analyses.

2. Explicitly separate and describe the political, social, and economic components that affect biological recovery goals, and indicate the extinction risk associated with each of these components.

3. Provide short-term recovery goals that combine biological, political, social, and economic components and establish a reasonable time for reassessment.

4. Require that recovery plans be written and approved within three years of listing.

5. Establish research priorities for gathering information specifically needed to assess whether recovery efforts are achieving the goal of establishing or maintaining viable populations in the ecosystems in which they occur. Attention should be paid to collecting information needed for PVAs, recognizing that, all else being equal, the probability of population survival is greater for larger populations than smaller populations, and that a species, survival probabilities are greater with multiple populations than a single population as long as population size of the subpopulations does not decline below viable levels.

6. In recovery planning modelling efforts, do not use the minimum population estimates from viability analyses. Instead, for population size and number estimates, use some measure of central tendency (mean, median, mode) and/or the upper confidence interval so that management strategies that use the data will incorporate a greater margin for error.

7. Continue to protect and recover populations and subspecies as well as full species.

8. Where possible, procure and protect suitable habitat for species, subspecies, and populations, recognizing that the probability of extinction is inversely related to the extent of a population's geographic distribution. The area of concern must be the species' historical range because its current range may be only a fraction of that which was historically occupied. In cases where suitable habitat is severely limited, habitat restorations should be included as a necessary component of recovery efforts to ensure that recovered populations can be viable.

9. If warranted, revise all existing unrevised recovery plans, emphasizing appropriate population goals in relation to current population size.

10. Avoid generalized recovery plans: clearly show how both proximate and ultimate limiting factors are addressed in the recovery effort, give specific target dates for recovery goals, and include a more formal public attitude assessment in conjunction with education component in recovery efforts.

11. Place more emphasis on multi-species and ecosystem-level recovery plans (especially plans with multiple taxonomic groups) as an interim step towards broader, ecosystem-level protection, recognizing that in some instances single-species recovery plans will still be necessary.

12. Recognize that our ability to save endangered species is limited; initiate an aggressive and proactive effort to save species while they are still common and a viable part of self-sustaining natural ecosystems.

APPENDIX 9

Cook, R. E. and Dixon, P., 1991. A review of recovery plans for threatened and endangered plant species. A Report for the World Wildlife Fund, Cornell Plantations. Cornell University, Ithaca. New York. 132 pp.

1. The U.S. Fish and Wildlife Service should write a primer of plant population biology for recovery plan writers.

   Since managers often lack basic knowledge about species and resources necessary for biological research, the primer should identify threats to plants, focus on population biology of plants, guide field investigations, and establish recovery priorities.

2. The U.S. Fish and Wildlife Service should provide guidelines for writing recovery plans for plants.

   New plans are too often modelled on old plans. Guidelines should discourage "boilerplate" language, specify methods for research, encourage setting of research priorities, and direct plans to focus on uniqueness of species.

3. Research recommendations in recovery plans should focus on basic species biology, simple observations of population behaviour, and the value of preliminary information to guide future research.

   The concepts of "limiting factors" and "habitat requirements" have little utility for management efforts towards reintroduction. So plans should discourage large-scale research and encourage general descriptions and field experience.

4. Recovery plans must address the special biological problems of plants.

   Population recruitment, loss of habitat through natural successional change, and minor taxonomic variations are some of these problems.

5. The U.S. Fish and Wildlife Service should produce an umbrella plan for certain elements that appear in all recovery plans.

   These include trade in endangered plants, strategy/policy to reduce removal from the wild, reintroductions, and public education.

6. Institutionalize "adaptive management" in plans.

   Recovery teams must demonstrate "prepared responsiveness." This would be fostered by common sense recovery plans, stressing the value of preliminary information and simple observations, modest, economical field studies, and ranking of research questions and management activities.

7. Review implementation of existing plans to see if progress has been made toward de-listing.

   Semi-annual or annual reports at regional or national levels or annual conferences to review progress would foster an adaptive management approach.

APPENDIX 10

Culbert, R. and Blair, R., 1989. Recovery planning and endangered species. *Endangered Species Update* 6: 2–8.

1. Recovery plans should identify the causes of a species' decline and specify actions needed to reverse the decline.

   Plans consist of three parts: (1) description of life history, status, and causes of decline, (2) outline of actions needed to alleviate threats and increase population sizes and numbers, and (3) implementation schedule by priority. There are also review, approval, updating, and revision processes. As of 1/1/89, only 284 of 531 listed species had approved recovery plans.

2. Selection and composition of recovery teams should be standardized.

   Political biases and conflicts of interest should be avoided, while non-agency experts should be included. The authors recommend a two-stage process for preparing plans: first, experts should address species' biology solely; and second, advisory team of commercial, conservation, and other interests should assist in making explicit decisions concerning tradeoffs.

3. Recovery plans should address specific ecological concerns.

   Many prblems exist. Plans often fail to consider restoration of previous habitat, population or taxonomic status, competition and predation, environmental variability, and movement between populations. Recovery objectives usually do not provide specific, concrete solutions to such problems.

4. Community or multiple species recovery plans could benefit more species at lower cost with simpler implementation and fewer conflicts.

5. Recovery plans should estimate population viability explicitly.

   "Population viability analysis" (PVA) is a useful systems approach to estimate probability of extinction of populations facing systematic and stochastic pressures. PVA can also provide useful knowledge about life history and environment, determine sensitivities to changes in reproduction, survival, or mortality, and suggest future research priorities and minimum areas.

6. Recovery plans should have adequate maps.

   Types of maps needed for recovery planning include: (1) habitat map of historic range and pertinent environmental variables, (2) land use map identifying precise reasons for habitat loss (past, present, and future), and (3) recovery plan map synthesizing analysis and specifying needed actions.

7. Base recovery objectives (numbers, sizes, and distributions of populations) on rigorous analysis.

   Too often, "plans provide for nothing more than token existence in a limited part of the species' range". Plans should recommend self-sustaining populations throughout the species' historic range (including environmentally-independent sites) and avoid simplistic rules of thumb (e.g., $N_e = 500$).

8. Recovery plans should provide a framework for implementation.

   Inter-agency co-operation can be enhanced by clearly stating management strategies and objectives and providing procedures to monitor co-operation; inter-agency task groups and regional protection plan committees are useful models for co-operative efforts. Recovery plan budgets should reflect stated priorities and estimate needs responsibly. Plans should provide a political feasibility analysis. Public education may generate community support.

APPENDIX 11

Beatley, T., 1994. Habitat conservation planning: endangered species and urban growth. University of Austin Press: Austin. 234 pp.

Habit conservation planning (HCP) is being used increasingly to resolve species protection/development conflicts, and because the conflicts will likely increase in the future, it is important to evaluate the success and desirability of the device and perhaps to offer suggestions for improving habitat conservation efforts.

Evaluation is difficult because the program is clearly a moving target, a large number are in progress, and few have been completed or fully settled in their implementation. Moreover, to judge certain important evaluative criteria, plans that result in the protection and recovery of listed species without loss of community prosperity, will require a number of years and extensive monitoring and study. This has not been done yet. Conclusions must be tentative and inconclusive.

Some successes to date have set aside considerable amounts of natural habitat (e.g., Coachella Valley HCP, California; 16 000 acres). Several areas of concern, questions, and future directions need attention:

1. Are the habitat conservation measures adequate?
2. Are the HCPs useful for the long-term viability of the species and its habitat?
3. Are current decisions based on limited biological knowledge adequate to protect species and habitats?
4. What are the potential benefits of HCP vs. comprehensive, multi-species approaches?
5. How can HCPs be integrated with land use planning?
6. How are HCPs to be funded over the long haul?
7. How will management of the HCP process in the future be carried out?

8. Will there be long-term developer satisfaction and support?
9. Will there be long-term community support and prosperity?

Recommendations include:

1. The HCP process should include representatives of all affected stakeholder groups.
2. The HCP should incorporate as complete and as thorough a biological and scientific information base as possible.
3. Integrate the HCP into local and regional long-range planning.
4. Develop an equitable, long-term funding program.
5. Protect habitat for multiple species simultaneously; seek to protect broader patterns of diversity.
6. Look for ways of dovetailing habitat conservation with other community goals.

APPENDIX 12

Some additional elements that should be included in all recovery plans

1. Need an evolutionary perspective to understand the problem and restore species.
2. Need an understanding of speciation theory and extinction theory (see Gilpin and Soulé 1986). Understand that extinction is a process usually requiring many years; it is not a single event.
3. Be clear about the unit of conservation (i.e., a population, subspecies, or species). Its taxonomy, both morphological and biochemical, should be worked out.
4. Use the extinction map (Fig. 1, this paper) as a basis for mapping the status of knowledge about the species, its habitat, and threatening processes.
5. A population viability analysis should be at the heart of the plan, as a basis of goal setting, planning, and evaluation. At the least, the PVA should be qualitative and verbal, at best, quantitative and mathematical.
6. Identify management levers and estimate (predict) outcome. Set up adaptive management responses. Monitor and adapt as needed.
7. Establish a program to carry out needed work in a timely way and on a scale that is needed.
8. The task should be seen as a social process challenge, not merely a technical biological task.

APPENDIX 13

Clark, T. W. and Cragun, J. R., 1994. Organization and management of endangered species restoration programs and recovery teams. Pp. 9–33 *in* Restoration of endangered species: conceptual issues, planning and implementation ed by M. L. Bowles and C. J. Whelan. Cambridge University Press: New York.

Analysing problems and developing action plans for endangered species recovery

A procedure useful in analysing a problem and developing an action plan for endangered species is outlined. Our 14-step procedure includes example work sheets to be filled out by you and your team. It can be applied to solving organizational as well as technical problems. It is self-explanatory.

Steps to follow in analysing a problem and developing an action plan

Problem identification

Step 1. Identify a problem you feel is important to the success of your organization over which you exercise some control.

Step 2. Translate the problem identified into an "objective" statement and write that objective on the *Problem Solving Worksheet*.

Remember: to increase the probability of success, objectives must be:

specific: having a special application or reference; specified; explicit, particular, precise, or definite

realistic: concerned with what is real or practical; representing things as they really are

attainable: that which can be reached, achieved, or accomplished by continued effort

measurable: that which can be measured, which is process of ascertaining extent, dimensions, quantity, etc.

acceptable: pleasing to the receiver, agreeable, welcome; worthy of being accepted, i.e., received with approval or favour

Step 3. Understand the problem(s) by answering the following:

1. I understand the problem *specifically* to be . . .
2. Examples of the problem that I have observed or experienced are . . .
3. The people (at all levels) directly involved in the problem are . . .
4. Those circumstances or conditions in the environment which contribute to the problem are . . .

Step 4. Assess the degree to which you can control or influence the problem and what the probabilities of success are.

Step 5. Restate your objective if necessary.

Step 6. Enter on your page 1 of the *Problem Solving Worksheet* those "facilitating forces" which you could use to help move toward reaching your objective. Enter also those "hindering forces" which stand in your way of reaching your objective.

Step 7. Evaluate each of these entries (from Step 6) in terms of how important they are. Identify those which are most important.

Development of alternative strategies

Step 8. Analyse your entries on page 1 and list on page 2 various strategies you could implement to reach your objective. Your strategies should consider both increasing the facilitating forces and decreasing the hindering forces.

Step 9. Select the one best strategy or combination of strategies which help meet your objective.

Development of your action plan

Step 10. Enter on page 3 of the *Problem Solving Worksheet* the specific tasks to be accomplished, who will do each task, and a date by which it should be completed. Also indicate in the section titled "Planning Notes" any details related to the various tasks, such as other people involved, arrangements to be made, etc.

Implementation and evaluation of action plan

Step 11. Specify those criteria that can be used to determine how effective you are in meeting your objective.

Step 12. Discuss with your supervisor your action plan and specifically consider how he/she could assist you in meeting the various tasks.

Step 13. Implement your plan, conduct an ongoing evaluation, and adjust as necessary.

Step 14. Evaluate the success of your efforts.

Problem Solving Worksheet

Page 1 asks participants to write down the "Objective" and has two columns for "Facilitating forces" and "Hindering forces".

Page 2 asks for "Strategies for Improvement" to be listed.

Page 3, on planning details, has five columns for "Task to be accomplished", "Who will do it", "When will it start", "When will it be done", and "Planning notes (who else is involved, what resources are required, etc.)".

Page 4, the action plan assessment form, lists the following criteria and asks whether the action plan meets each one, and if not, suggestions for improvement:

1. The plan logically leads to goal achievement.
2. Action steps are clear.
3. Actions steps are complete.
4. Actions are within the control and influence of the participant.
5. Time frames are real.
6. Plan is feasible: it makes sense, can be done.
7. Plan deals with concrete actions that can be monitored.
8. Plan as work-group commitment and support.
9. The responsibilities of people involved are clearly outlined.
10. Roadblocks have been identified and can be overcome.
11. Results can be measured and it will be known when task has been accomplished.

APPENDIX 14

Clark, T. W. and Westrum, R., 1989. High-performance teams in wildlife conservation: a species reintroduction and recovery example. *Environ. Manage.* **13**: 663–70.

1. Not every endangered species case is unique. A large number of common organizational issues are shared by all.
2. Certain organizational structures and modes of operation are much more likely to be effective than others. An effective organization is well matched to its task (i.e., its structure is appropriate to its function); it is well-staffed, led, and buffered from its political enviroment; and it processes information well, learns rapidly from its own mistakes, and is creative. A healthy team spends most of its energies solving external problems rather than working on itself.
3. Recovery programs generally consist of a *team* of people directly involved in the effort (requiring effective small-group dynamics) and a larger *network* of influential people and organizations (requiring maintenance of co-operation and commitment).
4. The goal in staffing the high-performance team is to get individuals who can handle the ambiguities and risks of highly uncertain situations without becoming rigid or "burning out".
5. Two important *cognitive* traits characterize the high-performance team: rapid decision making, and use of the entire team's cognitive skills.
6. Two *emotional* traits also characterize the high-performance team: internal support within the group, and buffering from the external environment.
7. The team leader must do two things: foster these cognitive and emotional characteristics and manage conflict within the team (i.e., steady the team and "explain" it to itself), and provide links to outside sources of power, influence, and knowledge (i.e., act as public relations arm and apologist for the team).

APPENDIX 15

Defenders of Wildlife, 1995. Saving America's wildlife: renewing the *Endangered Species Act.* Defenders of Wildlife: Washington. 63 pp.

1. Expand the role of science.
2. Create a National Commission on Species extinction.
3. Use science to set priorities.
4. Ensure adequate peer review.
5. Encourage work of the National Biological Service.
6. Create mechanisms for technical assistance.
7. Reduce the need for listings.
8. Provide incentives for private landowners.
9. Increase the role of states.
10. Promote regional ecosystem management planning.
11. Provide long-term funding.
12. End harmful subsidies.
13. Expand existing funding programs.
14. Close existing loopholes.

APPENDIX 16

Committee on Endangered Species, 1995. Strengthening the use of science in achieving the goals of the *Endangered Species Act:* an assessment. Ecological Society of America: Washington 21 pp.

1. The 1973 ESA is a powerful and sensible way to protect biological diversity; it contains the procedures and mechanisms by which to achieve this goal.
2. On the basis of science, the most important priorities for deciding which candidate species to list are: (1) number of other species that will benefit from the listing; (2) ecological role of species; (3) the organism's recovery potential; and (4) its taxonomic distinctness.
3. Formal Population Viability Analysis offers a method to identify how a species' survival potential can be maximized in the least controversial manner.
4. The likelihood of restoring the viability of an endangered species is enhanced when: (1) recovery plans seek to achieve a population distributed in suitable habitats across the landscape; and (2) these plans are developed and implemented expeditiously.
5. Additional programs for ecosystem-level protection that would complement existing legislation offer promise for a proactive approach that would effectively protect our nation's biological heritage at lower long-term costs.

REFERENCES

Administrative Conference of the United States, 1995. Proposed recommendations: improving decision-making under the *Endangered Species Act.* Administrative Conference of the US: Washington.

Alvarez, K., 1993. Twilight of the panther: biology, bureaucracy and failure in an endangered species program. Myakka River Publishing: Sarasota, Florida.

Anonymous, 1995. Making the ESA work better. *Endangered Species Bulletin* **XX**(3): 4–7.

Argyris, C., 1982. Reasoning, learning, and action: individual and organizational. Jossey-Bass: San Francisco.

Argyris, C., 1993. Knowledge for action: a guide to overcoming barriers to organizational change. Jossey-Bass: San Francisco.

Argyris, C. and Schön, D. A., 1978. Organizational learning: a theory of action perspective. Addison-Wesley: Reading, Massachusetts.

Ascher, W. and Healy, R., 1990. Natural resource policymaking in developing countries: environment, economic growth, and income distribution. Duke University Press: Durham, North Carolina.

Australian Nature Conservation Agency, 1993. On the brink! *Newslet. Endangered Species Prog.* **4**: 1–16.

Australian Nature Conservation Agency, 1994. *Endangered Species Protection Act* 1992: an outline. Commonwealth of Australia: Canberra.

Backhouse, G. N., 1992. Recovery plan for the Eastern Barred Bandicoot *Perameles gunnii.* Department of Conservation and Environment: Melbourne.

Backhouse, G. N., 1993. Recovery plan for the Eastern Barred Bandicoot *Perameles gunnii,* 1993–1995. Department of Conservation and Natural Resources: Melbourne.

Backhouse, G. N., Clark, T. W. and Reading, R. P., 1994a. Reintroductions for recovery of the Eastern Barred Bandicoot *Perameles gunnii* in Victoria, Australia. Pp. 209–18 *in* Reintroduction biology of Australian and New Zealand fauna ed by M. Serena. Surrey Beatty & Sons: Chipping Norton.

Backhouse, G. N., Clark, T. W. and Reading, R. P., 1994b. The Australian Eastern Barred Bandicoot program, evaluation and reorganization. Pp. 251–71 *in* Endangered species recovery: finding the lessons, improving the process ed by T. W. Clark, R. P. Reading and A. L. Clarke. Island Press: Washington.

Backhouse, G. N. and Clark, T. W., 1995. Case studies and policy initiatives in endangered species recovery in Australia: recommendations. Pp. 110–16 *in* People and nature conservation: perspectives on private land use and endangered species recovery ed by A. Bennett, G. Backhouse and T. Clark. Transactions of Royal Zoological Society of New South Wales. Surrey Beatty & Sons: Chipping Norton.

Bardach, E., 1977. The implementation game: what happens after a bill becomes law. MIT Press: Cambridge, Massachusetts.

Bean, M. J., Fitzgerald, S. G. and O'Connell, M. A., 1991. Reconciling conflicts under the *Endangered Species Act:* the habitat conservation planning experience. World Wildlife Fund-US: Washington.

Beatley, T., 1994. Habitat conservation planning: endangered species and urban growth. University of Texas Press: Austin.

Beedy, E. C., 1995. Ten ways to fix the *Endangered Species Act. Endangered Species Update* **12**(6): 12–14.

Bowles, M. L. and Whelan, C. J. (eds), 1994. Restoration of endangered species: conceptual issues, planning and implementation. Cambridge University Press: New York.

Brewer, G. D. and deLeon, P., 1983. The foundations of policy analysis. Dorsey Press: Homewood, Illinois.

Brunner, R. D., 1996. Policy and global change research: a modest proposal. *Climate Change* **32**: 121–36.

Butler, D. and Merton, D., 1992. The black robin: saving the world's most endangered bird. Oxford University Press: Auckland.

Caughley, G., 1994. Directions in conservation biology. *J. Anim. Ecol.* **63**: 215–44.

Cheney, K. A., 1993. A selective bibliography on the Endangered Species Act. *Nat. Resour. Environ.* **8**: 43–45, 77–79.

Clark, J. A., 1994. The Endangered Species Act: its history, provisions, and effectiveness. Pp. 19–45 *in* Endangered species recovery: finding the lessons, improving the process ed by T. W. Clark, R. P. Reading and A. L. Clarke. Island Press; Washington.

Clark, T. W., 1989. Conservation biology of the black-footed ferret, *Mustela nigripes.* Wildlife Preservation Trust Special Scientific Report No. 3.

Clark, T. W., 1992. Practicing natural resource management with a policy orientation. *Environ. Manage.* **16**: 423–33.

Clark, T. W., 1993. Creating and using knowledge for species and ecosystem management: science, organizations, and policy. *Perspectives in Biol. Med.* **36**: 497–525 + appendices.

Clark, T. W., in press. Averting extinction: reconstructing endangered species recovery. Yale University Press: New Haven, Connecticut.

Clark, T. W., 1996. Learning as a strategy for improving endangered species conservation. *Endangered Species Update* **13**(1 and 2): 5–6, 22–24.

Clark, T. W. and Westrum, R., 1987. Paradigms and ferrets. *Social Stud. Sci.* **3**: 3–33.

Clark, T. W., Crete, R. and Cada, J., 1989. Designing and managing successful endangered species recovery programs. *Environ. Manage.* **13**: 159–70.

Clark, T. W. and Westrum, R., 1989. High performance teams in wildlife conservation: a species reintroduction and recovery example. *Environ. Manage.* **13**: 663–70.

Clark, T. W., Warneke, R. M. and George, G. G., 1990. Management and conservation of small populations. Pp. 1–18 *in* Management and conservation of small populations ed by T. W. Clark and J. H. Seebeck. Chicago Zoological Society: Brookfield, Illinois.

Clark, T. W. and Seebeck, J. H. (eds), 1990. Management and conservation of small populations. Chicago Zoological Society: Brookfield, Illinois.

Clark, T. W. and Harvey, A. H., 1991. Implementing recovery policy: learning as we go? Pp. 147–63 *in* Balancing on the brink of extinction: the *Endangered Species Act* and lessons for the future ed by K. A. Kohm. Island Press: Washington.

Clark, T. W. and Cragun, J. R., 1994. Organization and management of endangered species restoration programs and recovery teams. Pp. 9–33 *in* Restoration of endangered species: conceptual issues, planning and implementation ed by M. L. Bowles and C. J. Whelan. Cambridge University Press: New York.

Clark, T. W. and Reading, R. P., 1994. A professional perspective: improving problem-solving, communication, and effectiveness. Pp. 351–70 *in* Endangered species recovery: finding the lessons, improving the process ed by T. W. Clark, R. P. Reading and A. L. Clarke. Island Press: Washington.

Clark, T. W., Reading, R. P. and Clarke, A. L. (eds), 1994. Endangered species recovery: finding the lessons, improving the process. Island Press: Washington.

Clark, T. W., Backhouse, G. N. and Reading, R. P., 1995a. Prototyping in endangered species recovery programs: the Eastern Barred Bandicoot experience. Pp. 50–62 *in* People and nature conservation: perspectives on private land use and endangered species recovery ed by A. Bennett, G. Backhouse and T. Clark. Transactions of the Royal Zoological Society of New South Wales. Surrey Beatty & Sons: Chipping Norton.

Clark, T. W., Gibbs, J. P. and Goldstraw, P. W., 1995b. Some demographics of the extirpation from the wild of Eastern Barred Bandicoots *(Perameles gunnii)* in 1988–91, near Hamilton, Victoria, Australia. *Wildl. Res.* **22**: 289–97.

Clark, T. W., Wallace, R. L. and Kellert, S. R., unpublished. The Endangered Species Act: lessons from Balancing on the Brink of Extinction and a framework for analyzing and improving performance.

Comer, R. D. (ed), 1993. Endangered species protection. *Nat. Resour. Environ.* **8**: 1–88.

Committee on Endangered Species, 1995. Strengthening the use of science in achieving the goals of the *Endangered Species Act:* an assessment. Ecological Society of America: Washington.

Cook, R. E. and Dixon, P., 1991. A review of recovery plans for threatened and endangered plant species. A report for the World Wildlife Fund, Cornell Plantations. Cornell University: Ithaca, New York.

Culbert, R. and Blair, R., 1989. Recovery planning and endangered species. *Endangered Species Update* **6**: 2–8.

Davis, G., Wanna, J., Warhurst, J. and Weller, P., 1993. Public policy in Australia. Allen and Unwin: St. Leonards, New South Wales.

Defenders of Wildlife, 1995. Saving America's wildlife: renewing the Endangered Species Act. Defenders of Wildlife: Washington.

Dixon, P. M. and Cook, R. E., 1989. Science, planning, and the recovery of endangered plants. *Endangered Species Update* **6**: 9–14.

Easter-Pilcher, A. L., 1993. Analysis of the listing of species as endangered or threatened under the *Endangered Species Act.* Ph.D. Dissertation, Montana State University, Bozeman.

Ehrenfeld, D. M., 1976. The conservation of non-resources. *Am. Sci.* **64**: 648–56.

Eisner, T., Lubchenco, J., Wilson, E. O., Wilcove, D. S. and Bean, M. J., 1995. Building a scientifically sound policy for protecting endangered species. *Science* **268**: 1231–232.

Endangered Species Unit, 1994. Recovery plan and funding proposal guidelines for endangered and vulnerable species and ecological communities. Australian Nature Conservation Agency: Canberra.

Etheredge, L. S., 1985. Can governments learn? Pergamon Press: New York.

Etheredge, L. S. and Short, J., 1983. Thinking about government learning. *J. Manage. Stud.* **20**: 41–58.

Falk, D. A., 1987. Integrated conservation strategies for endangered plants. *Nat. Areas J.* **7**: 118–23.

Falk, D. A., 1990. Endangered forest resources in the U.S.: integrated strategies for conservation of rare species and genetic diversity. *Forest Ecol. Manage.* **35**: 91–117.

Falk, D. A. and Holsinger, K. E. (eds), 1992. Genetics and conservation of rare plants. Oxford University Press: Oxford.

Federal Register, 1983. Fish and Wildlife Service: endangered and threatened species listing and recovery priority guidelines. *Federal Register* **48**(184): 43098–3015.

Flannery, T. F., 1994. Future eaters: an ecological history of the Australasian lands and people. Reed Books: Chatswood.

Gilpin, M. and Soulé, M., 1986. Minimum viable populations: processes of species extinction. Pp. 19–34 *in* Conservation biology ed by M. E. Soulé. Sinauer Associates: Sunderland, Massachusetts.

Groves, C. R., 1994. Candidate and sensitive species programs: lessons for cost-effective conservation. Pp. 227–50 *in* Endangered species recovery: finding the lessons, improving the process ed by T. W. Clark, R. P. Reading and A. L. Clarke. Island Press: Washington.

Houck, O. A., 1993. The Endangered Species Act and its implementation by the U.S. Departments of Interior and Commerce. *Uni. Colorado Law Rev.* **64**: 277–370.

Humphries, R. K. and Seebeck, J. H., 1995. Conservation of the Eastern Barred Bandicoot *Perameles gunnii* on private land in Victoria. Pp. 156–62 *in* People and nature conservation: perspectives on private land use and endangered species recovery ed by A. Bennett, G. Backhouse and T. Clark. Transactions of the Royal Zoological Society of New South Wales. Surrey Beatty & Sons: Chipping Norton.

Jordan, W. R., III, Gilpin, M. E. and Aber, J. D., 1987. Restoration ecology: a synthetic approach to ecological research. Cambridge University Press: New York.

Kellert, S. R., 1985. Social and perceptual factors in endangered species management. *J. Wildl. Manage.* **49**: 528–36.

Kellert, S. R., 1994. A sociological perspective: valuational, socioeconomic, and organizational factors. Pp. 371–90 *in* Endangered species recovery: finding the lessons, improving the process ed by T. W. Clark, R. P. Reading and A. L. Clarke. Island Press: Washington.

Kennedy, M. (ed), 1991. Australia's endangered species: the extinction dilemma. Simon and Schuster: Brookvale.

Kohm, K. A. (ed), 1991. Balancing on the brink of extinction: the Endangered Species Act and lessons for the future. Island Press: Washington.

Lacy, R. C. and Clark, T. W., 1990. Population viability assessment of Eastern Barred Bandicoots in Victoria. Pp. 131–46 *in* Management and conservation of small populations ed by T. W. Clark and J. H. Seebeck. Chicago Zoological Society: Brookfield, Illinois.

Lasswell, H. D., 1963. The future of political science. Prentice-Hall: New York.

Lasswell, H. D., 1971a. From fragmentation to configuration. *Policy Sci.* **2**: 439–446.

Lasswell, H. D., 1971b. A pre-view of policy sciences. American Elesvier: New York.

Lasswell, H. D. and McDougal, M. S., 1992. Jurisprudence for a free society: studies in law, science, and policy. 2 v. New Haven Press: New Haven, Connecticut.

Lawton, J. H. and May, R. M. (eds), 1995. Extinction rates. Oxford University Press: New York.

Leeuw, F. L., Rist, R. C. and Sonnichsen, R. C. (eds), 1994. Can governments learn? Comparative perspectives on evaluation and organizational learning. Transaction Publishers: New Brunswick.

Lindenmayer, D., Clark, T. W., Thomas, V. and Lacy, R. C., 1993. Population viability as a tool in wildlife conservation policy: with reference to Australia. *Environ. Manage.* **17**: 745–58.

Lovejoy, T., 1976. We must decide which species will go forever. *Smithsonian* **4**: 52–59.

Mace, G. M. and Lande, R., 1991. Assessing extinction threats: towards a re-evaluation of IUCN threatened species categories. *Conserv. Biol.* **5**: 148–57.

Maguire, L. A., Lacy, R. C., Begg, R. J. and Clark, T. W., 1990. An analysis of alternative strategies for recovering the Eastern Barred Bandicoot in Victoria. Pp. 147–64 *in* Management and conservation of small populations ed by T. W. Clark and J. H. Seebeck. Chicago Zoological Society: Brookfield, Illinois.

Mansergh, I., Jelinek, A. and Clunie, P., 1995. A review of the Action Statement process under the Victorian *Flora and Fauna Guarantee Act* 1988. Pp. 68–77 *in* People and nature conservation: perspectives on private land use and endangered species recovery ed by A. Bennett, G. Backhouse and T. Clark. Transactions of the Royal Zoological Society of New South Wales. Surrey Beatty & Sons: Chipping Norton.

Mattson, D. J. and Craighead, J. J., 1994. The Yellowstone grizzly bear recovery program: uncertain information, uncertain policy. Pp. 101–30 *in* Endangered species recovery: finding the lessons, improving the process ed by T. W. Clark, R. P. Reading and A. L. Clarke. Island Press: Washington.

Maunder, M., 1992. Plant reintroduction: an overview. *Biodiv. Conserv.* **1**: 51–61.

McFarlane, R. W., 1992. A stillness in the pines: the ecology of the Red-cockaded Woodpecker. W. W. Norton and Co.: New York.

Michael, D. N., 1995. Barriers and bridges to learning in a turbulent human ecology. Pp. 461–88 *in* Barriers and bridges to the renewal of ecosystems and institutions ed by L. H. Gunderson, C. S. Holling and S. S. Light. Columbia University Press: New York.

Miller, B. J., Reading, R. P. and Forrest, S. C., 1996. Prairie night: black-footed ferrets and the recovery of engangered species. Smithsonian Press: Washington.

Miller, R., Reading, R., Jackson, J. A., Snyder, N., Forrest, S. and Frazier, J., 1994. A model for improving endangered species recovery programs. *Environ. Manage.* **18**: 637–45.

Morgan, G., 1986. Images of organization. Sage Publications: Beverley Hills.

Murphy, D., 1991. Invertebrate conservation. Pp. 181–98 *in* Balancing on the brink of extinction: the *Endangered Species Act* and lessons for the future ed by K.A. Kohm. Island Press: Washington.

Myroniuk, P., 1995. Captive management of the threatened Eastern Barred Bandicoot: zoos and co-operative conservation. Pp. 63–67 *in* People and nature conservation: perspectives on private land use and endangered species recovery ed by A. Bennett, G. Backhouse and T. Clark. Transactions of the Royal Zoological Society of New South Wales. Surrey Beatty & Sons: Chipping Norton.

Nakamura, R. T. and Smallwood, F., 1980. The politics of policy implementation. St. Martin's Press: New York.

National Research Council, 1995. Science and the *Endangered Species Act*. Committee on Scientific Issues in the *Endangered Species Act*: Washington.

New, T. R.; 1984. Insect conservation: an Australian perspective. Junk: Dordrecht.

New, T. R., 1991. Butterfly conservation. Oxford University Press: Melbourne.

New, T. R. and Yen, A. L., 1995. Species management and recovery plans for butterflies (Insecta: Lepidoptera) in Australia. Pp. 15–21 *in* People and nature conservation: perspectives on private land use and endangered species recovery ed by A. Bennett, G. Backhouse and T. Clark. Transactions of the Royal Zoological Society of New South Wales. Surrey Beatty & Sons: Chipping Norton.

Nilsson, G., 1986. The endangered species handbook. Animal Welfare Institute: Washington.

Pressman, J. L. and Wildavsky, A. 1978. Implementation. University of California Press: Berkeley.

Reading, R. P., 1993. Toward an endangered species reintroduction paradigm: A case study of the Black-footed Ferret. Ph. D. Dissertation, Yale University, New Haven, Connecticut.

Reading, R. P., Clark, T. W. and Kellert, S. R., 1991. Towards an endangered species reintroduction paradigm. *Endangered Species Update* **8**(11): 1–4.

Reading, R. P. and Miller, B. J., 1994. The Black-footed Ferret recovery program: unmasking professional and organizational weaknesses. Pp. 73–100 *in* Endangered species recovery: fnding the lessons, improving the process ed by T. W. Clark, R. P. Reading and A. L. Clarke. Island Press: Washington.

Reading, R. P., Clark, T. W. and Arnold, A., 1995. Attitudes toward the endangered Eastern Barred Bandicoot. *Anthrozöos* **VII**(4): 255–69.

Reading, R. P., Clark, T. W., Seebeck, J. H. and Pearce, J., 1996. Habitat suitability index model for the Eastern Barred Bandicoot *(Perameles gunnii)*. *Wildl. Res.* **23**: 221–35.

Reading, R. P., Clark, T. W. and Griffith, B., in press. The influence of valuational and organizational considerations on the success of rare species translocations. *Biol. Conserv.*

Robertson, H. A., Hay, J. R., Saul, E. K. and McCormack, G. V., 1994. Recovery of the Kakerori: an endangered forest bird of the Cook Islands. *Conserv. Biol.* **8**: 1078–086.

Robinson, N. A., 1995. Implications from mitochondrial DNA for management to conserve the Eastern Barred Bandicoot *(Perameles gunnii)*. *Conserv. Biol.* **9**: 114–25.

Robinson, N. A., Murray, N. D. and Sherwin, W. B., 1993. VNTR loci reveal differentiation between and within structure of populations of the Eastern Barred Bandicoot *Perameles gunnii*. *Molecular Ecol.* **2**: 195–207.

Rohlf, D. J., 1989. The Endangered Species Act: a guide to its protection and implementation. Stanford Environmental Law Society: Stanford, California.

Rose, R., 1993. Lesson-drawing in public policy: a guide to learning across time and space. Chatham House Publications: Chatham, New Jersey.

Schön, D. A., 1983. The reflective practitioner: how professionals think in action. Basic Books: New York.

Schön, D. A., 1987. Educating the reflective practitioner. Jossey-Bass: San Francisco.

Schön, D. A. and Rein, M., 1994. Frame reflection: toward the resolution of intractable policy controversies. Basic Books: New York.

Scott, J. M., Davis, R., Csuti, B., Noss, R., Butterfield, B., Groes, C., Anderson, H., Caicco, S., D'Erchia, F., Edwards, T. C., Ulliman, J. and Wright, G. R., 1993. Gap analysis: a geographic approach to protection of biological diversity. *Wildl. Monogr.* **123**: 1–41.

Scott, J. M., Tear, T. H. and Mills, L. S., 1995. Socioeconomics and the recovery of endangered species: biological assessments in a political world. *Conserv. Biol.* **9**: 214–16.

Seebeck, J. H. and Patrick, C., in press. An annotated bibliography of the Eastern Barred Bandicoot. Department of Natural Resources and Environment: Melbourne.

Senge, P. M., 1990. The fifth discipline: the art and practice of the learning organization. Doubleday: New York.

Tear, T. H., Scott, J. M., Hayward, P. H. and Griffith, B., 1993. Status and prospects for success of the *Endangered Species Act*: a look at recovery plans. *Science* **262**: 976–77.

Tear, T. H., Scott, J. M., Hayward, P. H. and Griffith, B., 1995. Recovery plans and the *Endangered Species Act*: are criticisms supported by data? *Conserv. Biol.* **9**: 182–95.

Thibodeau, R. F., 1983. Endangered species: deciding which species to save. *Environ. Manage.* **7**: 101–07.

Tobin, R., 1990. The expendable future: U.S. politics and the protection of biological diversity. Duke University Press: Durham, North Carolina.

Toyne, P., 1994. The reluctant nation: environment, law and politics in Australia. ABC Books: Sydney.

U.S. Fish and Wildlife Service, 1990. Policy and guidelines for planning and co-ordinating recovery of endangered and threatened species. U.S. Department of Interior: Washington.

U.S. General Accounting Office, 1981. Congressional guidance and better federal coordination would improve marine mammal management. U.S. General Accounting Office: Washington.

U.S. General Accounting Office, 1988. Endangered species: management could enhance recovery program. U.S. General Accounting Office: Washington.

U.S. General Accounting Office, 1992. Types and numbers of implementing actions. U.S. General Accounting Office: Washington.

U.S. General Accounting Office, 1993a. Endangered species: factors associated with delayed listing decisions. U.S. General Accounting Office: Washington.

U.S. General Accounting Office, 1993b. Protected species: marine mammals predation on varieties of fish. U.S. General Accounting Office: Washington.

Wallace, R. L., 1994. The Florida Manatee recovery program: organizational learning and a model for improving recovery programs. Pp. 131–56 *in* Endangered species recovery: finding the lessons, improving the process ed by T. W. Clark, R. P. Reading and A. L. Clarke. Island Press: Washington.

Walters, M. J., 1992. A shadow and a song. Chelsea Green Publishing: Post Mills, Vermont.

Warshall, P., 1994. The biopolitics of the Mt. Graham Red Squirrel *(Tamiasciuris hudsonicus grahamensis)*. *Conserv. Biol.* **8**: 977–88.

Westrum, R., 1994. An organizational perspective: designing recovery teams from the inside out. Pp. 327–49 *in* Endangered species recovery: finding the lessons, improving the process ed by T. W. Clark, R. P. Reading and A. L. Clarke. Island Press: Washington.

Wilcove, D. S., McMillan, M. and Winston, K. C., 1993. What exactly is an endangered species? An analysis of the U.S. Endangered Species List 1985–1991. *Conserv. Biol.* **7**: 87–93.

Wilson, B. and Clark, T. W., 1995. The Victorian *Flora and Fauna Guarantee Act* 1988: a five year review of its implementation. Pp. 87–103 *in* People and nature conservation: perspectives on private land use and endangered species recovery ed by A. Bennett, G. Backhouse and T. W. Clark. Transactions of the Royal Zoological Society of New South Wales. Surrey Beatty & Sons: Chipping Norton.

Wondolleck, J. M., Yaffee, S. L. and Crowfoot, J. E., 1994. A conflict management perspective: applying the principles of alternative dispute resolution. Pp. 305–26 *in* Endangered species recovery: finding the lessons, improving the process ed by T. W. Clark, R. P. Reading and A. L. Clarke. Island Press: Washington.

World Conservation Monitoring Centre, 1992. Global biodiversity: status of the Earth's living resources. Chapman and Hill: New York.

Yaffee, S. L., 1982. Prohibitive policy: implementing the federal *Endangered Species Act*. MIT Press: Cambridge.

Yaffee, S. L., 1994a. The wisdom of the Spotted Owl: policy lessons for a new century. Island Press: Washington.

Yaffee, S. L., 1994b. The Northern Spotted Owl: an indicator of the importance of sociopolitical context. Pp. 47–72 *in* Endangered species recovery: finding the lessons, improving the process ed by T. W. Clark, R. P. Reading and A. L. Clarke. Island Press: Washington.

Yen, A. L., New, T. R., Van Praagh, B. D. and Vaughan, P. J., 1990. Invertebrate conservation: three case studies in southeastern Australia. Pp. 207–24 *in* Management and conservation of small populations ed by T. W. Clark and J. H. Seebeck. Chicago Zoological Society: Brookfield.

Yen, A. and New, T., 1995. Some problems affecting implementation of recovery programs for threatened Australian non-marine invertebrate species. Pp. 9–14 *in* People and nature conservation: perspectives on private land use and endangered species recovery ed by A. Bennett, G. Backhouse and T. Clark. Transactions of the Royal Zoological Society of New South Wales. Surrey Beatty & Sons: Chipping Norton.

Recovery of Australian threatened species — a national perspective

Bruce Male[1]

The Commonwealth Endangered Species Program was established in 1989 to prevent further extinctions of Australian flora and fauna and restore those species and ecological communities currently threatened with extinction to a secure status in the wild. The Commonwealth *Endangered Species Protection Act 1992* provides the legislative framework for the program and places specific statutory obligations on the Commonwealth Government with regard to its own lands and waters and its decisions and actions.

The approach adopted by this program, including action plans, recovery plans, amelioration of threatening processes and education and public awareness campaigns and the legislative linkage, is described and explained.

INTRODUCTION

THE international context for threatened species conservation in Australia is established by the Convention on Biological Diversity. Australian obligations under this convention are supported by the draft document, The Conservation of Threatened Species and Ecological Communities threatened with Extinction: the National Strategy (Anon. 1994a), which has received support from all Australian Governments. While this strategy is not legally binding, it does place moral obligations on governments and provides a framework within which government actions can be judged. The primary goal of the strategy is:

> *To ensure that endangered and vulnerable species and ecological communities can survive and flourish, retain their genetic diversity and potential for evolutionary development in their natural habitat, and to prevent further species and ecological communities from becoming endangered.*

Threatened species issues were given public prominence in Australia in 1989 when a 10-year Endangered Species Program was announced in a Prime Ministerial Statement on the Environment (Hawke 1989). However, it is important to recognize that threatened species conservation action had been undertaken for many years prior to this important initiative. Most threatened species conservation action is implemented by the Australian State and Territory government conservation agencies, very often with assistance from the Commonwealth. Early conservation actions on this front were often characterized by a lack of consistency, co-ordination and national perspective.

Under the Constitution, the Commonwealth has complete authority for foreign policy. Where Australia is party to international agreements and treaties, such as the Convention on the Conservation of Migratory Species of Wild Animals (Bonn) and the Convention on Biological Diversity, the Commonwealth may use its external affairs powers to override the jurisdiction of the States. To date this has not been necessary to effect Australian obligations with regard to threatened species matters. All States and Territories have nature conservation agencies with long-standing records in wildlife management. Almost without exception, all native flora and fauna are protected in all jurisdictions and all governments either have or intend to introduce legislation to conserve threatened species.

As wildlife and natural ecosystems recognize no legal boundaries, the effective national conservation measures that can be implemented in Australia are co-operative ones between the Commonwealth and State governments. The Australian National Parks and Wildlife Service, now the Australian Nature Conservation Agency (ANCA), was created in 1975. The agency's corporate purpose is the conservation and appreciation of Australia's natural and associated cultural heritage. Its primary role is to provide leadership, develop national nature conservation policy, and implement co-operative national programs for the conservation and restoration of biodiversity (Anon. 1994b).

HISTORY, DEVELOPMENT AND PHILOSOPHY OF THE ENDANGERED SPECIES PROGRAM

In 1988 the Commonwealth Minister for the Environment established a committee of conservation practitioners, scientists and conservation campaigners to provide advice on threatened species issues. This *ad hoc* ministerial committee was the first Endangered Species Advisory Committee known as ESAC. The first draft national strategy for endangered species was developed by this group (Anon. 1989). One of the early recommendations to the Minister for the Environment was that a program be established by the Commonwealth Government to halt and reverse the decline of threatened species. In mid-1989 the government established Australia's first national program for threatened species with a funding

[1]Endangered Species Unit, Australian Nature Conservation Agency, GPO Box 636, Canberra, Australian Capital Territory, Australia 2601.
Pages 23–27 *in* BACK FROM THE BRINK: REFINING THE THREATENED SPECIES RECOVERY PROCESS ed by S. Stephens and S. Maxwell. Surrey Beatty & Sons, 1996.

base of $2 million per annum (Hawke 1989) and expenditure on threatened species research and management commenced in early 1990.

Initially project proposals were called for by nationwide public announcement and selected from those received. While promoting access to the limited resources available under the program, this practice had a number of drawbacks, the principal one being that the choice was limited to those projects proposed by others. In addition, the wide and prominent public advertisement of funding led to vastly more proposals being submitted (total value of approximately $8 million) than could be supported from the available $2 million. This approach, if continued, may have lead to widespread dissatisfaction within the scientific and conservation community. It would also have left the program staggering from one group of conservation crises to another, while allowing only limited control over its direction and outcomes.

The Endangered Species Advisory Committee and officers of the Australian National Parks and Wildlife Service responsible for administering the program recognized that a better strategic base was required that would enable proposals to be solicited and selected according to national priorities (Dr Barry Reville, pers. comm.). The recovery plan concept that had already been used extensively in the United States and New Zealand was adopted. This approach was already being used for one Australian bird species. In 1984 the South Australian, Tasmanian, Victorian and Commonwealth Governments co-operated to develop and implement a recovery plan (Brown and Wilson 1984) for the endangered Orange-bellied Parrot.

During 1990 the first Endangered Species Program Recovery Plan Guidelines were rapidly developed by the Endangered Species Unit in close co-operation with the State and Territory conservation agencies and the members of the Endangered Species Advisory Committee. By late 1990 there was in place for the first time national recovery plan guidelines that had the support of all Australian government conservation agencies. All these agencies commenced preparation of recovery plans, supported by Endangered Species Program funding, based on these guidelines. In 1991 the first of these plans received funding for implementation from the program and a strong strategic focus based on medium to long-term planning had been established for recovery action for threatened species.

The recovery plan process continued to be developed progressively in consultation with States and Territories and has undergone further refinement based on experience and feedback ever since.

The philosophy of the program was, and remains, one that is directed at identified problems, follows a structured methodology and is focused on achieving positive and practical conservation outcomes.

ENDANGERED SPECIES PROGRAM GOALS AND OBJECTIVES

The goal of the program is twofold. Firstly to prevent further extinctions of Australian flora and fauna, and secondly to restore threatened species and ecological communities to a secure status in the wild.

The objectives are to promote the recovery of threatened species and ecological communities and to promote greater community understanding and awareness of the need to conserve threatened flora and fauna.

The implementation of the program has a multi-stranded structure consisting of related themes. These are:

1. research and management action for threatened species and ecological communities;
2. research and management action to control key threatening processes; and
3. education, community involvement and public awareness.

The first of these themes is dealt with by the development, publication and revision of action plans that provide an overview of the conservation status of major groups of taxa and provide information to guide the development of detailed recovery plans for priority species. This is followed by the preparation and implementation of these recovery plans.

The program addresses the second theme through the funding of strategic research to assist the understanding of threatening processes followed by the preparation and implementation of national threat abatement plans for key threatening processes.

Action for the education, involvement and awareness theme consists of the preparation and distribution of interpretive and educational materials through a range of media, from traditional print based brochures to innovative interactive presentations on the ANCA Internet World Wide Web server. Community involvement is fostered through the funding of selected non-government organizations that actively promote and achieve community participation in recovery plan conservation actions. One example of this is the Threatened Bird Network run by the Royal Australasian Ornithologists Union.

COMMONWEALTH PROGRAM MANAGEMENT AND AUDIT REQUIREMENTS

The Department of Finance imposes some basic requirements on all Commonwealth programs (Anon. 1992a). These program budgeting guidelines include the "up-front" definition of the programs objectives in the new policy proposal at the time the proposal is being considered for funding. The program objective given must mirror the objective in the original and subsequent proposals for funding from various Commonwealth budgets. In addition, the policy proposals must include performance indicators for the program. All Commonwealth programs are subjected to periodic review of their performance against their objectives, usually at intervals of five years.

The philosophy of the program as described previously is one that underpins all sound management practices irrespective of the discipline to which they are applied. It also forms the basis of the Commonwealth Public Service program budgeting guidelines.

The Endangered Species Program is subject also to both internal and external audit and must satisfy the requirements of Commonwealth audit legislation. It must also meet any guidelines or directions of the Auditor-General, such as the best practice guidelines for the administration of funding to external bodies (Anon. 1994c). Such audits are not confined to financial aspects, accounting and expenditure but also examine questions of program access, equity and probity. A "cradle to grave" approach is adopted that examines project selection through implementation, review and completion taking account of the project objectives and using its performance indicators.

These Commonwealth requirements and practices have been reflected in the design and structure of the Endangered Species Program so far described.

ENDANGERED SPECIES PROTECTION ACT

The Endangered Species Program receives its legislative mandate from the *Endangered Species Protection Act 1992* that came into force on 30 April 1993. Legislation by its nature is the highest form of government policy. It is legally binding and is not lightly or readily amended. Given that it is vital that legislation is effective and efficient in its practical operation from its enactment, it is significant that the legislation followed the development and implementation of the program. The recovery plan provisions of the Endangered Species Protection Act were tested by the program over a number of years before being enshrined in legislation. Accordingly, these provisions have benefited from this experience and should not require amendment for some time.

The objects of the Act are to:

- promote the recovery of threatened species and ecological communities;
- prevent other species and communities from becoming threatened;
- reduce conflict in land management through readily understood mechanisms for the conservation of threatened species and communities;
- provide for public involvement in, and promote public understanding of threatened species and communities conservation; and
- encourage co-operative management for the conservation of threatened species and communities (Anon. 1992b).

The objects of the Act do not include ecologically sustainable development (ESD), however, a number of provisions oblige the Commonwealth to have regard to the principles of ESD when framing various actions under the legislation (Anon 1992b). The principles of ESD include the protection of biodiversity, ergo the conservation of threatened species, as a fundamental principle (Anon. 1992c).

The Act applies to:

- land owned or leased by the Commonwealth Government or its agencies;
- Australia's external territories;
- the coastal sea, except areas vested in a State or Territory;
- the Australian continental shelf (sea-bed and waters);
- the Australian Fishing Zone;
- areas declared under the *National Parks and Wildlife Conservation Act 1975;* and
- the actions of Commonwealth Ministers, their agents, departments, bodies and companies (Anon. 1992b).

While the Act does not usually apply to land and waters under State and Territory jurisdiction, where it does it can only constrain Commonwealth activity in these jurisdictions. Only by invoking the foreign affairs power [under the Constitution] can the Act override State and Territory law.

The Act lists threatened ecological units (species and communities) and key threatening processes and obliges the Commonwealth to prepare recovery and threat abatement plans, as appropriate, as well as conferring protected status on listed species and ecological communities. There are provisions for the application of conservation orders restraining activities that threaten listed species and communities on Commonwealth land and in Commonwealth waters.

Two Ministerial statutory advisory bodies to the Minister for the Environment are established under the provisions of the Act. The Endangered Species Advisory Committee (ESAC) is appointed by the Minister and its membership is drawn from the conservation, scientific, rural and business communities as well as Commonwealth and State and Territory Governments. It is charged with a number of advisory responsibilities, but the most significant of these with respect to program delivery are:

- advising the Minister on the priorities for recovery and threat abatement plans;
- advising the Minister on the priorities for the preparation of recovery and threat abatement plans;
- commenting on draft recovery and threat abatement plans; and
- advising the Minister on the preparation and adoption of recovery and threat abatement plans (Anon. 1992b).

The Endangered Species Scientific Subcommittee (ESSS) consists of the scientific members of the Endangered Species Advisory Committee, but it has quite separate functions and reports to the Minister independently of the Endangered Species Advisory Committee. Its primary functions are to advise the Minister on the listing of species, ecological communities and key threatening processes on the Schedules to the Act (Anon. 1992b).

RECOVERY PLAN PROCESS

The most significant provisions of the Act, with far reaching implications, are those that define the content of recovery and threat abatement plans.

A recovery plan prepared to meet the requirements of the Act must have the following features:

- an objective to be achieved over the life of the plan;
- criteria against which the objective is measured;
- costed and timed actions needed to satisfy the criteria; and
- actions to protect critical habitats.

In addition, the plan must also identify interests or parties affected by the plan, those involved in performance evaluation, and specify any major benefits to non-target species or ecological communities (Anon. 1992b).

RECOVERY TEAMS AND THEIR ROLES

The nationally agreed recovery plan process establishes recovery teams as the mechanism for guiding the development and managing the implementation of recovery plans. The guidelines (Anon. 1995) describe the responsibilities of these teams and include:

- supporting the lead conservation agency responsible for the implementation of the plan with the writing and updating of the plan;
- monitoring and assessing the progress of the actions prescribed in the plan;
- advising parties to the plan whether the implementation of recovery actions is consistent with the plan's objectives;
- commenting on the drafts of progress and final reports that may be required by funding bodies; and
- monitoring press releases and other publicity to ensure accuracy and appropriate acknowledgment of the team's involvement.

In addition, the recovery team is also a forum for reporting on plan implementation, refining actions, discussing problems and recommending appropriate solutions.

Under the guidelines recovery teams are usually chaired by the lead agency implementing the plan and may include among their membership:

- persons undertaking research or management actions;
- specialist biologists (e.g., conservation agency, university, CSIRO, consultant);
- an Endangered Species Unit representative;
- a World Wide Fund for Nature representative (if a funding agency);
- a representative from other major funding agencies if appropriate;
- a representative from captive-breeding institution if appropriate;
- a business representative (if a funding agency or significant land manager);
- a local government representative (if a funding agency or significant land manager);
- other government agency representatives (e.g., forestry, mining), if such bodies are significant land managers or funding agencies; and
- community representatives if appropriate.

FUTURE DEVELOPMENT OF RECOVERY PLANNING

It is important to recognize the constraints that will affect of the future development and refinement of the recovery plan process under the auspices of the Endangered Species Program.

The most important and fundamental of these is that recovery plans are the legislative means of conserving threatened species and ecological communities and that the basic elements of these plans are defined in the Commonwealth's *Endangered Species Protection Act 1992*. As such, recovery plans are likely to remain as the primary delivery mechanism for the Commonwealth's Endangered Species Program. Other constraints such as those imposed by the Commonwealth's budgetary, financial and program management, and audit requirements have already been described.

There is, however, a substantial flexibility within this basic framework. The recovery planning process has been, and should continue to be, a practice-based learning experience that draws on the collective wisdom and knowledge of all its participants. The *Endangered Species Protection Act 1992* is no more prescriptive in its definition of a recovery plan than has been described in this paper. Within the Endangered Species Program recovery plan guidelines there exists considerable scope for the program practitioners and managers to expand on and define mechanisms to enhance their efficiency and effectiveness. As such, these guidelines are to remain a working set that evolves to meet changing circumstances, be these changes administrative or scientific in nature.

The experience of the Endangered Species Program managers so far has raised some important questions that must be considered.

For example, how specific should recovery plan objectives be? General objectives require less revision throughout the life of the plan and on one level are likely to remain valid. However, it is more difficult to critically assess whether general objectives have been met. Specific and detailed objectives are easier to assess but are far more likely to require amendment during the life of the plan. There are risks to the credibility of the process associated with continuous or frequent amendment to objectives. The perception could be that the "goal posts" are being moved and the objectives are being tinkered with in order to make the plan appear more successful.

Another approach is to meld the general with the specific. In this approach a more general objective is set for the species which will necessarily extend beyond the three to five year life of the current plan, while specific goals are defined for that shorter period. This recognizes that most recoveries of threatened species will inevitably involve the sequential implementation of successive shorter plans and does not create a false expectation that the recovery of these species can be achieved by anything less than strategic long-term management.

In order to assess their objectives all plans must have robust and measurable criteria. To achieve this, such criteria must be carefully crafted, readily assessable and linked to the specific as well as the general objectives.

Practice has shown that the actions required in a recovery plan will almost certainly undergo some degree of amendment during the three to five year life of a plan. The greater the actions are detailed in the plan the greater the likelihood that they will need adjustment. This can be administratively tedious and will not necessarily add to the effectiveness of the recovery. Conversely, if the actions are not well detailed, the focus of the plan can be lost and the plan can stray from its primary tasks. This issue of specificity versus plasticity must be balanced in each plan to achieve the best practical outcomes for the species without the administrative overheads stifling responsiveness.

Although it is not explicitly stated in the legislation, it is clearly the intent of the provisions of the *Endangered Species Protection Act 1992* that there be strong and visible linkages between all of the fundamental components of recovery plans. Without these connections between objectives, criteria, actions and performance evaluation, the discipline and rigour of plans is lost. The program administrators have identified a lack of such linkages in varying degrees in many plans, and see it as a most serious issue affecting performance. This is an area which needs greater attention from those preparing and revising plans in the future.

There are many different mechanisms and philosophical approaches to performance evaluation. The most fundamental are that the processes used are transparent, open and consultative. This is an area to which all participants in the process must give greater attention, particularly as the Endangered Species Program approaches a major statutory review in 1998.

The success of a recovery plan is as dependent on the group of individuals that manage and implement it as on its contents. It is a truism that recovery teams, which are *ad hoc* non-statutory groups, can make or break a recovery plan. To be successful recovery teams must consist of willing, co-operative and trusting participants whose primary goal is the achievement of the plan's objectives. Recovery teams must pool their collective wisdom and knowledge to meet a variety of challenges. They should be the forums for the peer review of plans where as far as possible, the participants leave behind them the pursuit of individual or corporate agendas in order to achieve the common goal.

Aside from these considerations, there is great scope and flexibility available to recovery plan managers to develop new mechanisms for recovery teams. The format and ambit of recovery teams is not constrained by legislation, only imagination and innovation.

Finally, throughout its development and implementation the Endangered Species Program has strived to be management oriented and outcome focused with sound performance evaluation. This can be the only approach that will deliver its objectives and ensure that the best opportunities are taken to enable threatened species to survive and flourish and prevent others from becoming threatened.

REFERENCES

Anon., 1989. An Australian National Strategy for the Conservation of Species and Habitats Threatened with Extinction. Canberra, Australia, Endangered Species Advisory Committee, Australian National Parks and Wildlife Service.

Anon., 1992a. Commonwealth Financial Management Handbook. Canberra, Australia, Department of Finance, Australian Government Publishing Service.

Anon., 1992b. *Endangered Species Protection Act 1992*. Canberra, Australia, Australian Government Publishing Service.

Anon., 1992c. Draft National Strategy for Ecologically Sustainable Development. Canberra, Australia, Australian Government Publishing Service.

Anon., 1994a. The Conservation of Threatened Species and Ecological Communities threatened with Extinction: the National Strategy (draft). Canberra, Australia, Australian Nature Conservation Agency.

Anon., 1994b. The Australian Nature Conservation Agency Corporate Plan. Canberra, Australia, Australian Nature Conservation Agency.

Anon., 1994c. Best Practice Guidelines for the Administration of Grants. Canberra, Australia, Australian National Audit Office, Australian Government Publishing Service.

Anon., 1995. Recovery plan 1 guidelines for endangered and vulnerable species and endangered ecological communities. Canberra, Australia, Endangered Species Unit, Australian Nature Conservation Agency.

Brown, P. B. and Wilson, R. I., 1984. Orange-bellied Parrot Recovery Plan: Management recommendations to the governments of Tasmania, Victoria and South Australia. Hobart, Australia, National Parks and Wildlife Service of Tasmania.

Hawke, R. J. L., 1989. Our Country, Our Future: Statement on the Environment by the Prime Minister. Canberra, Australia, Australian Government Publishing Service.

ADDITIONAL READING

Anon., 1994. *The Endangered Species Protection Act 1992*: An Outline. Canberra, Australia, Australian Nature Conservation Agency.

Anon., 1996. ESAC priorities for the conservation action. Canberra, Australia, Endangered Species Advisory Committee, Australian Nature Conservation Agency.

Cogger, H. G., Cameron, E. E., Sadlier, R. A. and Eggler, P., 1993. The Action Plan for Australian Reptiles. Canberra, Australia, Australian Nature Conservation Agency.

Garnett, S., 1992. The Action Plan for Australian Birds. Canberra, Australia, Australian National Parks and Wildlife Service.

Lee, A. K., 1995. The Action Plan for Australian Rodents. Canberra, Australia, Australian Nature Conservation Agency.

Male, B., 1993. Action, Recovery and Research Plans: what are they? Pp. 6–7 *in* On the Brink! Volume 5. Canberra, Australia, Endangered Species Unit, Australian Nature Conservation Agency.

Male, B., 1996. Listing Process: Threatened Species Lists. Pp. 12–13 *in* On the Brink! Volume 8. Canberra, Australia, Endangered Species Unit, Australian Nature Conservation Agency.

Maxwell, S., Burbidge, A. A. and Morris, K. D. (Eds, in press). The 1996 Action Plan for Australian Marsupials and Monotremes. Canberra, Australia, Australian Nature Conservation Agency.

Richards, G. and Hall, L., (in prep). The Action Plan for Australian Bats. Canberra, Australia, Australian Nature Conservation Agency.

Tyler, M. (in prep). The Action Plan for Australian Amphibians. Canberra, Australia, Australian Nature Conservation Agency.

Wager, R. and Jackson, P., 1993. The Action Plan for Australian Freshwater Fishes. Canberra, Australia, Australian Nature Conservation Agency.

Commonwealth and State Government frameworks for recovery planning for threatened species and communities

David Papps[1]

The Commonwealth and State Governments of Australia share a commitment to the conservation of threatened species. There are differences in legislative, policy and practical approaches to recovery planning. Some of these differences reflect the different stages in the development and implementation of legislation or associated programs in States and the Commonwealth.

All States have had involvement in the preparation and implementation of recovery plans, often in co-operation with the Commonwealth's Endangered Species Program.

The framework for recovery planning in each jurisdiction is noted and problems identified.

Restoring threatened species and ecological communities to a secure status in the wild offers one of the greatest challenges in biodiversity conservation. Co-operative endeavour between the community and conservation agencies and across political boundaries, built on rigorous, critical and continuing review of policy and programs, is essential to success.

Key words: Recovery plans.

INTRODUCTION

THE State (including Territories) and Commonwealth Governments share a commitment to the conservation of threatened species. All have endorsed *Conservation of Species and Ecological Communities Threatened with Extinction: the National Strategy*.

This strategy was developed jointly by the Commonwealth and States and Territories through the relevant Ministerial Council (the Australian and New Zealand Environment and Conservation Council) from an earlier version prepared by the Commonwealth's Endangered Species Advisory Committee. Other relevant Ministerial Councils and key industry and conservation organizations were consulted.

In endorsing the strategy, governments committed themselves to "... the implementation of the strategy's objectives within the budgetary priorities and financial constraints of individual jurisdictions."

The strategy identifies 12 inter-related objectives. One of these is to:

> "... prepare and implement recovery plans for all endangered species and ecological communities, and where necessary, for vulnerable species."

This paper deals only with recovery planning. However, this focus does not imply a diminution of the importance of other objectives and programs necessary for the conservation of threatened species.

All Australian States and the Commonwealth have some form of legislation for the protection of threatened species. In those jurisdictions where recovery planning is not mandatory, it is still utilized as a conservation planning tool.

In this paper I provide brief summaries of the legislative basis for endangered species recovery planning in the States (in this paper a reference to the States also includes the Territories) and the Commonwealth; describe the programs and strategies in place for recovery planning in each jurisdiction; identify problems and describe how these problems have been addressed (if at all). This information is based solely on advice provided by the relevant government conservation agency in each jurisdiction. It is not an independent review.

TERMINOLOGY

This paper adopts the terminology used by the Australian Nature Conservation Agency except where otherwise noted (ANCA 1995).

A *recovery plan* is a comprehensive plan that details, schedules and costs all actions assessed as necessary to support the recovery of a species or ecological community.

A *research plan* includes research and survey actions which need to be carried out to provide adequate information for the preparation of a recovery plan.

[1]NSW National Parks and Wildlife Service, P.O. Box 1967, Hurstville, New South Wales, Australia 2220.
Pages 28–36 *in* BACK FROM THE BRINK: REFINING THE THREATENED SPECIES RECOVERY PROCESS ed by S. Stephens and S. Maxwell. Surrey Beatty & Sons, 1996.

A *recovery team* is a group of people with relevant expertise and responsibility charged with assisting the lead agency/agencies in writing, implementing and monitoring a recovery plan or where appropriate, a research plan.

In this paper "threatened" is taken to mean both endangered and vulnerable conservation categories. The precise definition of both terms may vary from state to state. Generally speaking, an endangered species is one in danger of extinction and whose survival is unlikely if the causal factors continue operating. Vulnerable species are those believed likely to move into the endangered category in the near future if causal factors continue operating (ESAC 1992).

Ecological communities are generally defined as integrated assemblages of species (ESAC 1992). Again, there are jurisdictional differences in definitions. For example, New South Wales' *Threatened Species Conservation Act 1995* defines an ecological community as "an assemblage of species occupying a particular area." The broadening of protection from species to ecological communities reflects the adoption of the concept of biodiversity conservation which embraces genetic diversity within a species through to the diversity across ecosystems. In arguing this case the Commonwealth's Endangered Species Advisory Committee (ESAC) said:

> "The best hope for conservation of the many species that have so far been undescribed or poorly studied lies in the conservation of the widest possible range of natural ecological communities, particularly endangered and vulnerable communities." (ESAC 1992).

LEGISLATIVE BASIS FOR RECOVERY PLANNING

Australian Capital Territory

In October 1994 the Australian Capital Territory Legislative Assembly amended the *Nature Conservation Act 1980* to provide for the identification and protection of native species threatened with extinction.

The amending Act established the Flora and Fauna Committee — an expert scientific body with responsibility to advise the Minister for the Environment, Land and Planning of native species and ecological communities that are threatened with extinction and processes that threaten or may threaten the survival, abundance or evolution of native flora and fauna, with a recommendation that they be declared accordingly.

The Flora and Fauna Committee assesses the conservation status of species and communities, and the ecological significance of processes against specified criteria on nature conservation grounds only and in a regional context.

The Minister's response to a Committee recommendation for declaration is discretionary. If the Minister accepts the Committee's recommendation and makes a declaration, the listing process is complete. A species may be declared vulnerable or endangered; an ecological community may be declared endangered; a process may be declared a threatening process.

The Conservator of Flora and Fauna is required to respond to each declaration with an action plan. In the case of a species declared as endangered, the Conservator is also required to declare the species as having special protection status — the highest level of statutory protection provided by the Nature Conservation Act.

Action plans are a management response to a declaration and are equivalent to a recovery plan in other jurisdictions. They have no statutory force in themselves. However, they can guide decisions by the Conservator in relation to other statutory responsibilities for nature conservation or propose statutory action or policy directly.

The Act specifies that action plans are to be prepared as "soon as practicable" for each declaration and "shall include proposals to ensure, as far as is practicable, the identification, protection and survival" of the declared species or ecological community, or "proposals to minimize the effect of any process" that is declared a threatening process.

There is no mandatory timeline or priority for action plan production, or requirement to address economic or social matters specifically. In practice, the Conservator's priority for action plan production is guided by an informal priority determined by the Flora and Fauna Committee as part of its assessment process.

At a policy level, guidelines for action plan content have been developed. Action plans are to include an outline of the information known about a species or ecological community, and to discuss the important conservation issues along with the current and proposed management actions which are necessary to enhance the conservation of the species or community, or management of the threatening process. Action plans are also to provide details about the conservation status of a species together with references for further information. The Flora and Fauna Committee are consulted routinely during action plan production.

Draft action plans are available for public comment for a minimum period of 21 days.

The Conservator is to consider any public comments lodged, and after consideration and effecting any appropriate revisions, prepare the action plan. Again, the action plan is available for public inspection for a minimum period of 21 days.

The action plan is then tabled in the Australian Capital Territory Legislative Assembly, as a disallowable instrument, meaning that it can be rejected within a period of 15 sitting days from the date of tabling.

The Act also enables the Conservator to prepare a draft variation of an action plan in the same manner as if it were a draft action plan.

South Australia

The legislative basis for threatened species conservation is the *National Parks and Wildlife Act 1972* which has schedules of endangered, vulnerable and rare species of vascular plants and vertebrates, except fish and frogs.

The *Native Vegetation Act 1991* and the *Pastoral Land Management and Conservation Act 1989* also contain objectives relating to the conservation of habitats of threatened species. Much of about 500 000 hectares conserved under South Australia's Heritage Agreements scheme was set aside because of its value as habitat of threatened species or communities.

The South Australian Government is considering a "Draft Threatened Species Strategy" which focuses on the management of threatening processes rather than species or communities. Recovery planning is not a formally recognized statutory process, but is used as a management tool.

Tasmania

The *Threatened Species Protection Act 1995* provides for the listing and protection of endangered, vulnerable and rare species of native Tasmanian flora and fauna (including invertebrates) in three schedules. Ecological communities are not included.

The legislation provides for the Director, with the Minister's approval, to make a recovery plan for any listed taxon of flora and fauna. There is no time frame for the making of a recovery plan after it is listed and it is not mandatory that a recovery plan is made for every listed species.

A recovery plan must specify:

1. the listed taxon to which it applies;
2. the objectives for the conservation and management of that taxon, including:
   a. the ways in which those objectives are to be achieved; and
   b. the resources required to achieve those objectives.
3. the date within five years when the recovery plan is to be reviewed.

In the making of a recovery plan, the Director must consider social and economic issues as well as matters relating to nature conservation.

When a draft recovery plan is made, copies must be provided to two committees which are established by the legislation, these being the Scientific Advisory Committee and the Community Review Committee. This last committee has direct access to the Minister and may address matters on behalf of interested landholders.

Draft recovery plans are subject to public comment. Their making must be advertised in the *Government Gazette* and in newspapers circulating in the relevant area of Tasmania. Public notification must be given within 30 days of publication of a draft recovery plan and must allow for a 30-day comment period. After consideration of any public comments, the Director may, with the Minister's approval, make a final recovery plan.

The same procedures apply when revoking and amending a recovery plan.

Similar conditions and arrangements apply to the making of a threat abatement plan. This also requires Ministerial approval.

Flowing on from recovery plans, the Director may also make a land management plan which applies to specific areas of land, be it a single property or a group of properties where a threatened species occurs. This would usually include an agreed strategy of actions for the management of an area of land for the period of the plan (usually five years) to achieve the required conservation objectives. This would usually lead to an agreement between the Director and the land holder to provide for the carrying out of works and may include payment for those works. The Director may cancel an agreement if it is no longer required.

Victoria

Victoria's *Flora and Fauna Guarantee Act 1988* protects threatened taxa and communities and seeks to manage threatening processes.

The Act provides mechanisms for a Scientific Advisory Committee (SAC) to evaluate all native species and list threatened taxa of flora and fauna, ecological communities and potentially threatening processes. A consequence of listing is the requirement for the Department of Natural Resources and Environment (DNRE) to produce an action statement "as soon as possible". These document high priority actions required to conserve and restore the item (and therefore clearly meet the definition of a recovery plan) or what actions are necessary to ameliorate the adverse effects of threatening processes.

The listing process (including the Minister who accepts or rejects recommendations of the SAC) is based only on conservation aspects whilst the action statements must consider the socio-economic effects of the actions. There is no public submission requirement under the Act although any parties expressing interest when the listings are advertised (for public comment) are given the opportunity to comment. The SAC and the Catchment and Land Protection Council are required to comment on the action statements prior to approval and publication by the DNRE.

Action statements are policy documents of the Department of Natural Resources and Environment and although there is no specific legal requirement to comply, other sections of the Act apply to action statements. These are that the Crown is bound; public authorities must have due regard to the objectives of the Act; and the Secretary (of DNRE) may require public authorities to undertake or refrain from actions that threaten species or communities. In Administrative Appeals Tribunal hearings, action statements have been taken as persuasive evidence.

A detailed review of the action statement process is provided by Mansergh *et al.* (1995). Action statements should be seen in the broader context of the Flora and Fauna Guarantee Act which is biodiversity conservation legislation that extends beyond "threatened species"; for example, the requirements for a Strategy, public authority Management Agreements, Protected Flora Controls and the like.

Western Australia

The *Wildlife Conservation Act 1950*, which is administered by the Department of Conservation and Land Management (CALM), provides for conservation of threatened species of flora and fauna. Threatened fauna and flora are declared via Ministerial notices in the *Government Gazette*. The Minister receives advice on changes to the lists from the Threatened Species Scientific Committee, which considers nominations for changes as well as reviewing existing lists.

At least every two years, all declared threatened species are ranked by the Threatened Species Scientific Committee for conservation action pursuant to CALM Policy Statement No. 50 *Setting priorities for the conservation of Western Australia's threatened flora and fauna*. This process results in taxa being allocated to the IUCN 1994 Red List categories of "Critically Endangered"; "Endangered"; and "Vulnerable". The Policy Statement provides that all taxa identified as critically endangered are conserved, through the preparation and implementation of interim recovery plans and that taxa identified as endangered and vulnerable are allocated research and management resources in priority order.

It also provides that conservation action for taxa identified as critically endangered commences as soon as possible and always within one year of endorsement of that rank by the Minister.

An important strategy that CALM is using to assist the listing, prioritizing and conservation management of vascular plants, is the development of Regional and District Threatened Flora Management Programs. These review the status of declared threatened flora and of priority taxa (taxa that are insufficiently known or which require monitoring) in a CALM administrative region or district, develop local priorities and outline management needs.

There is currently no equivalent legislation or formal policy covering threatened ecological communities. However, an informal, non-statutory process, including advice from a scientific advisory committee; the establishment of threatened ecological communities database; and steps for assigning ecological communities to categories of threat, is now in place.

CALM Policy Statement No. 44 *Wildlife Management Programs* lays down procedures for the development and implementation of recovery plans and interim recovery plans. The Policy also covers the setting up and operation of recovery teams. Recovery teams report annually to CALM's Corporate Executive and to funding agencies.

Northern Territory

The primary legislative instrument for the protection and management of wildlife (plants and animals) is the *Territory Parks and Wildlife Conservation Act 1980*. Currently all vertebrate animals are declared as protected, but there is no specific provision for threatened or endangered species. Instead there is a category of "specially protected" animals and plants to cover species that may be unusually vulnerable, either locally or more broadly, to habitat loss or harvesting.

Classification as specially protected imposes no additional obligations on government to actively manage species to improve their conservation status. Enhanced protection is achieved by constraining the range of actions that landholders may take that have the potential to affect populations directly, or through effects on habitat.

In the absence of legislative mandate for endangered species recovery programs, reliance is placed on the Act's provisions relating to the implementation of management programs for individual species. These programs may incorporate any steps necessary to secure the long-term conservation status of the species, and require the approval of the Administrator of the Northern Territory. Approval at this level provides the security of government support necessary to meet effectively the long-term obligations inherent in endangered species management.

Queensland

The *Nature Conservation Act 1992* contains schedules of threatened wildlife and provides for the preparation of recovery plans. Recovery plans may also be plans prepared or adopted by the Commonwealth under its own legislation.

New South Wales

The *Threatened Species Conservation Act 1995* covers all native animals and plants (except for fish and marine plants which will be protected under other statutes). Species and ecological communities can be listed by an independent Scientific Committee as threatened (i.e., endangered or vulnerable) and populations of a species which is not itself listed as endangered may be listed as endangered if that population meets specific criteria. There is no Ministerial discretion over this process.

The Director-General of the National Parks and Wildlife Service (NPWS) is obliged by the Act to prepare a recovery plan for each endangered species, population and ecological community and each vulnerable species. Recovery plans must be prepared as soon as practicable after listing but at least within the same time limits as those set out in the Commonwealth's *Endangered Species Protection Act 1992*.

The general purpose of recovery plans is "to promote the recovery of the species, population or ecological community to a position of viability in nature."

The Act provides guidelines for preparing recovery plans to which the Director-General must have regard. These are the objectives of the Act amongst others, (this includes the prevention of extinction and promotion of the recovery of threatened species, populations and ecological communities); the minimizing of any significant adverse social and economic consequences; and the most efficient and effective use of available resources. The Director-General is also obliged to consider (and implicitly include in the plan) any measures by which the public can co-operate in the conservation task.

The legislation also obliges the Director-General (after consultation with the independent Scientific Committee) to determine priorities for the preparation of recovery plans. Highest priority must be accorded to nationally endangered species, populations and ecological communities (the New South Wales legislation bases its eligibility for listing on the conservation status in New South Wales alone and so may include species and communities not threatened on an Australia-wide basis). Priority after this is determined on four criteria — likelihood of extinction; likelihood of recovery; and whether the species in question is a keystone (i.e., many other species dependent for survival on it) or indicator (i.e., where the relative abundance of the species is a measure of the overall health of its ecosystem) species.

The contents of each recovery plan are prescribed in some detail by the Act. This ranges from the obvious (identifying the species, population or community to which it applies; identifying critical habitat and threatening processes) to the difficult (identifying methods by which adverse social and economic consequences of the making of the plan can be minimized). A plan also must spell out the actions necessary to ensure recovery and to protect any critical habitat. It must document criteria against which progress can be assessed. Each recovery action must be assigned to a person(s) or public authority(ies) who is responsible for its implementation. Each plan must also include a date by which it would be subject to review by the Director-General.

Plans undergo a broad process of consultation before adoption. A draft recovery plan is prepared as a basis for this consultation. During its preparation, the relevant Ministers must consult over any action likely to be included for which a public authority is responsible. Such an action can not be included in a recovery plan without the approval of the relevant Minister.

The draft plan is publicly exhibited and made available to a number of key groups including the Scientific Committee and relevant public authorities.

The Director-General must consider all written submissions and may amend the draft plan accordingly. The draft is then forwarded to the Minister for the Environment along with the public submissions and summaries of other advice. The Minister only has three options: approve the plan with or without amendment; or refuse to approve the plan (on the basis of its likely social or economic consequences); or refer it back to the Director-General for further consideration. However, if a plan is approved in a form different to that of the draft, the plan must state the reasons for any amendments.

Approved plans are published and each recovery plan is reviewed by its identified date, if not before.

This process is similar to the long-standing process for preparation and adoption of national park plans of management in New South Wales.

The Act also contains provisions governing the implementation of recovery plans. The principal of these obliges Ministers and public authorities (including the NPWS) to take "any appropriate action available to them to implement those measures included in a recovery plan for which they are responsible and must not make decisions that are inconsistent with the provisions of a recovery plan". However, this obligation is fettered. The Act specifically allows that a recovery plan does not affect the statutory discretion of a Minister or public authority. In these circumstances, the Minister or authority need only take the recovery plan into account.

The Act does ensure that such departures from a recovery plan are both transparent and subject to negotiation. If a public authority proposes to exercise a function in a manner that is inconsistent with a recovery plan (and obviously can contemplate this because it is a statutory discretion) it must give written notice to the Director-General (in the unlikely event that the public authority is the National Parks and Wildlife Service, notification is to the Minister for the Environment).

Such mandatory notification leads to an assessment by the Director-General as to whether exercise of the function in the manner proposed is likely to jeopardize the effective implementation of the plan. A "jeopardy opinion" leads to further mandatory consultation between the NPWS and the public authority in an attempt to resolve the matter (by modification of the proposed action or other mutually acceptable means).

Failure to achieve resolution at this level then involves referral to relevant Ministers. If the Ministers also cannot agree, the Premier must make a decision which is binding on all parties.

Public authorities must report annually to Parliament on their progress in implementing measures in relevant recovery plans.

Commonwealth

The *Endangered Species Protection Act 1992* provides the statutory basis for recovery plans for species in Commonwealth areas (i.e., land and water owned or managed by the Commonwealth).

The Commonwealth is required to prepare and implement recovery plans for species and communities found in Commonwealth areas. Perhaps more importantly, the Commonwealth seeks the co-operation of the States in which those species or communities occur outside of a Commonwealth area, with view to joint preparation and implementation of a recovery plan. The Endangered Species Protection Act also provides for adoption by the Commonwealth of State plans, and for assistance to the States for the preparation and implementation of recovery plans.

CURRENT RECOVERY PROGRAMS AND STRATEGIES

Australian Capital Territory

The Australian Capital Territory has not yet prepared any action plans but it is implementing part of the Commonwealth's recovery plan for lowland native grasslands (a threatened ecological community).

Research and survey programs have been established for several species and ecological communities, which are widely acknowledged as being under threat.

South Australia

The South Australian Department of Environment and Natural Resources (DENR) has contributed to recovery outlines for national action plans and has prepared and implemented research and/or recovery plans for the most relevant taxa. A threatened plant population database has been established.

Tasmania

Tasmania's Parks and Wildlife Service has produced booklets on threatened plants, vertebrates and invertebrates in Tasmania and has established supporting committees to advise on each of these groups. The Service is also writing and implementing recovery plans for several threatened species.

A "Directory of Conservation Priorities", and as required by the new Act, a State-wide Threatened Species Strategy are being prepared.

Victoria

Since 1991 over 300 items have been listed under the Flora and Fauna Guarantee Act. Seventy action statements (covering 84 items) have been published with a further 30 in advanced draft form (as at June 1996). A systematic monitoring program is in place to evaluate progress (e.g., change in populations' size and numbers of populations known). Other aspects of the Act are also being implemented in co-ordination with many other programs; for example, Land for Wildlife and the ANCA Endangered Species and Invasive Species Programs. The statutory Draft Flora and Fauna Guarantee Strategy has been published for public comment and the final will be published soon. Victoria's ten new Catchment and Land Protection Boards have biodiversity objectives linked to and complementing the Flora and Fauna Guarantee Act.

Western Australia

CALM established the Western Australian Threatened Species and Communities Unit in 1992. Its role is to co-ordinate, assist with, and promote the conservation of threatened species and communities. Its recovery process involves the setting up of recovery teams and the preparation of costed recovery plans. There are currently 23 recovery teams and more are being set up.

The recovery of threatened flora in the diverse south-west of the state is built around District and Regional Flora Recovery Teams and Plans, not around the single species approach. This work is the subject of another paper in this volume.

CALM also concentrates much of its research and management effort on combating threatening processes, for example, the control of foxes.

Queensland

Queensland's Department of Environment (DoE) has been involved in the preparation of recovery plans for fauna since 1991. Most of the 18 prepared to date have been funded through the Commonwealth's Endangered Species Program. A number of plans have been prepared for plants, however there has been little implementation of these yet.

New South Wales

Notwithstanding the very recent nature of statutory recovery plans, New South Wales has had a program for the preparation of recovery plans (earlier versions were referred to as species management plans) for at least a decade. Many have been funded through the Commonwealth's Endangered Species Program. Few have been fully implemented and none have yet been through the statutory process of adoption.

Commonwealth

To date the Commonwealth has not approved any recovery plans of its own.

PROBLEMS AND SOLUTIONS

Australian Capital Territory

The Australian Capital Territory has very limited experience but notes the following problems:

- differing priorities for dealing with threatened species across political boundaries;
- lack of regional data on threatened species and communities; and
- potential conflicts between economic activity and the conservation of threatened species and communities.

The Australian Capital Territory is active in participating with other relevant jurisdictions (especially New South Wales and the Commonwealth) and is attempting to better integrate threatened species recovery into land use planning.

South Australia

South Australia's DENR identifies two major problems:

- funding gap between preparation and implementation of recovery plans; and
- occasional lack of communication/co-ordination between State agencies over threatened species which cross jurisdictional boundaries.

DENR suggests there are no simple solutions to problems of this nature.

Tasmania

The following problems are evident in Tasmania:

- conserving habitats of threatened species on private land (and this is, in part, a lack of funding to support voluntary conservation agreements); and
- a lack of staff resources and expertise to manage recovery programs.

Victoria

The DNRE identifies five major problems:

- the publication of action statements can create unrealistic expectations in the community (i.e., the problem is now solved);
- inadequate resourcing;
- strong focus on charismatic vertebrates, rather than effort on dealing with threatening processes;
- public also do not understand (and so do not support) conservation effort on populations and communities; and
- listing process is drawn out.

DNRE is responding by:

- adoption of a State-wide strategy;
- producing more action statements on ecological communities and threatening processes;
- encouraging public authorities to build threatened species conservation into their policies, programs and decision-making;
- building real and measurable performance indicators into all recovery programs; and
- abandoning historically mis-directed programs (e.g., DNRE's involvement in captive breeding).

Western Australia

CALM notes two major problems:

- lack of information; and
- lack of resources.

Western Australia has about 45% of all Australia's threatened taxa and the vast majority do not occur in other States. Conversely, while Western Australia covers 33% of Australia, it has only about 10% of the population (with the corresponding levels of public funding).

In response CALM claims the largest and most focused threatened species scientific research group in Australia. Financial problems are more intractable. CALM's threatened species programs now depend to a significant extent on money from the Commonwealth's Endangered Species and Invasive Species Programs. There are also local fund-raising initiatives.

CALM also notes it is taking a proactive (although not legislatively-based) approach to the conservation of threatened ecological communities.

Queensland

The DoE notes that the recovery planning process is relatively recent. Problems identified to date are:

- existing staff and resources are spread thinly to cover the number of species;
- there is a lack of good information on threatening processes; and
- plans have predominantly focused on defining the problems rather than implementing solutions.

DoE is proposing to establish a threatened species unit to better co-ordinate and drive the recovery program.

New South Wales

The National Parks and Wildlife Service (NPWS) notes the following problems:

- recovery plan preparation often reflects available in-house expertise/interest rather than conservation priorities;
- do not know if recovery plans are working (most have yet to be implemented);
- inadequate funding for preparation and implementation;
- lack of interest/support from other public agencies and private landowners;
- differing expectations for recovery plans, with some interest groups regarding them simply as instruments to regulate the activities of government departments (e.g., the State's forestry agency) or individuals;
- ineffective recovery teams (high conflict, poor focus);
- poor information base;
- mistaken belief that recovery of threatened species is simply a scientific problem (rather than having organizational, policy, social, economic, and political components);
- too much focus on charismatic mega-vertebrates; and
- not enough effort mitigating threatening processes or managing/planning across a bioregion.

The State Government has allocated about $4 million in its latest budget to fund the implementation of the Act. Further specific funding for recovery planning may be considered.

The NPWS has recently restructured to better focus resources on priority programs, and specialist threatened species groups have been established in head office and key regional centres.

Threatened species conservation will be better integrated into the Service's other programs, for example, its pest and weed control programs will focus far more on projects which will contribute to threatened species conservation.

The NPWS is also increasing its efforts in community-based conservation work, including an in-principle commitment to establishing a program similar to Victoria's Land for Wildlife to complement its own voluntary Conservation Agreement scheme. This type of program is an essential mechanism for private sector involvement in nature conservation, especially threatened species recovery.

Commonwealth

The Commonwealth advises of no particular problems in its own program but its experience with the States has reinforced the importance of forming a properly representative recovery team early in the project.

DISCUSSION

Adoption of sensible principles in the nationally endorsed strategy and the enactment of comprehensive threatened species legislation within each jurisdiction are the first steps in building a framework for threatened species recovery. Unless such principles and laws are given effect to in a way that delivers meaningful outcomes (i.e., species are saved from extinction) they remain no more than statements of good intentions.

There is a range of legislative, policy and practical approaches to recovery planning across the States and the Commonwealth. Some of these differences simply reflect different stages in the development of essentially the same process. It appears that many of these differences are at the level of detail rather than being fundamental in nature.

While this paper simply reports agency views on recovery planning, it is possible to make a number of general observations.

Most agency responses lacked detailed analysis of the problems encountered in the recovery planning process. There could be a number of explanations for this. However, if it reflects inadequate resources for agencies to undertake critical self-review then this could be seen as a serious problem. Experience suggests that this type of feedback loop is essential, both in assessing the effectiveness of individual plans and the performance of the agency in delivering the programs.

There are only five "problems" identified by more than one agency:

- lack of information/knowledge;
- lack of funding/resources/skills;
- focus on just species rather than processes/communities;
- lack of public support;
- ineffective recovery teams.

Of these, the first two were most commonly identified (with lack of funding the major problem). Obviously, species recovery cannot occur without funding but sufficient resourcing alone will not guarantee success. Agencies must exercise care in their analysis of poor performance in recovery planning. It is possible that inadequate resourcing masks other program failings.

Comparison of these issues with those identified in recovery planning programs elsewhere may be useful for Australian agencies. Some commentary on both the New Zealand and the American experience appear elsewhere in this volume. There has been considerable review, in particular, of the recovery planning processes undertaken as a result of the United States *Endangered Species Act* (e.g., Clark *et al.* 1994). Clark and others took account of considerable first-hand expert experience in the administration of the American legislation as well as input from a range of relevant professionals to produce a set of "meta-lessons" which they suggest encapsulate the most important guidelines for successful implementation of recovery planning programs. Their views on improving the effectiveness of recovery teams and government agencies involved in species recovery appear particularly pertinent.

Experience to date in New South Wales (and many of these observations also reflect the collective views of agencies previously outlined) suggests some other areas to which agency attention might be increasingly directed.

First, recovery planning for threatened species/communities is a task with scientific, economic, social, political and organizational components. This is a position strongly held by Clark and others who note:

> "Attempting to restore species by ignoring everything but the species' biology invites failure" (Clark *et al.* 1994).

Second, Australia is a federation of States; for recovery planning to work efficiently there must be co-operation and co-ordination across jurisdictions that reflects biological realities but also accepts political realities (e.g., that the great bulk of law and on-ground programs are the province of State governments).

Third, most land in Australia is owned or managed by private individuals. Conservation of threatened species cannot ever be achieved solely by reservation of land in a national park system. Therefore, successful outcomes are dependent on co-operative endeavour between the community and conservation agencies.

Fourth, saving species costs money. Conservation agencies will need to spend more time with their treasury officials and the private sector convincing them of the importance of saving threatened species.

Fifth, conservation of biodiversity, and particularly threatened species, must be integrated into the mainstream of political and economic decision making.

Sixth, recovery plans need to be supported by effective recovery teams. Plans must also be subject to rigorous, critical and continuing review of their effectiveness. Measures of effectiveness must reflect real conservation outcomes. Mistakes must be acknowledged and rectified without punishment. This too reflects one of the "meta-lessons" of Clark *et al.* (1994).

Seventh, more effort should go into threatened species conservation across bioregions and into threat abatement itself. Exclusive focus on single species is not cost-effective.

Finally, legislation governing threatened species conservation should encourage public participation and ensure independent scrutiny of agency decisions.

ACKNOWLEDGEMENTS

The summaries of the State and Commonwealth frameworks for recovery planning were supplied by the relevant government agency. Apart from minor editorial changes I have repeated that information.

Elsewhere the views expressed are mine and are not necessarily those of either the NSW NPWS or the New South Wales Government.

The paper was improved by the comments of Ian Mansergh and an anonymous reviewer.

REFERENCES

Australian Nature Conservation Agency, 1995. Recovery Plan Guidelines for Endangered and Vulnerable Species and Endangered Ecological Communities. ANCA: Canberra.

Clark, T. W., Reading, R. P. and Clarke, A. L., 1994. Synthesis in Endangered Species Recovery. Finding the Lessons, Improving the Process ed by T. W. Clark, R. P. Reading and A. L. Clarke. Island Press: Washington DC.

Endangered Species Advisory Committee, 1992. An Australian National Strategy for the Conservation of Australian Species and Communities Threatened with Extinction. ANPWS: Canberra.

Mansergh, I., Jelinek, A. and Clunie, P., 1995. A review of the Action Statement process under the *Victorian Flora and Fauna Guarantee Act* 1988. *In* People and nature conservation. Perspectives on private land use and endangered species recovery ed by A. Bennett, G. Backhouse and T. W. Clark. Transactions of the Royal Zoological Society of New South Wales. Surrey Beatty & Sons: Chipping Norton.

The New Zealand experience: legislation, priorities and recovery planning

Ian Govey[1] and Janet Owen[1]

The Department of Conservation is responsible for drafting and implementing recovery plans for New Zealand's indigenous vertebrates, invertebrates and plants. Recovery plans are not a statutory requirement in New Zealand. They are intended to provide concise statements of proposed management actions. They cover a defined time span — usually five years. The Department of Conservation has developed a ranking system to identify species in most need of conservation action, which takes account of their status, the threats they face, their vulnerability and distinctiveness. Recovery plans complement this priority ranking system. The purpose of recovery plans is to promote public involvement and support, identify and prioritize actions needed for recovery of the species, and provide the basis for work plans and funding requirements. Recovery groups are established to oversee implementation of the plans. They include people with specific expertise on the species, and the function of the group is to resolve technical issues and make recommendations on the detail of conservation activities. Recovery plans are useful in describing the Department's management intentions, allowing the support of private landowners to be elicited in making land management changes for the benefit of threatened species. The public, and non-governmental organizations are encouraged to participate in the decision-making process. Recovery plans are also valuable in the Department's business planning. Work plans can be developed around the intended outcomes listed in the plan, and the outcomes can be measured. With a published list of objectives, funding can be sought from sponsors to assist with specific funding projects.

INTRODUCTION

THE Department of Conservation (hereafter referred to as the Department) is responsible for drafting and implementing recovery plans for New Zealand's indigenous vertebrates, invertebrates and plants. The Department administers several different statutes which protect:

- all marine mammals;
- all but one indigenous bird;
- all endemic reptiles and amphibians; and
- some terrestrial invertebrates.

Responsibilities have also recently been transferred from another agency, allowing the Department to protect corals and marine fish.

Plants are a noteworthy omission in this list. In New Zealand there is no effective overall legislation to protect plants. Protection of rare plants depends on co-operation from the landowner, or takes place on land which is administered by the Department. Fortunately the Department has some degree of administration or control over approximately one-third of the country's land area, which contains habitats for many rare plants.

Another noteworthy point about New Zealand legislation is that it does not require recovery plans to be compiled. There is also no administrative requirement for recovery plans. In New Zealand recovery plans are produced on a voluntary basis for the advantages which will be discussed.

SETTING PRIORITIES FOR RECOVERY PLANNING

As there are no legislative criteria which identify when, and for which species a recovery plan should be produced, the choice is made by the Department according to an assessment of priorities; which species have the greatest need. In recent years the Department has given a good deal of attention to priority setting in many aspects of its operations.

To assist with threatened species management, it was decided for a number of reasons that some relatively objective criteria were needed to more accurately assess the conservation needs of different species. Among these reasons were:

- The same criteria cannot be used to judge the conservation status and recovery requirements for different groups; for example, invertebrates and plants. Comparisons must be based as objectively as possible, having regard to species behaviour such as mobility, biological factors such as the number of offspring produced, and spatial factors such as the area of habitat required to support a viable population.

[1]Department of Conservation Head Office, P.O. Box 10420, Wellington, New Zealand.

Pages 37–39 *in* BACK FROM THE BRINK: REFINING THE THREATENED SPECIES RECOVERY PROCESS ed by S. Stephens and S. Maxwell. Surrey Beatty & Sons, 1996.

- Because resources are limited, and the processes which threaten species tend to be long-term ones, the most needy species require action first.
- There is a need to continually evaluate priorities in light of new information, and not assume that earlier conclusions still remain valid.
- There is a danger of building on conservation successes, and not coming properly to grips with the harder tasks.

With those points in mind, the Department has designed a ranking system, identifying those threatened species in most need of conservation action. Each species is scored according to 17 criteria, which assess five factors; taxonomic distinctiveness, status, threats, vulnerability and human values. Panels of experts were established to carry out the scoring process for each major taxonomic group to ensure that the most recent information was considered. The document, "Setting Priorities for the Conservation of New Zealand's Threatened Plants and Animals" has been published (Molloy and Davis 1994).

The ranking system is designed to provide a basis for assessing species for conservation action, rather than a strict list of "most endangered". Where a descending order of priority is used, there is a danger of exaggerating the importance of a minor difference in scores. For this reason the scores are treated as indicative, rather than as strict priorities. In order to make this distinction clear, actual scores are not included in the published document, and the species are grouped into three categories which reflect the total scored in the ranking process. The expert panels were consulted in setting the cut-off point between the three categories. The document is now in its second edition, and the Department is confident that the scoring and review by expert panels has provided a robust basis for the priority rankings.

Recovery plans are complementary to the priority ranking system, being one way in which the priorities identified can be actioned. This document is used as the first point in deciding which species are in need of a recovery plan, but there are a host of other logistic, financial, technical and political factors which also need to be considered before a plan can be drafted.

Among these factors are the complexities of the issues faced by managers in each case. The adequacy of the information base in both research and management must be borne in mind when the achievability of the project is considered. The factors which limit populations or threaten the species must be identified with sufficient confidence that logistically possible and technically feasible remedies can be proposed. The issue of cost must have a bearing on the decision, and also whether the work to be undertaken has other benefits that might flow on to other species or the ecological community as a whole.

It is in considering these factors that recovery planning is just as important in the process as it is in the product.

PREPARING RECOVERY PLANS

The Department of Conservation identifies three purposes for its recovery plans:

- To identify and prioritize actions needed for the recovery of the species.
- To promote public involvement and support.
- To provide the basis for work plans and funding requirements.

Once the Department has identified through the priority rankings that a recovery plan is needed for a particular species, that sufficient information is available to identify the problems the species faces, and that remedial work is likely to be achievable, stakeholders are brought together to form a recovery group to develop the plan in detail. The primary purposes of the recovery group are to provide technical advice to the Department, and while the plan is being drafted, to act as a communication channel between Departmental and non-departmental managers, scientists, Departmental staff responsible for liaising with the media and local landowners, and other interested groups.

Once the plan has been drafted, the group has an important role in reviewing the progress of the recovery actions being undertaken, and in ensuring that the priority actions recommended in the plan have been included in the Department's business planning.

Recovery plans are seen as concise statements of management intentions, and have a life-span appropriate to the recovery strategy being recommended. In most cases this is likely to be between five and ten years. Because public involvement is sought in the recovery planning process, the plan must be produced in a format suitable for wide readership. The documents contain enough supporting information to explain the decisions reached, but not a comprehensive detailing of the species' ecology and status. It is also important that a degree of flexibility is written into the plan to cope with changing circumstances, by allowing for the periodic evaluation of outcomes and management activities, and assessment of their successes in contributing to the larger aim of the plan.

EXPERIENCE WITH RECOVERY PLANNING

One of the important constraints on threatened species management is funding. The Department's overall budget for protected species management includes the preparation and implementation of recovery plans, predator control and eradication on offshore islands and mainland areas. The budget remained relatively static in real terms from its establishment in 1987 to the 1994–95 year, at approximately $13.6 million. (Department of Conservation Annual Report 1995). It was estimated that within that figure, 60% was likely to have been spent on threatened species recovery, a total of somewhat less than $8.2 million. This funding was supplemented by a contribution of $620,000 per year in major sponsorship for four threatened species programs.

Beginning in the 1995–96 financial year the budget was increased by $3 million per annum, for a period of three years. In 1996 the government announced a further significant increase in funding. An additional $18 million has been made available, to be spread over three years beginning from the 1996–97 financial year. The new funding has been targeted to threatened species recovery programs, including pest and predator control. The increased funding will allow active management to be undertaken on species currently receiving little attention, and for several new projects to be initiated in the intensive management of forest ecosystems, building on the successful results achieved in trial programs.

These funds were made available following the government's recognition of the worsening problems facing threatened species. The Department's second edition of its publication "Setting Priorities for the Conservation of New Zealand's Threatened Plants and Animals" (Molloy and Davis 1994) played a valuable role in outlining the scale of those problems. The new edition contained an additional 113 species in the threatened category, including 34 species of liverworts and mosses, a taxonomic group that was ranked for the first time.

The New Zealand experience with the development of recovery plans has identified several advantages:

- **Established goals and priorities**

  The essence of an effective recovery plan is in its setting of a series of clear goals and priorities, and a plan of action for achieving them. While some notable successes were achieved in earlier days with a more *ad hoc* approach, modern concepts of accountability and effectiveness in the public sector require closer monitoring of resources. While it is important that goals remain flexible, clear priorities assist managers, field workers and stakeholders.

- **Work planning and budgeting**

  Business planning is carried out by the Department on an annual basis. Work plans can be developed with confidence around the intended outcomes set out in the plan, avoiding any requirement for sudden additional resourcing, to the detriment of other planned work.

- **Performance monitoring**

  Performance can be measured against the goals that have been set. Continuous evaluation is a very important part of the recovery process, given that programs are, of necessity, undertaken with imperfect information. As experience is gained in each program, goals may need to be reviewed, but the evaluation process gives managers confidence that funding and work programs are being applied effectively.

- **Agreement between stakeholders**

  The process of establishing a recovery group and its contributing to the development of a recovery plan provides opportunities for interested individuals to contribute to the recovery of threatened species. One of the keys to successful species recovery planning is consultation. Without effective communication between the Department and other stakeholders, it is unlikely that conservation benefits will be sustained in the longer term. Where private landowners need to be involved, recovery plans have a vital function as an advocacy tool.

  Where these stakeholders are involved, the process also promotes agreement which allows for co-ordinated action on agreed objectives and goals.

- **Funding and sponsorship opportunities**

  With work plans and expected expenditure established, sponsors can be sought to assist with funding specific projects. Work planning offers advantages for the Department, by allowing a sponsor's involvement to be predicted and applied to specific areas. Management on a more *ad hoc* basis carries the danger of sponsors possibly pulling out at a crucial stage, leaving a funding deficit.

  Sponsors also gain confidence in contributing to actions published in a recovery plan, as they have the assurance of knowing that the Department has clearly defined its objectives and work plans, and has stakeholders committed to the success of the project. Contributions are not sought on the basis of vague plans or unproven ideas.

CONCLUSIONS

Recovery plans are a relatively new development in New Zealand's efforts to protect and preserve threatened species. They have proven their usefulness as a planning tool, when used together with initiatives to prioritize aspects of the Department of Conservation's work. The challenge for the future will be to improve on the effectiveness of both planning and management techniques, in the knowledge that threats to indigenous biodiversity will intensify.

REFERENCES

Department of Conservation Annual Report for the year ended 30 June 1995. Department of Conservation: Wellington.

Molloy, J. and Davis, A., 1994. Setting Priorities for the Conservation of New Zealand's Threatened Plants and Animals (second edition). Department of Conservation: Wellington.

Determining priorities for plant conservation: an economic necessity

Katrina Jensz[1]

A national system to prioritize threatened plant species for conservation action is necessary to assess those species that are critically endangered and need urgent attention, to determine whether conservation action for those species is feasible, to ensure that maximum benefit is obtained from the limited resources available, and to enable agencies to work co-operatively on an agreed list of priorities.

The conventional approach for determining priorities for conservation action has been to only assess the degree to which species are threatened, with priority given to the most critically endangered species. Such systems have not been helpful when applied to assess funding priorities for plant conservation given the large number of threatened plants and the inadequate fiscal resources available.

This paper describes a system which prioritizes threatened plant species for conservation action by considering their *potential for recovery* as well as their state of endangerment. The system is currently being used to assess all species listed as nationally endangered. The priority list will be used for future funding decisions under the Endangered Species Program.

INTRODUCTION

ONE of the central problems in threatened species conservation is deciding how to allocate limited resources to saving the many species that are threatened with extinction. Ideally we would like to protect all threatened species but, as this is impossible at the present level of funding, we need to develop ranking systems (Given and Norton 1993).

The conventional approach for determining priorities for conservation action has been to only assess the degree to which species are threatened, with priority given to the most critically endangered species. This approach has been widely applied within Australia, and is typified by the stated policy of the Western Australian Department of Conservation and Land Management (WA CALM) that conservation resources should be allocated on a priority basis to the most threatened taxa of plants and animals (Burbidge and Brown 1993).

Mace and Lande (1991) identified the need to clearly separate the process of assessing threat from the process of setting priorities for conservation action. They believe a system for setting priorities should recognize the likelihood of extinction, but embrace numerous other factors. Embracing these views, systems such as WA CALM's *Criteria for Determining Priorities for Recovery Action* (Burbidge and Brown 1993) have been developed in Australia for ranking species already listed under threat categories. Such systems, however, continue to place priority largely on the *degree of threat,* and have not been helpful when applied to assess funding priorities against the scenario of severely inadequate fiscal resources for plant conservation.

Despite the number of threatened plants vastly exceeding the number of threatened vertebrates, funding continues to be applied disproportionately to these groups in favour of vertebrates. By way of example, there are over 1 000 species of vascular plants recognized as nationally threatened in Australia (endangered or vulnerable). In contrast, there are 143 species of vertebrates listed as nationally threatened. In 1995–96 the Endangered Species Program contributed approximately $4.5 million towards projects concerning the recovery of threatened species, of which $2.7 million (60%) went towards recovery projects for vertebrate species and only $1.7 million (38%) towards vascular plants.

Given that this level of disproportionate funding was likely to continue for the foreseeable future, it was clearly necessary to develop a system to prioritize threatened plant species to assist in decisions on allocation of funds. In 1994, the Australian Nature Conservation Agency (ANCA) undertook to develop a new approach to this problem. This paper reports on how this process was undertaken and describes a system which prioritizes threatened plant species for conservation action by considering their potential for recovery as well as their state of endangerment.

[1]Endangered Species Unit, Australian Nature Conservation Agency, P.O. Box 636, Canberra, Australian Capital Territory, Australia 2601.
Pages 40–49 *in* BACK FROM THE BRINK: REFINING THE THREATENED SPECIES RECOVERY PROCESS ed by S. Stephens and S. Maxwell. Surrey Beatty & Sons, 1996.

Fig. 1. Criteria used in threatened flora ranking scheme.

Degree of threat and Potential for recovery ↓	The two criteria, *degree of threat* and *potential for recovery,* are assessed concurrently
Benefits to habitat/ecosystem ↓	
Genetic distinctiveness ↓ Socio-economic values	The species may then be assessed against the other criteria in a hierarchial manner

NATIONAL SYSTEM FOR RANKING THREATENED FLORA FOR CONSERVATION ACTION

In 1994 it was recognized by ANCA, the Australian and New Zealand Environment and Conservation Council Endangered Flora Network (EFN) and the Commonwealth Endangered Species Advisory Committee (ESAC) that a need existed to rank plants on a continental scale to assist in decisions relating to conservation action. Such a system would assess those species that are critically endangered and in need of urgent attention, determine whether conservation action for those species was feasible, ensure that maximum benefit was obtained from the resources available and enable agencies to work co-operatively on an agreed list of priorities.

ANCA facilitated a series of workshops to address this need. The first, an EFN workshop, held in November 1994, aimed to develop criteria for consideration in preparing a national system. Participants at this workshop agreed that the following criteria should be considered:

- degree of threat;
- potential for recovery;
- benefits to habitat/community;
- genetic distinctiveness; and
- socio-economic/cultural value.

It was agreed that the categories of *degree of threat* and *potential for recovery* should have the highest priority. The other three categories should be discretionary and only considered, if required, such as to separate the ranking of two species that rank equally under the two main criteria.

Following the EFN workshop, ESAC suggested that the concepts developed should be further advanced. A second workshop was held in August 1995 with EFN and ESAC representation. The outcome of the workshop was a system which ranked threatened flora by first assessing both the *degree of threat* and the *potential for recovery,* and then using the criteria *benefits to habitat/ecosystem, genetic distinctiveness,* and *socio-economic values,* in that order, if necessary (Fig. 1).

A series of questions were developed against each of the five criteria, which permits a score to be calculated for each species assessed. These questions, and other points for consideration when assessing plants against each criterion, are described below and included in full in Appendix 1. Another criterion, *resources (or cost) of recovery,* was considered but rejected as not being appropriate for consideration when prioritizing species for conservation. However, this is useful information about a species and questions against this criterion are included in the system, although they are not scored and do not contribute towards the final ranking.

1. Degree of threat

It is important to note that this ranking system does not propose to re-assign threat categories but to merely clarify the relative level of threat each species is subject to within the existing threat categories. *Degree of threat* is an important consideration as the first step towards ranking for priority for conservation action.

The questions developed are based on criteria used to assess threat status in the IUCN Red List Categories (IUCN 1994). They are:

- total number of individuals;
- total number of populations;
- geographical distribution;
- area of occupancy;
- stability of populations (rate of decline);
- legal protection of land/water; and
- seriousness of threats (and threats to each population).

The seriousness of threats to each population is an additional factor to the IUCN system. This allows consideration of both the percentage of the population affected by different threats, and the severity of threats. A species with all populations affected by a serious threat would be considered more threatened than a species with only some populations affected by a threat or with populations affected by less serious threats.

Legal protection of land/water on which the populations occur, also supplementary to IUCN criteria, is considered under this criterion. For example, if populations occur on land that has a secure tenure (such as a conservation reserve) the species is recognized to be less threatened. Land tenure of the habitat is important with regard to protection against further habitat loss.

2. Potential for recovery

Once the degree of threat has been determined, *potential for recovery* is considered to be the next most important criterion to use in setting priorities for conservation action. An understanding of the potential for the species to recover is important to ensure that resources may be directed to actions that maximize the chances of survival of the species in the wild and to identify those species which will benefit most from recovery actions. Under this criterion the following factors are considered:

- specificity of habitat requirements (availability of suitable habitat and potential for recovery of degraded habitat);
- protective measures (whether negotiation of legal protection of land/water would be beneficial);
- reproductive issues (regeneration capacity);
- extent of biological knowledge on species and similar species (sufficient information to allow recovery action to proceed);
- integrated conservation potential (if *ex situ* propagation is necessary and possible, and the potential use of germplasm banks); and
- threatening processes (ability to implement action on known identified threats and the degree to which these threats are manageable).

3. Benefits to habitat/ecosystem

Assessment against this criterion considers the following:

- whether recovery actions on the species concerned are beneficial to the general habitat in which the species occurs;
- whether the recovery actions are beneficial to other species of significance (both flora and fauna) or to threatened communities; and
- whether new techniques will be developed for the recovery of the species.

A species occurring in the same habitat as other threatened species, for which the restoration of degraded habitat would assist their recovery, would score highly under this criterion.

4. Genetic distinctiveness

Genetic distinctiveness is considered to be fairly broad, taking into account the following:

- whether the family or genus is monotypic; and
- whether the whole family, whole genus or whole species is threatened nationally.

Species which do not have any close relatives would score higher under this criterion than those that have close relatives.

If taxonomic distinctiveness were more heavily weighted, conservation dollars may be allocated to taxonomically distinct species not in any immediate danger of extinction, rather than species which are critically endangered (New Zealand Department of Conservation 1994). This criterion is not given a high priority in this system as it is generally felt that all taxa should be given equal weighting.

5. Socio-economic values

This criterion is given a low priority as social and economic considerations are generally not significant for plant conservation efforts. Factors considered include:

- the cultural importance of a species;
- educational value;
- scientific interest; and
- commercial uses.

Consideration of this criterion should be minimal. The value humans place on a species should not distract from conserving other species facing imminent extinction (New Zealand Department of Conservation 1994).

Examples

Table 1 shows scores for *degree of threat* and *potential for recovery* for a sample of species assessed using this priority system. It is interesting to note that an exercise assessing species against the IUCN criteria showed that "Critically Endangered" species generally scored 17–18 or higher for *degree of threat* and "Endangered" species scored less.

Table 1 also shows those species for which recovery projects are currently being funded by the Endangered Species Program (ESP). Decisions to fund projects on those species were made over the last six years. Each year the process and criteria has become more stringent with fewer funds available for new projects (Endangered Species Unit 1995).

The scores in this table indicate how funding of some species may be justified over the funding of others. Conversely, funding allocated to other projects may have been better justified towards funding projects for species with higher scores. For example, funding recovery projects for *Acacia cretacea, Acacia whibleyana* and *Zieria prostrata* could well be justified as they score 18 for *degree of threat* (critically endangered) and the *potential for recovery* scores are also high.

The projects for species with lower *degree of threat* scores, such as *Pimelea spicata* and *Pterostylis arenicola,* will be looked at closely with regard to the commitment of future funding. In the case of *Pt. arenicola*, surveys have revealed that this species is more widespread than originally thought and the status of the species has consequently been downgraded to "Rare" in the recently revised ROTAP list. The *Pt. arenicola* Recovery Team has recommended that ESP funds be withdrawn from the project for future years.

For *Pi. spicata*, the Recovery Team has recognized that particular actions, such as genetic research, are not necessary for the species and that other actions have a higher priority. Changes have been made to the priorities of the recovery actions for this project and funds have been redistributed accordingly. Management actions such as liaison with land owners and managers for the protection of *Pi. spicata* populations has proved very beneficial to the conservation and recovery of the species and would have assisted in reducing the score for *degree of threat*. The current commitment for ESP funds for this project ceases by the end of 1997 and it is unlikely that further funds will be allocated to continuing the project.

DISCUSSION

The system described in this paper will first be used to rank those species listed as nationally endangered (approximately 300 species). State and Territory conservation agencies are currently using the system to assess the endangered species that occur in their State

Table 1. Scores for a sample of species assessed and suggested order of priority for recovery action. The total score for each species is shown for comparative purposes only, and has not been used for ranking.

Species	Priority	Degree of Threat (Maximum score 25)	Potential for Recovery (Maximum score 25)	Total	ESP currently funding conservation action
Grevillea mccutcheonii	1	25	13	38	No
Agrostis limitanea	2	22	13	35	No
Acacia cretacea	3	18	14	32	1993–1997
Acacia whibleyana	4	18	13	31	1995–1998
Phebalium daviesii	5	21	12	33	1996–1999
Drakea isolata	6	23	11	34	No
Zieria prostrata	7	18	12	30	1992–1996
Pterostylis gibbosa	8	17	12	29	1996–2000
Caladenia robinsonii	9	24	10	34	1996–1997
Allocasuarina portuensis	10	20	9	29	No
Fontainea oraria	11	23	8	31	1992–1999
Prasophyllum affine	12	22	8.5	30.5	No
Zieria baeuerlenii	13	20	7.5	27.5	No
Prostanthera euryboides	15	12	13	25	1993–1997
Pterostylis arenicola	14	10	13	23	1993–1997
Rulingia prostrata	16	12	12	24	No
Pimelea spicata	17	11	11.5	22.5	1993–1997
Ptychosperma bleeseri	18	14	9.25	23.25	1993–1997
Eucalyptus copulans	19	25	5	30	No
Alectryon ramiflorus	20	21	5	26	1996–1999
Pomaderris cotoneaster	21	8	4	12	No

or Territory and the scores against each species are being compiled to develop a list of priority species. A substantial number of assessments have already been received. The list of priority species is to be used as a guide in the allocation of Endangered Species Program funds to State and Territory agencies for recovery projects on threatened plants.

This list is not considered absolute as some species are yet to be assessed for inclusion, and the system itself may be subject to modification which may result in some adjustment to the scores and therefore a change in the ranking order. It is important to note that there is not, and probably never will be, a perfect system for ranking threatened taxa (Burbidge and Brown 1993) and application of this system has identified some limitations. It is recognized that there is scope for improvement to the system.

First it should be noted that the factors scored under each criterion have not been weighted intentionally and are therefore weighted implicitly (i.e., equal weights to all attributes). Appropriate weighting may be given further consideration where some attributes are thought to be more important.

Also, there is a possibility of varied interpretation of some questions as the system does have a degree of subjectivity. A comparison is currently being done to determine the extent of variability between assessments made by different people. Recommendations may be made on how this could be reduced, perhaps through more stringent definitions. It should be recognized that it is difficult to avoid a degree of subjectivity in any system.

Several comments have been received by those using the system in relation to a bias towards species for which data on threats are already known (T. Auld, D. Keith, pers. comm.). To overcome this, a precautionary principle could be included in the system, that is, "do not know" could receive a score. Assessment of species using the system may identify some species that would be more appropriately placed in the "K" category (insufficiently known or "Data Deficient"). For example, the assessment of *Pomaderris cotoneaster* identified that further surveys are required and that the species should probably be given "K" status rather than be listed as endangered.

The system may also help in identifying species for which minimal management actions are needed to assist their recovery and where a full national recovery plan may not be necessary. *Allocasuarina portuensis*, which scores highly, has only one population *in situ* consisting of one original individual and several plants re-introduced from *ex situ* propagation. The population occurs in a national park and, as protective measures have been established, it would be difficult to justify funding the implementation of a national recovery plan for this species.

Threatened species recovery proposals submitted to ANCA for funding in the past have been assessed against the Endangered Species Program priorities, also taking into account the State and Territory agency priorities and the quality of the proposals. Rather than seeking proposals for species which have been identified as priority species, ANCA has selected proposals from those submitted by State and Territory agencies which best met the priorities of the program. These were sometimes projects on species that are now recognized as having a lower priority.

In future, proposals will be sought for projects for those species that rank highest in the priorities determined for the Endangered Species Program. In the case of plant species, those that rank highest using the system described in this paper will be considered for funding. When assessing proposals for funding, consideration will still be given to the quality and merits of the proposal.

Obviously, many less than ideal decisions have to be made now if we are to stem the tidal wave of extinctions that threaten to engulf many species (Frankel *et al.* 1994). The system described in this paper may be considered controversial in that it takes a pragmatic approach to endangered species conservation. This approach is necessary due to the limited resources available. The concept of ranking a highly endangered plant as a lower conservation priority than a less endangered species because of their *potential for recovery* may not be readily accepted by many people. However, making hard decisions like this may be the only feasible approach in dealing with the problem that confronts us.

ACKNOWLEDGEMENTS

I thank John Briggs for his contribution and time invested in the development of the ranking system. A large number of people have been involved in using the system to assess listed endangered species and in particular I thank Tony Auld, Ken Atkins, Peter Lang, David Keith and Steve Harris for their efforts and helpful comments. Also, thanks to Jamie Kirkpatrick and Jamie Pook for their comments in reviewing this paper, and to Barry Baker for his advice and help in editing this paper.

REFERENCES

Atkins, K., 1993. Priority Ranking of Threatened Flora. Western Australia Department of Conservation and Land Management.

Briggs, J. D. and Leigh J. H., 1988. Rare or Threatened Australian Plants. Australian National Parks and Wildlife Service.

Burbidge, A. and Brown, A., 1993. Setting Priorities for the Conservation of Western Australia's Threatened Flora and Fauna. CALM Policy Statement No. 50. Western Australia Department of Conservation and Land Management.

Chalson, J. M. and Keith, D. A., 1995. A Risk Assessment Scheme for Vascular Plants: Pilot Application to the Flora of New South Wales. Project No. 450. Commonwealth Endangered Species Program.

Endangered Species Unit, 1995. Recovery Plan Guidelines for Endangered and Vulnerable Species and Endangered Ecological Communities. Australian Nature Conservation Agency.

Frankel, Brown, and Burdon, 1995. The Conservation of Plant Biodiversity. Cambridge University Press.

Given, D. R. and Norton, D. A., 1993. A Multivariate Approach to Assessing Threat and for Priority Setting in Threatened Species Conservation. *Biol. Conserv.* **64**.

Griffith, S. J., 1992. *Zieria prostrata ms* Species Recovery Plan. NSW National Parks and Wildlife Service.

IUCN, 1994. IUCN Red List Categories. International Union for the Conservation of Nature Species Survival Commission: Gland.

Jusaitis, M., 1994. *Pterostylis arenicola* Recovery Plan. South Australia Black Hill Flora Centre.

Jusaitis, M., 1994. Research Plans — *Brachycome muelleri* and *Acacia whibleyana*. South Australia Black Hill Flora Centre.

Liddle, D., Wightman, G. and Taylor, S., 1992. Recovery Plan for the Palm *Ptychosperma bleeseri*. Conservation Commission of the Northern Territory.

Mace, G. and Lande, R., 1991. Assessing Extinction Threats: Toward a Reevaluation of IUCN Threatened Species Categories. *Conserv. Biol.* **5**: 2.

Mace, G., Collar, N., Cooke, J., Gaston, K., Ginsberg, J., Leader Williams, N., Maunder, M. and Milner-Gulland, E. J., 1993. The development of New Criteria for Listing Species on the IUCN List. *Species* **19**.

Millsap, B. A., Gore, J. A., Runde, D. E. and Cerulean, S. I., 1990. Setting Priorities for the Conservation of Fish and Wildlife Species in Florida. *Wildl. Monogr.* **111**.

Nash, S. and Matthes, M., 1994. *Allocasuarina portuensis,* Conservation Research Statement and Species Recovery Plan. New South Wales National Parks and Wildlife Service.

Nash, S. and Matthes, M., 1993. *Pimelea spicata,* Conservation Research Statement and Species Recovery Plan. New South Wales National Parks and Wildlife Service.

New Zealand Department of Conservation, 1994. Setting Priorities for the Conservation of New Zealand's Threatened Plants and Animals. Second edition. Collated by C. Tisdall.

Robinson, C. J. and Coates, D. J., 1995. Declared Rare and Poorly Known Flora in the Albany District, Western Australia Wildlife Management Program No 20. Western Australia Department of Conservation and Land Management.

APPENDIX 1

Proposed national system for ranking threatened flora for conservation action

This system is based on the criteria and factors proposed at the EFN plant priorities workshop and has been developed by the ESAC plant working group.

Species are first ranked according to the *Degree of Threat* and a score is calculated against each species using the attached questions.

Species are then assessed under the criteria for *Potential for Recovery* and a score is calculated against each species under this category using the attached questions.

Under this system each taxon would be allocated two scores, one for *Degree of Threat* and one for *Potential for Recovery*, and would then be grouped in a grid arrangement using these scores as follows:

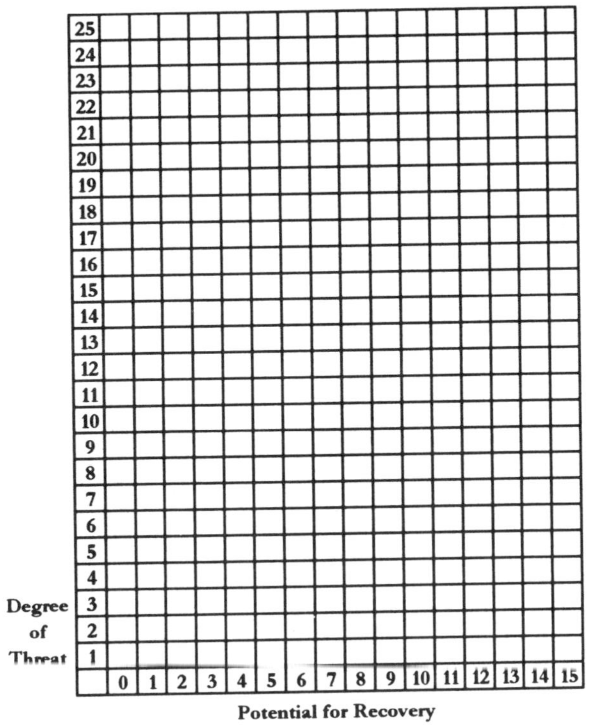

Priority will be given to those groups with the highest Degree of Threat and the highest Potential for Recovery.

The scores against the other criteria, *Benefits to Habitat/Ecosystem*, *Genetic Distinctiveness* and *Socio-economic values*, will be considered only where it is necessary to split species within a group.

Additional information is requested under *Resources for Recovery*. This is useful information about the species but it is NOT scored and does NOT contribute towards ranking the species in the priority system.

QUESTIONS AGAINST CRITERIA

The following questions are to be used to allocate scores for each taxon against each criterion.

It is important to answer **ALL** questions.

E.g:

Number of individuals		
Are there >50 mature individuals?	Yes ☐	No ☐ score 1
Are there >250 mature individuals?	Yes ☐	No ☐ score 1
Are there >2 500 mature individuals?	Yes ☐	No ☐ score 1

For a species with less than 50 mature individuals, tick **No** to all 3 questions. The subtotal under this category would be 3.

For a species with less than 250 mature individuals, tick **Yes** to the first question and **No** to the next 2 questions. The subtotal under this category would be 2.

For a species with less than 2 500 mature individuals, tick **Yes** to the first 2 questions and **No** to the last question. The subtotal under this category would be 1.

For a species with greater than 2 500 mature individuals, tick **Yes** to all 3 questions. The subtotal under this category would be 0.

Name of species being assessed:

....

State/Territory: ....

Name of Assessor: ....

DEGREE OF THREAT

For species for which there are currently recovery actions being undertaken, if those actions are not guaranteed to continue (i.e., if funding is not guaranteed), assess these questions as if the actions will not continue.

	Non-scoring column	**Scoring column**
Current geographical distribution		
Is the range >5 km?	Yes ☐	No ☐ score 1
Is the range >50 km?	Yes ☐	No ☐ score 1
Is the range >100 km?	Yes ☐	No ☐ score 1
Number of populations		
Is there only one known population?	No ☐	Yes ☐ score 1
Are there >5 known populations?	Yes ☐	No ☐ score 1
Are there >10 known populations?	Yes ☐	No ☐ score 1
Number of individuals		
Are there >50 mature individuals?	Yes ☐	No ☐ score 1
Are there >250 mature individuals?	Yes ☐	No ☐ score 1
Are there >2 500 mature individuals?	Yes ☐	No ☐ score 1

Degree of Threat — *continued*

Area of occupancy	Non-scoring column	Scoring column
Is the area of occupancy >0.01 km² (1 ha)?	Yes □	No □ score 1
Is the area of occupancy >0.1 km² (10 ha)?	Yes □	No □ score 1
Is the area of occupancy >1.0 km² (100 ha)?	Yes □	No □ score 1

Sub-total □

Populations: Geographically or otherwise distinct group of individuals between which there is little genetic exchange.

Mature individuals: Individuals known, estimated or inferred to be capable of reproduction. In the case of taxa that naturally lose all or a subset of mature individuals at some point in their life cycle, the estimate should be made at the appropriate time, when mature individuals are available for reproduction.

Rate of decline (observed, inferred or projected)	Non-scoring column	Scoring column
Is there a continuing decline in the: • extent of occurrence, and/or • area of occupancy, and/or • quality of habitat, and/or • number of locations or populations, and/or • number of mature individuals?	No □	Yes □ score 1 Don't know □ score 1
Is there an estimated continuing decline of mature individuals of at least 25% over 3 years?	No □ Don't know □	Yes □ score 1
Is there an estimated continuing decline of mature individuals of at least 20% over 5 years?	No □ Don't know □	Yes □ score 1
Is there an estimated continuing decline of mature individuals of at least 10% over 10 years?	No □ Don't know □	Yes □ score 1

If you ticked **Yes** to any questions above please provide details to support this:

Legal protection of land/water (conservation agreements, reserves, etc.)	Non-scoring column	Scoring column
Do any mature individuals occur on sites with legal protection?	Yes □	No □ score 1
Do <25% mature individuals occur on sites with legal protection?	No □	Yes □ score 1
Do <50% mature individuals occur on sites with legal protection?	No □	Yes □ score 1

Sub-total □

Extent of occurrence: Area contained within the shortest continuous imaginary boundary which can be drawn to encompass all the known populations of the taxon.

Area of occupancy: Area within its "extent of occurrence" which is occupied by a taxon.

Checklist of *current* and *potential* threats

Do threats to populations have High, Medium or Low probability of occurring and affecting **one or more** populations? (tick those that apply)

Threats	High	Medium	Low
Plant pathogens	□	□	□
Invasive species (native, feral)	□	□	□
Inappropriate fire regimes	□	□	□
Inappropriate grazing regimes (feral, native, domestic)	□	□	□
Land clearance (management activities, roadworks, development, mining, reclamation, inundation etc.)	□	□	□
Changes in hydrology	□	□	□
Collecting/harvesting (inc. forestry)	□	□	□
Pollution	□	□	□
Other, please specify	□	□	□
Other, please specify	□	□	□

If there is one or more **High** probability threats; **Score 3**

If there are no High probability threats but there are one or more **Medium** probability threats; **Score 2**

If there are no High or Medium probability threats but there are one or more **Low** probability threats; **Score 1**

Sub-total □

(Highest possible score is 3)

	Non-scoring column	Scoring column
Are **all** populations affected by a high probability threat?	No □	Yes □ score 1
Is every population affected by at least a medium probability threat?	No □	Yes □ score 1
Is every population affected by at least a low probability threat?	No □	Yes □ score 1

Sub-total □

Note: "Low numbers of individuals" or "low genetic variability" is not scored as a threat in this section as it has already been scored in this criterion under N*umber of populations* and N*umber of individuals.*

POTENTIAL FOR RECOVERY

	Non-scoring column	Scoring column
Specificity of habitat requirements		
Is there adequate habitat/potential habitat for recovery of the species?	No ☐	Yes ☐ score 1
Is there potential for management actions to be effective for recovery of degraded habitat?	No ☐	Yes ☐ score 1

	Non-scoring column	Scoring column
Protective measures (management, land tenure)		
Would successful negotiation of legal protection of land/water (e.g., conservation agreements, protected areas, etc.) significantly benefit recovery of the species?	No ☐	Yes ☐ score 1

	Non-scoring column	Scoring column
Reproductive issues (regeneration capacity)		
Does the species have the capacity to multiply in the wild successfully:		
1. by seed?	No ☐ Don't know ☐	Yes ☐ score 1
2. by vegetative means?	No ☐ Don't know ☐	Yes ☐ score 1
Does the species regenerate successfully in the wild, or can known management actions effect successful regeneration?	No ☐ Don't know ☐	Yes ☐ score 1

	Non-scoring column	Scoring column
Extent of biological knowledge relevant to the species		
Is there sufficient biological knowledge to allow recovery action to proceed?	No ☐ Don't know ☐	Yes ☐ score 1

Sub-total ☐

Potential habitat: Suitable habitat for translocation (need not be former habitat)

	Non-scoring column	Scoring column
Integrated conservation potential		
Is the species known to be successfully propagated *ex-situ*?	No ☐ Don't know ☐	Yes ☐ score 2
If Not::		
Does it have the potential to be successfully propagated *ex-situ*?	No ☐ Don't know ☐	Yes ☐ score 1
Is the species known to be, or does it have the potential to be, successfully propagated from material stored in germplasm banks?	No ☐ Don't know ☐	Yes ☐ score 1

Sub-total ☐

Germplasm banks: Ex-situ collections of reproductive material such as seed banks or cryogenic storage.

Material: Includes seed, propagules, etc.

Potential for Recovery — *continued*

Major Threats (Ability to implement action on known identified threats)

Are any high probability threats identified under *Degree of Threat* unmanageable?	Yes ☐	No ☐ score 1

How many **high probability** threats were identified under *Degree of Threat?*	A ☐	No score

Of those **high probability** threats identified under *Degree of Threat* what is the degree of manageability for each? score for High, Medium or Low manageability (tick only those that apply)

Threats	High	Medium	Low
Plant pathogens	☐ score 3	☐ score 1	☐ score 0
Invasive species (native, feral)	☐ score 3	☐ score 1	☐ score 0
Inappropriate fire regimes	☐ score 3	☐ score 1	☐ score 0
Inappropriate grazing regimes (feral, native, domestic)	☐ score 3	☐ score 1	☐ score 0
Land clearance (management activities, roadworks, development, mining, reclamation, inundation)	☐ score 3	☐ score 1	☐ score 0
Changes in hydrology	☐ score 3	☐ score 1	☐ score 0
Collecting/harvesting (inc. forestry)	☐ score 3	☐ score 1	☐ score 0
Pollution	☐ score 3	☐ score 1	☐ score 0
Other, please specify	☐ score 3	☐ score 1	☐ score 0
Other, please specify	☐ score 3	☐ score 1	☐ score 0

Total of High Medium and Low manageability:	B ☐ No score	Divide B (manageability) by A (number of high probability threats):	☐ score

Sub-total ☐

Degree of Manageability:

High — Management action will effectively control threat

Medium — Management action will partly control threat

Low — Little or no known management action will control threat

BENEFITS TO HABITAT/COMMUNITIES

	Non-scoring column	Scoring column
Would recovery actions disadvantage other flora and fauna species in the community?	Yes ☐	No ☐ score 1
Are new techniques to be developed for the recovery of the species?	No ☐	Yes ☐ score 1
Would recovery actions benefit other rare or threatened communities or species?	No ☐	Yes ☐ score 1

Sub-total ☐

GENETIC DISTINCTIVENESS

Is the family monotypic?	No ☐	Yes ☐ score 1
Is the whole family threatened nationally?	No ☐	Yes ☐ score 1
Is the genus monotypic?	No ☐	Yes ☐ score 1
Is the whole genus threatened nationally?	No ☐	Yes ☐ score 1
Is the whole species threatened nationally (including all sub-species, varieties etc.)?	No ☐	Yes ☐ score 1

Sub-total ☐

SOCIO-ECONOMIC/CULTURAL VALUE

Is the species of cultural importance?	No ☐	Yes ☐ score 1
Is the species of educational value or of scientific interest?	No ☐	Yes ☐ score 1
Is the species used commercially (e.g., timber, horticulture), or	No ☐	Yes ☐ score 2
Does the species have high potential to be used commercially?	No ☐	Yes ☐ score 1
Please specify commercial use:		

Sub-total ☐

ADDITIONAL INFORMATION
(Non-scoring)

Please provide as much information as possible.

RESOURCES FOR RECOVERY

Resources for recovery		
Is land purchase essential to ensure protection and recovery of the species?	Yes ☐ Don't know ☐	No ☐
• If Yes, what is the estimated cost of land purchase? (provide details in space below)		
• If No, what other options are available to ensure protection and recovery of the species, and what resources would be required? (provide details in space below)		
Is a community group likely to become actively involved in the recovery project?	No ☐ Don't know ☐	Yes ☐
Are necessary management actions for recovery likely to be supported by land managers?	No ☐ Don't know ☐	Yes ☐
Are there options for integrating research/recovery actions with work on other species, thereby reducing resources required per species?	No ☐ Don't know ☐	Yes ☐

Details:

Conservation overviews for invertebrates and non-vascular plants and fungi: setting the scene for recovery plans

Tim R. New[1]

The problems of defining conservation priorities among very diverse groups of organisms, exemplified by invertebrates and non-vascular plants and fungi, are formidable. National conservation overviews for these taxa reveal many problems for the development of practical conservation management programs, collectively reflecting their high diversity, poorly understood taxonomy (even at the levels of genera and families in some groups), lack of biological and distributional knowledge, and lack of public sympathy and of resources available for conservation.

INTRODUCTION

It may appear unusual to consider biota as disparate as invertebrates, non-vascular plants and fungi in a single paper on appraisal of conservation priorities and needs for recovery plans. However, these diverse and poorly documented groups share many features which hamper conservation assessment, and which have led to their neglect on practical conservation agendas through being considered either unimportant and tangential to mainstream conservation need or "too hard to deal with".

The development and execution of a recovery plan for any species presupposes certain knowledge of that species, particularly of its status (distribution, abundance, history of decline, vulnerability to current or anticipated threats: see New 1996) and of the kinds of practical management needed. Management may be orientated to maintain the current status (for example of a naturally and stably rare species, perhaps by obviating further threat if some decline has occurred) or be more aggressively aimed to build up numbers and augment diminished populations, establish additional populations within the species' historical range, or to monitor and refine management as needs arise. This procedure may be feasible for large, relatively well-known, conspicuous and easily detected organisms (such as many vertebrates) but there are considerable difficulties in defining the optimal needs of many invertebrates, non-vascular plants and fungi, even though the cost of formulating and undertaking a recovery plan may be fully equivalent to that for a "typical" vertebrate or vascular plant. The central problem of defining management needs has been clarified for only very few species. Many invertebrates and non-vascular plants naturally have very small distributions, many may always have been highly localized and in very small numbers and, despite their apparent rarity, may not be under any tangible threat. In many instances "doing nothing" may be the best current management decision.

Yen and New's (1995) comment that "Few recovery plans for Australian invertebrates have been prepared and fewer have been implemented" transfers easily to non-vascular plants and we face three major problems:

1. Knowing taxa well enough to appraise their conservation need.
2. Setting realistic and justifiable priorities among a multitude of localized and rare taxa which may need management to assure their sustainability.
3. Seeking viable alternatives to species-level conservation management.

The last becomes important in trying to employ limited funds and capability as effectively as possible. Many deserving taxa could be nominated as targets for individual conservation — but the major practical consequence in Australia will be a list so lengthy that funding will not be available for most of the listed taxa, and hence the legislation (or, more precisely, the commitment to management of listed species) is not able to be implemented. In 1990, 58 species (10% of those listed) under the United States Endangered Species Act accounted for 90% of available funding (La Roe 1993), and many species had no funding allocated. In Australia the commitment already exceeds the resources likely to become available. Formal listing, nevertheless, implies a societal recognition that action to conserve a species is needed, and the fact that resources may not be available to implement this should not detract from defining the need as fully as possible. Listing of invertebrates and non-vascular plants will help to assess the scale of conservation activity which will be needed in Australia.

---

[1]School of Zoology, La Trobe University, Bundoora, Victoria, Australia 3083.

Pages 50–54 *in* BACK FROM THE BRINK: REFINING THE THREATENED SPECIES RECOVERY PROCESS ed by S. Stephens and S. Maxwell. Surrey Beatty & Sons, 1996.

Objective evaluation of status becomes vital (New 1995a, 1996; Taylor 1993) in assuring that the "most deserving" cases/taxa are funded and to counter zealous over-statement of conservation need founded on inadequate information: as Taylor (1993) commented, in many cases "we do not understand the true relative importance of various species so we are afraid to assign any a low value", in part for fear of confusing objectivity with more subjective value judgments and condemning species to extinction.

Problems of setting priorities are recognized widely, and one consequence has been to hope that invertebrates (in particular) can be protected adequately under the umbrella of managing the more charismatic organisms which are easier to define and fund. However, this option is not universally viable, and the listing of selected taxa can provide "flagships" to indicate the conservation needs of invertebrates and non-vascular plants, and to communicate the intricacies of their biology to managers and funding agencies. The impetus to invertebrate conservation gained in Victoria, for example, by using the Eltham Copper Butterfly *Paralucia pyrodiscus lucida*, the Hemiphlebia Damselfly *Hemiphlebia mirabilis* and the Giant Gippsland Earthworm *Megascolides australis* in this way has been of inestimable value in promoting the needs of invertebrates to a wide and increasingly sympathetic audience (Yen *et al.* 1990). One benefit is that a greater number of informed and interested practitioners are recognizing the need to develop standard and/or replicable protocols for status evaluation and recovery plan (more broadly, management plan) design.

CONSERVATION OVERVIEWS

Two documents commissioned through the Australian Nature Conservation Agency (ANCA) are designed to help define priorities for these less "fashionable" groups and to fulfil Australia's broad obligations to conserve our unique biota. "Conservation overviews" for non-marine invertebrates (Yen and Butcher 1994) and non-vascular plants and fungi (Scott 1995): both set out to define and document major conservation needs, discuss major impediments to conservation and how these may be overcome, identify major gaps in knowledge, and suggest taxa or groups of priority concern. For invertebrates particularly, a number of taxon-based or State-based overview documents are also available on the status and relative priorities of particular taxonomic groups. As examples of these, a list of threatened Tasmanian invertebrates (Invertebrate Advisory Committee 1994), Tasmanian cave invertebrates (Eberhard *et al.* 1991), reviews of Australian butterflies (Kitching *et al.* 1996), freshwater crustaceans (Horwitz 1990) and hydrobiid molluscs (Ponder and Colgan 1992), raise various overlapping concerns. However, major sections of Australia's biota, such as the marine invertebrates and marine non-vascular plants, have not yet become the focus of any major review or synthesis emphasizing their conservation needs.

The overviews are of great relevance to development of the "recovery plan process" for invertebrates and non-vascular plants and fungi. More generally, the impetus for their development reflects the increasing and responsible need to consider complex biota fully and, for invertebrates, is helping to fulfil the intent of such documents as the Council of Nature Conservation Ministers' Australian Statement on Invertebrates (CONCOM 1989) and the International Union for the Conservation of Nature (IUCN) resolution 18/41 (1990), which emphasized the needs to consider invertebrates as vital constituents of a great variety of natural ecosystems.

PROBLEMS FOR PRACTICAL CONSERVATION

Many of the problems of conserving invertebrates, non-vascular plants and fungi are by no means limited to these groups, but are perhaps most universally applicable to them. Invertebrates, non-vascular plants and fungi share the following features:

1. Massive diversity, which is largely undocumented, so that for many groups even the approximate species number is unknown. For the totality of these groups, even the nearest order of magnitude is debatable. Non-marine invertebrates number several hundred thousand species in Australia. A figure of 250 000 species of fungi in Australia is accepted as reasonable (Scott 1995). In each of these groups, only a small proportion of species has been diagnosed formally and named.

2. Taxonomic problems include numerous undescribed and undiagnosed taxa, as well as much undetected synonymy, and many "phantom" taxa have not been seen since they were described. These may be extinct, elusive, or widespread. Levels of extinction are essentially unknown. Carlton (1993) discussed possible reasons for lack of documented marine invertebrate extinctions in recent years, and his arguments are of much wider application. As Gaston and Mound (1994) also noted, there has been some tendency to assume that many taxa not seen since they were described a century or more ago still exist and that failure to detect them reflects inadequacies of biological inventory: many such taxa may indeed be extinct, although not formally acknowledged as such.

3. As there are few experts on any group, and many of the groups are large, taxonomic progress to set solid baseline data for appraisal of assemblages and overcoming the "taxonomic impediment" is slow and uneven. At present, only small proportions of taxa in ecological surveys of most invertebrate and non-vascular plant groups can be identified reliably to species level. Uncertainties over species limits, incidence of cryptic taxa, and unknown patterns of variation all confound efforts at basic documentation.

4. Many taxa are small, inconspicuous (many invertebrates have complex life cycles and may be apparent for only days or weeks each year; many also have larval forms grossly different in appearance from the corresponding adults and the various life stages cannot be associated; alternation of generations in some non-vascular plants provides similar confusion; fungi may be recognizable only by their highly seasonal fruiting bodies), and species are difficult to discriminate. For many groups,

recognition even to order or family level necessitates dissection and/or microscopical examination and practical identification guides do not exist.

5. Biological and distributional knowledge for most species is fragmentary or non-existent. Factors governing the abundance of most taxa are unknown, and historical information on distribution is usually entirely absent except for sporadic label data on museum specimens. Many taxa are known only from their type localities, and even these may be documented imprecisely, simply by country or other major geographical region. Definition of status is difficult and approaches such as population and habitat viability analysis are simply not feasible for most taxa because of lack of life-table data. The vast ecological heterogeneity of invertebrates renders extrapolation from information available on closely related species unwise or, even, grossly misleading. Estimation of risk of extinction is therefore difficult.

6. Lack of resources for study and clarification of status other than in individual cases, and lack of widespread "precedent expertise" for undertaking management programs for most taxa.

7. Lack of public interest and sympathy. With few exceptions (such as butterflies and some molluscs among the invertebrates) the organisms are neglected or considered trivial even by many experienced conservation practitioners. Both overviews emphasize the need for greater public and scientific education on these groups. To date only two invertebrates and no non-vascular plants are listed on the *Endangered Species Protection Act 1992*. This reflects lack of knowledge and concern, and certainly does not reflect that no other species merit listing. The conservation overviews list many taxa which are likely candidates for future listing, and for which recovery or other management plans will consequently be needed.

DISCUSSION

Despite such formidable difficulties, it is not sufficient simply to lump all bryophytes, beetles, fungi or worms together for conservation. Each species has its biological peculiarities and without knowledge of these any species-focused management plan is likely to be inadequate. Both overviews nominate taxa for priority concern (Table 1), either as endangered or in need of priority evaluation of conservation status: but these listings are indicative rather than comprehensive. Some 560 non-vascular plants (excluding fungi) are noted by Scott and his colleagues, and the 422 invertebrates noted by Yen and Butcher are heavily biased toward insects. The additional — and at present, inevitable — problem may be exemplified by the insect taxa, which include 92 Lepidoptera (79 of them butterflies), and the 65 Diptera (which include 56 Drosophilidae), figures which reflect, respectively, the group with best-defined data and wide interest, and the zeal of specialists. The listings are therefore heavily biased towards a few groups of insects, and are obviously incomplete. It would be very misleading to claim that the 20 000 or so species of Australian moths contain only some 13 threatened taxa when around 20% of the butterflies have conservation significance, or that the other 97 families of Diptera represented in Australia are of no conservation interest.

Table 1. Numbers of taxa noted as having conservation significance in conservation overviews of Australian non-vascular plants and fungi (Scott 1995) and non-marine invertebrates (Yen and Butcher 1994).

a. Non-vascular plants (potentially vulnerable — extinct)	
lichens	206
bryophytes	149
algae	205
(fungi)	(unknown no.)
b. Invertebrates	
Platyhelminthes	1
Nemertea	5
Mollusca	14
Annelida	1
Onychophora	5
Crustacea	73
Arachnida	14
Collembola	9
(Insecta)	
Lepidoptera	92
Diptera	65
Coleoptera	31
Hymenoptera	4
others	106

Many phyla of invertebrates, and most groups of non-vascular plants, have had no constructive, focused work on conservation evaluation in Australia. On schemes for priority based on phylogenetic isolation, many Australian lineages would rank highly on any global priority system (Vane-Wright 1994). The needs, therefore, are to focus efforts rather differently, recognizing that conservation methods applied conventionally to other taxa are not always appropriate, because concepts of "rarity" and "threatened taxa" are applied more tentatively and are difficult to equate realistically with those. Concepts of rare and threatened species, and even of "an individual plant" may be inappropriate for many non-vascular plants (Scott 1995). Schemes for recording conservation status of plants have generally been designed for vascular plants. However, there is indeed sufficient consensus over the needs of particular taxa (especially of invertebrates) to suggest positive ways forward, once we allow for unknown levels of extinction, unknown status of most taxa in relation to setting action priorities, and whether we are indeed focusing on the optimal taxa. For invertebrates, there has been much discussion in the last few years on "optimal targets", and any groups which can be used also as "tools" in helping to monitor environmental health, serve as indicators of broader diversity, or protect largely undocumented assemblages, are considered especially important. Scott and his colleagues noted that "flagship taxa" are scarcely appropriate for non-vascular plants, and fungi, but there is some likelihood that "flagship groups" (rather than individual species) will become more important in overcoming the species-level taxonomic impediment. With fungi, it is not possible to select taxa that are known to be endangered or vulnerable, so possible flagship taxa may be selected to demonstrate their variety and ecological roles (Scott 1995). As one invertebrate example, Onychophora ("velvetworms") are unmistakable in general appearance although their species-level taxonomy is complex, confusing and incomplete. Presence of any species of Onychophora may be significant in denoting conservation values for the site/habitat where it is fund (New 1995b) and the species' name is largely irrelevant in conveying the need for site conservation. More precise taxonomic focus is needed for most other groups; for example it is not unusual for subspecies of butterfly to be conservation targets, and these must therefore be differentiated clearly from closely similar taxa.

In general, the impediments to pursuing species-focused conservation of most invertebrates and non-vascular plants render it difficult (and unwise) to use this approach as the major thrust of practical conservation. It is extremely difficult to define their real needs in relation to the massive number of taxa, and the overall lack of funds to pursue such expensive studies. Other options involve concentrating on higher level conservation — such as assemblage, community or habitat and controlling at least the more discernible threats, in conjunction with augmenting the base-line information so badly needed. Ecological emphasis thus tends to supplement and supplant taxon-focused conservation management. The overviews both stress the need for massively increased capability for inventories (as provisional checklists) of sites; completion of relevant volumes of the Flora of Australia and Zoological Catalogue/Fauna of Australia; registration of sites supporting notable taxa, centres of endemism or high diversity as a system equivalent to the British SSSI scheme; initiation of databasing schemes for selected groups; harmonizing of State and Federal legislation, and routine inclusion of selected groups in biological surveys. There is universal need for greater objective documentation of taxon diversity and biological correlates of this as an avenue towards understanding the roles of invertebrates and non-vascular plants in monitoring studies, including estimating the influences of future disturbances (such as climatic changes), for which long-term studies of faunal change at fixed sites along ecological gradients may be needed.

Universal emphasis on habitat security recognizes that a secure environment which can support its characteristic community is the most vital need for any taxon/taxa and that, as an interim step pending capability for more finely-focused management, provision of these as refuges and as reservoirs of biodiversity for the future is vital. For invertebrates and non-vascular plants, focusing on habitats or particular sites has the following practical conservation problems:

1. Although this broader conservation approach is likely to sustain many of the species which live there, it does not obviate the need for management to maintain habitat diversity, such as retaining early successional stages, of the habitat.
2. Difficulties in monitoring and documenting "what is there". In Australia, and elsewhere, a systematic program to document the biota of protected areas would be of immense help in setting priorities for the groups treated in this paper, simply in telling us what is represented in nominally reserved areas. No such program currently exists, and its establishment is a high priority.
3. Difficulties of security, because it may be difficult to define the boundaries of protected areas or sites and to buffer these adequately from external influences such as invasive exotic weeds. The recent controversy over permitting intrusion of apiculture into national parks in southeastern Australia is one example of this uncertainty, reflecting lack of knowledge of the potential interactions between native and introduced species (New 1994).
4. Future reserves will be nominated, almost inevitably, in a climate of multiple and diverse land use needs and priorities, but any comprehensive ranking or selection system is severely impoverished if the major invertebrate and non-vascular plant components are excluded.

Management or recovery plans are therefore possible in broad terms and at the "group level" where the problems common to many species can be defined, even in the most general terms. More detailed plans usually command much original research. Recovery plans for several butterflies have been designed and are being refined progressively (reviewed by New and Yen 1995) and other taxa investigated (particularly under the auspices of Victoria's *Flora and Fauna Guarantee Act 1988*) also provide valuable leads to the extent that the ultimate aim of de-listing taxa formerly supposed to be endangered is already occurring in some cases.

Recovery plans are the central theme of this volume. The overviews discussed here recognize that nominating particular species of invertebrates, fungi or non-vascular plants for need and design of management programmes is only a first step in the massive conservation investigation we face, and that bases for action are commonly flimsy. The positive steps needed to redress this involve science, education and pragmatism, with emphasis on increasing knowledge and the capability to acquire and interpret that information. It is indeed encouraging that these groups are now being considered on conservation agendas — not simply as tokenism but as a real acknowledgement of their massive importance in the natural world. But, as Yen and New (1995) emphasized for invertebrates, specialist knowledge of the groups is rare among wildlife and conservation biologists, and there is a critical need for informed personnel to facilitate the development and implementation of recovery plans as well as to provide the basic documentation needed to define conservation status of taxa adequately. Scott (1995) recommended establishment of at least two positions for taxonomists working on non-vascular plants (one position to be mycological) in each of the major State and Federal herbaria to help improve taxonomic knowledge, and Yen and Butcher (1994) foresaw the need for an invertebrate specialist in the Department of Conservation (or equivalent) in each State. The need is similar for non-vascular plants, and education to appreciate more fully the practical problems of documenting and conserving them is a key need in advancing practical conservation.

ACKNOWLEDGEMENTS

I appreciate the privilege of reading the two unpublished conservation overviews which form the basis for this paper, and of discussing aspects of them with the senior authors. I hope that I have not misrepresented their intentions unduly, and opinions expressed are my responsibility. Financial support from ANCA to attend the "Back from the Brink" conference is gratefully acknowledged, as is advice from Sally Stephens, and the thoughtful comments of two referees, Drs Margaret Brock and Pierre Horwitz.

REFERENCES

Carlton, J. T., 1993. Neoextinctions of marine invertebrates. *Am. Zool.* **33**: 499–509.

CONCOM (Council of Nature Conservation Ministers), 1989. Australian Statement on Invertebrates. Canberra.

Eberhard, S. M., Richardson, A. M. M. and Swain, R., 1991. The Invertebrate cave fauna of Tasmania. University of Tasmania: Hobart.

Gaston, K. J. and Mound, L. A., 1994. Conservation and systematics — the agony and the ecstasy. Pp. 185–95 *in* Perspectives on insect conservation ed by K. J. Gaston, T. R. New and M. J. Samways. Intercept: Andover.

Horwitz, P., 1990. The conservation status of Australian freshwater Crustacea. Report No. 14. Australian National Parks and Wildlife Service: Canberra.

Invertebrate Advisory Committee, 1994. Interim list of native invertebrates which are rare or threatened in Tasmania. Parks and Wildlife Service: Tasmania.

IUCN, 1994. The International Union for the Conservation of Nature Red List of Threatened Animals. Gland.

Kitching, R. L., Dunn, K. L. and Dexter, E., 1996. Conservation status of the Australian Butterflies. Australian Nature Conservation Agency (unpublished report).

LaRoe, E. T., 1993. Implementation of an ecosystem approach to endangered species conservation. *Endangered Species Update* **10**: 3–6.

New, T. R., 1994. Exotic Insects in Australia. Gleneagles Press: Adelaide.

New, T. R., 1995a. Introduction to Invertebrate Conservation Biology. Oxford University Press: Oxford.

New, T. R., 1995b. Onychophora in invertebrate conservation: priorities, practice and prospects. *Zool. J. Linn. Soc.* **114**: 77–89.

New, T. R., 1996. Evaluating the status of butterflies for conservation. Pp. 4–21 *in* Decline and Conservation of Butterflies in Japan III ed by S. A. Ae, T. Hirowatari, M. Ishii and L. P. Brower. Lepidopterological Society of Japan: Osaka.

New, T. R. and Yen, A. L., 1995. Species management and recovery plans for butterflies (Insecta: Lepidoptera) in Australia. Pp. 15–21 *in* People and nature conservation ed by A. Bennett, G. Backhouse and T. Clark. Surrey Beatty & Sons: Chipping Norton.

Ponder, W. F. and Colgan, D. J., 1992. Conservation of aquatic invertebrates with restricted distributions with special reference to hydrobiid gastropods. Australian National Parks and Wildlife Service (unpublished report).

Scott, G. A. M. (ed), 1995. Overview of the conservation of non-marine lichens, bryophytes, algae and fungi in Australia. Australian Nature Conservation Agency (unpublished report).

Taylor, S., 1993. Practical ecosystem management for plants and animals. *Endangered Species Update* **10**: 26–29.

Vane-Wright, R. I., 1994. Systematics and the conservation of biodiversity: global, national and local perspectives. Pp. 197–211 *in* Perspectives on insect conservation ed by K. J. Gaston, T. R. New and M. J. Samways. Intercept: Andover.

Yen, A. L. and Butcher, R. J., 1994. An overview of the conservation status of non-marine invertebrates in Australia. Australian Nature Conservation Agency (unpublished report).

Yen, A. L. and New, T. R., 1995. Some problems affecting implementation of recovery programs for threatened Australian non-marine invertebrate species. Pp. 9–14 *in* People and nature conservation ed by A. Bennett, G. Backhouse and T. Clark. Surrey Beatty & Sons: Chipping Norton.

Yen, A. L., New, T. R., van Praagh, B. D. and Vaughan, P. J., 1990. Invertebrate conservation: three case studies in south-eastern Australia. Pp. 207–24 *in* Management and conservation of small populations ed by T. W. Clark and J. H. Seebeck. Chicago Zoological Society, Brookfield: Illinois.

Essentials of a good recovery plan

Andrew A. Burbidge[1]

Recovery plans are best prepared by a recovery team and should be outcome-oriented, practical, easily understood documents containing recovery actions that are achievable with a reasonable input of resources. The most important parts of a plan are the recovery objectives, criteria for success and failure, and recovery actions. Costings should be clear and accurate, and detailed records should be kept of how costs were developed. Lack of biological or other information on a species or community should not delay the writing or commencement of implementation of a recovery plan. Where lack of information prevents the production of a full recovery plan, or where urgency dictates that recovery actions can not await a full plan, an interim recovery plan should be prepared, especially for critically endangered and endangered taxa or communities. An interim recovery plan prescribes work over a three year period, including any research needed to more clearly define recovery actions. Pitfalls for recovery plan writers include too much detail in introductory sections, wasting time and space on minor details that have little or no relevance to outcomes, and delaying the commencement of recovery actions because of delays in preparation of the plan. Recovery plans require commitment of staff and resources for their implementation and should be officially adopted by relevant Ministers and conservation agencies. The only real test of recovery plans is whether they successfully guide the recovery of a taxon or ecological community, so they need to be regularly reviewed and flexible, and be amended as knowledge improves or conditions change. Four recovery plans written to guide threatened species and ecological community recovery in Western Australia are reviewed for their adequacy and ease of subsequent implementation.

Key words: Recovery plans, interim recovery plans, threatened species, threatened ecological community.

INTRODUCTION

RECOVERY plans, documents that guide the "recovery" of threatened taxa (not only preventing species' extinctions but also implementing actions that lead to sufficient increases in range and abundance that they are no longer threatened with extinction), originated in the United States of America, where they are prepared according to the provisions of the United States Endangered Species Act (see Tear *et al.* 1995; Clark, this volume).

Recovery plans have become rapidly and widely adopted in Australia as the main guiding documents for the recovery of threatened taxa and are starting to be used in a similar way for recovering threatened ecological communities. In Australia, recovery plans have usually followed guidelines prepared by the Australian Nature Conservation Agency (ANCA), which were based in part on the United States experience.

In this paper I discuss priorities for writing recovery plans, dealing with a lack of information and for whom recovery plans are written, as well as providing guidelines for recovery planning. Four case studies from Western Australia are presented to illustrate some of the good points and pitfalls of plans currently being implemented.

Priorities for writing recovery plans

Many critically endangered and endangered Australian taxa are not the subject of recovery plans. Accordingly, no or few recovery actions are being undertaken to prevent their extinction. There are four main reasons for this:

1. Numerous taxa are threatened.
2. Conservation agencies have been slow to prioritize threatened taxa in terms of degree of threat.
3. "Flagship" taxa, or taxa for which research data are available, tend to have been treated first.
4. There are insufficient data on the limiting factors for many taxa and defining recovery actions is often difficult or can not be done with any degree of certainty.

Because there are so many taxa that need addressing, people working in the area can become overwhelmed by the size of the problem. Having hundreds of recovery plans and hundreds of recovery teams is not possible with present or anticipated resources. Anyway, it would not be cost-effective and should not be contemplated. A better method is setting up a recovery team to deal with many species suffering from similar threats within a specified geographic area. This method is being used in Western Australia to conserve

[1]Western Australian Department of Conservation and Land Management, Threatened Species and Communities Unit, P.O. Box 51, Wanneroo, Western Australia, Australia 6065.

Pages 55–62 *in* BACK FROM THE BRINK: REFINING THE THREATENED SPECIES RECOVERY PROCESS ed by S. Stephens and S. Maxwell.
Surrey Beatty and Sons, 1996.

threatened plants in the bio-diverse South-West Botanical Province and is the subject of a paper (Brown *et al.,* this volume).

Setting priorities is not difficult if carried out dispassionately. Setting priorities can now be based on the 1994 International Union for the Conservation of Nature Red List Criteria (IUCN 1994), which are much more quantitative than the old criteria. Some State agencies have already established their priorities, and priorities at the national level are currently being developed by the Endangered Species Advisory Committee with the assistance of Australian and New Zealand Environment and Conservation Council (ANZECC) networks.

Research plans

One way of dealing with the lack of information has been to prepare research plans. These have been treated by ANCA separately from recovery plans. Some are expensive and lack of funding has often delayed their implementation. Some research work proposed has been aimed at more clearly delineating conservation status. Lack of funding for research plans may have lead to delays in recovery actions for taxa or communities that are critically endangered.

Even where implemented, research plans tend to concentrate on research without considering important and sometimes relatively simple and cheap recovery actions that should have been implemented immediately. Research is one possible recovery action and in my view research should be part of a recovery plan, not a separate process. Where research is included in a recovery plan, along with other actions, limited funding may allow urgent recovery work to commence before research commences. The inclusion of research in a recovery plan will certainly promote the integration of research and management and allow a more experimental approach to management. I propose that recovery plans be the sole document used for all work relating to the recovery of threatened taxa and ecological communities in Australia.

Interim recovery plans

Recovery plans tend to be detailed documents and their writing consumes time and resources. Often, writing has been delayed because of lack of sufficient information to treat the taxon in detail. Where there are insufficient data available to clearly define all necessary recovery actions, or urgency dictates that recovery actions can not await the preparation of a full recovery plan, I recommend the preparation of interim recovery plans that prescribe work over a three year period, including any necessary scientific research.

Identification of taxa as "Critically Endangered" (CR), "Endangered" (EN) and "Vulnerable" (VU) under the new IUCN criteria is the best method of driving threatened species conservation priorities and should work just as effectively, with appropriate criteria, for ecological communities. In Western Australia, the Department of Conservation and Land Management (CALM) has adopted the policy that all CR taxa will have recovery plans or interim recovery plans and that the implementation of recovery actions will commence within one year of the endorsement of CR status by the Minister. The preparation and implementation of recovery plans for EN taxa will follow the completion of those for CR taxa, as resources for their conservation become available.

Why write a recovery plan and for whom are they written?

These questions are basic to any discussion of the essentials of a good recovery plan and are worth considering carefully before writing commences. Usually, a recovery plan is written so conservation work related to species or communities can be guided by accurate information, objectives, actions and detailed forward estimates of costs. Recovery plans should maximize the probability of recovery and minimize the probability of errors, including errors of omission, that might lead to a species becoming extinct or a community collapsing or disappearing. They also allow the public to know what is being done to save threatened species and communities.

Recovery plans have three main audiences — those who have a legal responsibility for nature conservation, those who will be funding and implementing the plan, and the interested public who want to know what is being done. All groups will include people with limited biological and ecological knowledge. Plans should be tailored to their audiences; something that some recovery plan writers sometimes seem to have forgotten.

Writers of recovery plans should be aware that recovery is defined in terms of de-listing, not in terms of a return to the former total range or levels of former abundance. Plans for the management of species on a local or regional level that are not threatened throughout their range should not be termed recovery plans. Recovery plans should not address recovery within a part of a species' or community's current geographic range — artificial boundaries, including political boundaries, are not biologically meaningful and "recovering" species within one area when they are not threatened overall is a waste of scarce conservation dollars.

Adoption of recovery plans

Recovery plans are prescriptive and require commitment of staff and resources for their implementation. It is of little value writing such plans if their implementation is not supported by and integrated into the existing legal, administrative and resourcing framework for nature conservation. Thus recovery plans need to be endorsed by the relevant State and Territory Ministers and conservation agencies and as well by the Federal Minister for the Environment under the Commonwealth *Endangered Species Protection Act 1992* (ESP Act).

GUIDELINES FOR RECOVERY PLANNING

There is some guiding legislation for recovery plan writers. Section 32(1) of the Commonwealth ESP Act states that recovery plans "must provide for the research and management actions necessary to stop the decline of, and support the recovery of, the species or community so that its chance of long-term survival in nature are maximized". Section 32(2) goes on to list matters that recovery plans must contain. Acts in the States and Territories also have provisions dealing with threatened taxa. Recovery plan writers should refer to those as necessary.

Guiding legislation can be helpful, but inspection of some of the recovery plans that have been prepared over the past few years shows a variety of documents of varying quality. This is not necessarily a bad thing, however, we should be able to learn from what has been done so far and make recovery plan writing more efficient and effective.

The following guidelines for those writing recovery plans are provided in the hope that they will be of assistance. They are based mainly on my experience of recovery planning in Western Australia and are not necessarily applicable to all jurisdictions or situations. They should be read in conjunction with relevant legislation and organizational policies.

1. Recovery plans should be written by or under the guidance of a recovery team

There are two issues here — ownership and accuracy. No one likes to be dictated to from outside. A recovery plan that is written by someone outside the relevant recovery team is less likely to be readily accepted by the team set up to implement it. As well, accuracy of information and the practicality of recovery actions are likely to be enhanced in a plan that is prepared by or in interaction with a recovery team that includes people who are to be involved in implementation. The preparation of a recovery plan by a person who has no stake in its successful implementation should be avoided.

2. Background sections should be brief and to the point

Recovery plans should contain background information on the description and history of discovery of the species or community, its conservation status and decline, its biology, ecology and limiting factors, and reasons for its decline. These sections need to be brief and to the point and easily understood by lay people. Too often recovery plans contain large amounts of detailed biological information that are not relevant to the outcomes of the plan. Not only does this unnecessarily complicate the plan from the reader's point of view, it can also delay its finalization while the writer takes time to ensure that every detail is accurate and up to date.

If detailed technical information is considered important, it can be placed in an appendix, as was done for information about effective population size in the Noisy Scrub-bird Recovery Plan (Danks *et al.* 1996).

3. Objective(s) should be clearly stated and achievable in the time frame adopted

The objective to be achieved is central to a recovery plan. It should be simple and clear. Objectives often are stated in terms of the species' or community's conservation status, for example, upgrading from endangered to vulnerable, or de-listing from a national threatened species list. For organisms where such a change in status is not achievable in the period of the plan (e.g., vertebrates with a long generation time), the objective should be stated in terms that relate to a longer term goal of upgrading or de-listing.

Recovery plans should not have objectives that are "wishes" or "stabs in the dark" and are not reasonably achievable within the time frame specified. Ambitious objectives are satisfactory as long as they can be achieved.

4. Criteria should be measurable

Setting criteria against which the success or failure of the recovery plans can be measured is of critical importance. These criteria must be measurable so that anyone (including the recovery team) can, when necessary, evaluate progress with the implementation of a recovery plan. Setting down measurable criteria for populations of threatened species or for an ecological community is not always easy; however, it is vital that this is done. For species, factors such as the number of mature individuals, the number of populations of a certain size in the wild and the number of individuals maintained in captivity or in propagation can be used. For communities, measurements of their area, species composition, community structure, and related abiotic factors, can be used.

Setting criteria for the failure of the recovery plan is of equal importance. Such criteria can be defined in terms of a continuing decline of a taxon towards extinction or of a community towards collapse or destruction. They can also be defined in terms of failure to carry out urgent recovery actions within a certain time frame.

5. Actions should be simple, achievable and accurately costed

The prescribed recovery actions are the most important part of a recovery plan — the success or failure of the plan will depend largely on whether the actions specified in it are appropriate and able to be implemented. It has been pointed out several times that it is implementation — what we do to and for a threatened species on the ground — that ultimately determines the survival of species (e.g., Backhouse and Clark 1995). No doubt the same applies to threatened ecological communities.

Recovery actions should be broken down into easily understandable and manageable segments. Each action should be defined by a clear, concise statement that describes what is to be done and by whom. Identification of the person or organizational unit that is responsible for implementing the action is vital, otherwise there can be no real accountability. This should be followed by a summary of costs by budgeting period, usually a year. Many recovery plans do not contain sufficient detail of how costs were developed, making it difficult for those implementing them a few years down the track to know exactly how the money was to be spent. It is very important to ensure that sufficient details are kept and either published as an appendix or retained on a file available to future managers.

Once costs are established, the source of funds needs to be identified. In some cases, future funding will be clear and can be listed in the plan. In many cases, however, at least a proportion of the necessary funds will need to be raised. In such cases, the plan should identify that amount of required funds that is committed (and by whom) and that amount that still needs to be found.

Funding for the implementation of recovery plans for threatened species and communities is in short supply. Therefore, it is important that recovery plan writers do not seek more funds than are absolutely necessary for the recovery of the species or community concerned. Expensive recovery plans will be implemented at the cost of recovering species and communities deemed of lower priority — therefore, savings identified in one recovery plan will allow others to be funded. This is a difficult area and I am not suggesting that a recovery plan should not include actions that are expensive if these are really necessary. What I am saying is that writers of recovery plans should be very careful not to include costs that are not necessary, for example, to clarify issues of scientific interest that are not necessary for the species' or community's recovery.

In Australia at present, few recovery plans are adequately refereed, especially in the sense of checking whether all recovery actions are necessary and whether the budget has been accurately calculated, before funding is sought or allocated. Section 32(3) of the ESP Act states that "in preparing a recovery plan, regard must be given to . . . (b) the most efficient and effective use of the resources that are allocated for conservation of species and ecological communities". I have to wonder whether this is actually being observed.

The current ANCA guidelines for the preparation of recovery plans suggest that costs should be shown in three places — against each action, in an implementation schedule and in the summary. The costs are compiled from the actions into the implementation schedule and summary in different ways, making arithmetic errors likely and increasing the possibility of mistakes when changes to costs are made. This can be overcome by not writing the implementation schedule and summary until the last moment, or by using a spreadsheet.

One of the most important factors that separates recovery plans from many other plans written by conservation agencies in Australia, including many management plans for national parks and other conservation reserves, is that recovery plans do not contain uncosted "wish lists". This is a most significant point and provides a good deal of credibility to the recovery process. Inherent in this, is the rule that *all* costs must be identified, including recurrent agency costs such as salaries and administrative overheads. This will preclude criticism that those working on the recovery of threatened species and communities are not realistic as to costs. Indeed, the argument can often be turned on those criticizing the process — how many other conservation programs are so well and so openly costed?

6. Recovery plans should be flexible and regularly reviewed, and be amended as knowledge improves or conditions change

Recovery plans are not documents that should be "set in concrete". Knowledge about species and communities continually changes, as more research is conducted and as recovery plans are implemented. Flexibility is important. The possible need for changes should be noted in the plan. The foreword to Western Australian recovery plans includes the statement: "Recovery plans delineate, justify and schedule management actions necessary to support the recovery of an endangered or vulnerable species or ecological community. The attainment of objectives and the provision of funds is subject to budgetary and other constraints affecting the parties involved, as well as the need to address other priorities . . . Approved recovery plans are subject to modification as dictated by new findings, changes in species' status and completion of recovery actions."

With the need for flexibility goes the need for regular reviews of the successes, failures and problems encountered during the implementation of a recovery plan. Reviews should be a regular feature of recovery team meetings. The recovery plan should include statements as to how and when reviews will conducted and also provide for annual reporting mechanisms. Major reviews should be undertaken at least every five years.

7. Recovery plans should promote and facilitate appropriate public participation

Public participation should occur during plan preparation, during a public review period and during implementation. Ideally, valid public concerns about the effectiveness or effects of recovery actions in the plan will be dealt with during the preparation of the plan itself. While socio-economic concerns are not taken into account during the listing of threatened species in Australia, they must be considered during recovery — failure to do so may lead to the failure of the plan.

Public participation in the preparation of a recovery plan is best facilitated by having appropriate community representatives among the members of the recovery team. Such people can include representatives of land owners and land managers involved in or affected by the recovery actions, representatives of local government, and representatives of local groups, including "Friends of" groups, who can assist with the plan's implementation. This is not a recommendation for large recovery teams — good planning, publicity, consultation and liaison should allow small and effective teams. It also is not a recommendation for team members to include those opposed to the recovery of the species or community, or to include lobbyists whose main aim is to use Team meetings to criticize government policies. Recovery teams are action-oriented, not policy-oriented. Other mechanisms can be used to ensure that such people can participate in policy development.

Financial and staff resources are limited in government conservation agencies and in most jurisdictions money is currently becoming scarcer. Involvement of volunteers in the implementation of recovery plans contributes considerable money and time and encourages people outside government to share responsibility for recovery.

Some legislation provides for draft recovery plans to be subject to a public comment. Having provisions for public comments on draft plans is now a common procedure in many areas of endeavour and should generally be promoted for recovery plans as well. However, public consultation can be expensive and time-consuming and considerations of the cost and staff time involved may not allow as much public involvement at this stage of the recovery process as may be desirable. Spending resources on the process of plan preparation rather than on recovery actions may not always be the best use of scarce conservation dollars — recovery plan writers may have to make a judgement on the level of public review a case by case basis.

The participation of local groups in the implementation of recovery plans should be strongly supported and promoted. Local "ownership" of recovery plans will enhance the chances of success and should also enable more to be done for less money. Many species will only be secure when local communities and private landholders take direct responsibility for ensuring their conservation on lands they manage.

8. Recovery plans should include sections on education and publicity

Preparation and implementation of recovery plans costs money — money that mostly, in Australia, comes from the taxpayer. Public support for the recovery process is vital if it is to be adequately funded from the public purse. Good publicity is also vital to obtaining sponsorships and ensuring that sponsors are adequately acknowledged. Education and publicity programs should not be left to chance but be written and costed into the recovery plan.

CASE STUDIES

How well are we doing? Recovery planning is a relatively new process in Australia, and one aim of the workshop was to evaluate successes and failures and learn from them. Examining case studies is one way of achieving this aim. The case studies below relate to selected Western Australian recovery plans with which I am familiar.

Western Swamp Tortoise *Pseudemydura umbrina*

The Western Swamp Tortoise is critically endangered and is the world's most endangered tortoise or turtle. Research studies and management actions aimed at its conservation have been under way since the 1960s. Work up to 1989 was summarized in a management program (Burbidge *et al.* 1990) and this formed the basis of a full recovery plan for the species for the period 1992–2002 (Burbidge and Kuchling 1994), which was written by recovery team members and reviewed by the whole recovery team. By the time that the recovery plan was written, the biology and ecology of the species were sufficiently well known to prescribe recovery actions with some confidence — many of the problems (e.g., captive breeding) had already been solved and recovery actions could be prescribed with confidence.

Recovery actions centred on monitoring of the single wild population, management of the remaining habitat in two small nature reserves, purchase of additional habitat to add to one of the reserves, captive breeding, translocation to establish a second population, and education and publicity. Funding for the plan was obtained from a variety of sources, including CALM, ANCA's Endangered Species Program, Perth Zoo, the Western Australian Water Authority, the World Wide Fund for Nature, and international herpetological societies, and the actions have been implemented according to schedule. Problems have been relatively minor:

- captive breeding has been more successful than anticipated leading to additional younger and smaller animals being translocated than was planned because of lack of holding space at Perth Zoo (the success of this strategy is still being evaluated);
- the recovery plan prescribed translocation to one site; the success of captive breeding now suggests that this did not allow enough flexibility and was insufficiently ambitious; a search for additional recovery sites commenced in late 1995;
- financial problems associated with cuts to the CALM budget have made it difficult (but not impossible) for that agency to meet its financial commitments;
- inadequate record keeping when costs were calculated has made the job of those administering the plan more difficult than should have been the case.

The conservation of the Western Swamp Tortoise has had a high public profile in Western Australia and is often used as an example of a successful plan, even though it has been expensive. National support has been attained through fund raising programs by the World Wide Fund for Nature Australia. It is now clear that the recovery plan will need some amendments after five years of operation; however, this should be no surprise to anyone — all recovery plans should have such review.

Lessons from the Western Swamp Tortoise Recovery Plan are:

1. Recovery plans are relatively easy to write when most biological problems have been solved.
2. Flexibility to allow for unanticipated successes is important.
3. Planned education and publicity programs can benefit the conservation of the species and threatened species conservation in general.
4. Good record keeping when calculating costs makes implementation easier.

The Woylie *Bettongia penicillata*

The Woylie (or Brush-tailed Bettong) is listed as endangered by ANZECC and in the ESP Act. A recovery plan for the Woylie was first prepared in 1990 (Hall *et al.* 1990). It included recovery actions in two states,

Western Australia and South Australia, with the South Australian component of the plan being based upon an existing management plan for establishment of the species in that State (Nelson *et al.* 1992). Funding for the recovery plan was obtained from ANCA, the Western Australian Department of Conservation and Land Management (CALM) and the South Australian Department of Environment and Natural Resources (SADENR) in 1991, and a recovery team was set up in 1992 to co-ordinate the recovery actions.

Very soon it became obvious that the plan was not adequate. There were several reasons for this:

- the first draft of the plan was written in a fairly short time frame by a consultant, who did not take account of all the facts and issues;
- allocation of responsibilities was to organizations, not individuals or positions, resulting in some actions being delayed;
- the plan did not cover all the populations of the species that occurred at the time;
- recovery actions already under way in Western Australia at the time of writing, mainly fox control via baiting, led to an unanticipated rapid increase in Woylie numbers;
- translocations of Woylies in Western Australia, in the presence of fox control, proved to be relatively easy and translocations commenced outside those prescribed in the plan;
- the costs of the South Australian component of the plan (which in retrospect was probably unnecessary for the recovery of the species in terms of de-listing, but not to the return of the species to more of its former range) were significantly under-estimated, mainly because the financial situation of the South Australian conservation agency changed;
- populations of Woylies in South Australia (which were derived from a small captive colony of Western Australian animals) proved to have lost a significant component of their genetic diversity through having too small a founder number.

In response to these changed circumstances, the recovery team quickly decided to write a new recovery plan. This was for two years only (1994–1995, reducing the total life of the recovery plan to five years), since the team considered that the species could be de-listed by the end of 1995 (Start *et al.* 1995) and that available funding should be re-directed towards more threatened taxa. Funding arrangements for the new plan were quickly finalized between the three agencies concerned, and the implementation of the new plan has proceeded without any significant problems. Woylies were categorized as "Lower Risk (Conservation Dependent)" under the 1994 IUCN Red List Categories by the Western Australia Scientific Ranking Panel for Threatened Species in mid-1995, and nominations for de-listing under Western Australian and Commonwealth legislation took place in late 1995[1].

[1]The Woylie was removed from Schedule 1 of the Endangered Species Protection Act and from the list of threatened fauna in Western Australia in April 1996.

Lessons from the Woylie Recovery Plan are:

1. Recovery plans should be written by the recovery team, not an outsider.
2. Costing must be realistic and contingency plans for budget cuts should be in place.
3. Allocation of responsibility for actions should be specific to a person or position in an organization and not be vague.
4. Recovery plans should not be too prescriptive about translocation sites; however, review of translocations should still be carried out rigorously before approval.
5. An inadequate recovery plan or one overtaken by events should be re-written as soon as possible.

Rose Mallee *Eucalyptus rhodantha*

The Rose Mallee has very large, pendulous, red flowers and silver-blue leaves. Once more widespread, it has become highly fragmented by agricultural land clearing. The species is listed as threatened pursuant to State legislation and as endangered under the ESP Act. At the time of plan writing the species was known from a total of about 500 mature plants in 11 fragmented populations. None occurred within conservation reserves.

It was a good subject for a recovery plan in the early days of plan writing because its biology and population genetics were well understood, having been the subject of a Ph.D. study, and because its spectacular flowers and popularity as a horticultural specimen ensured public support for its conservation. Since implementation commenced in 1992, there has been significant progress in implementing the recovery actions.

Issues include:

- the recovery plan was written by a consultant who worked closely with CALM staff and stayed on to help implement the plan;
- the recovery team included local farmers who were enthusiastic supporters of the recovery of the species and who were able to influence other farmers to support the actions;
- implementation of some actions has been delayed because the local CALM office had other priorities and insufficient staff to carry them out. Funds were not the issue, as a budget had been allocated to each recovery action.

Lessons from the Rose Mallee Recovery Plan are:

1. Strong local support enhances the probability of success, especially when a species is restricted to private land.
2. Conservation agency staff resources may be a limiting factor to success and need to be locked in to the plan or augmented.
3. Somebody in authority in the lead agency must be given the responsibility for ensuring that management actions are carried out.

Toolibin Lake

Freshwater wetlands supporting thickets of *Casuarina obesa* and sometimes *Melaleuca* species used to be widespread throughout the agricultural areas of southwestern Australia, before clearing for agriculture resulted in the salination of most wheat-belt wetlands and the death of emergent vegetation. Toolibin Lake, near the town of Narrogin, is the last significant example of perched freshwater wetlands with *Casuarina/Melaleuca* woodland across their bed. Concern about the possible salination of the lake has been long standing, with the first committee being set up to address the concerns as long ago as 1976. As well as nature conservation issues, landholders in the catchment of the lake were concerned that the same hydrological events were degrading their agricultural land.

After considerable research over many years a recovery plan was prepared for CALM in 1992 by engineering consultants (Bowman Bishaw Gorham *et al.* 1992). A recovery team, supported by a technical advisory group, was set up, with strong representation from landowners in the lake's catchment, and some of the necessary money has been raised from a number of sources: Alcoa of Australia, ANCA, the Western Australian Department of Conservation and Land Management, Agriculture WA, the Commonwealth Farm Forestry Program, the Toolibin Catchment Group and the Wickepin Land Conservation District Committee, landholders of the Toolibin catchment and the Water Authority of Western Australia.

The recovery plan stresses the fact that the conservation of Toolibin Lake is designed to achieve much more than saving a lake of high significance to nature conservation. It identifies five principal goals:

1. To conserve Lake Toolibin and its associated wildlife as a freshwater habitat.
2. To increase the economic production from agricultural lands within the Toolibin catchment in a manner which:
   - is sustainable in the long term (over 100 years);
   - reduces the current area of degraded land;
   - favours the conservation of local wildlife.
3. To demonstrate that, within a large catchment, it is practicable and profitable to stabilize and reverse hydrological changes that threaten both agricultural and nature conservation lands.
4. To show other land managers in Australia how to protect their agricultural and nature conservation lands.
5. To develop means for resolving divisive land conservation issues, such as drainage and disposal of groundwater.

The plan sets out recovery actions that are both ambitious and expensive. Nevertheless, much progress has been made under very difficult circumstances, although it is still too early to say whether the plan will be successful. To a significant degree, progress has been achieved because CALM was strongly committed to the plan, both at executive and Regional Manager level.

Lessons from the Toolibin Lake Recovery Plan are:

1. The plan integrates nature conservation goals into those of the wider human community of the area. Without this integration it would have had no chance of success.
2. Consultation with and co-operation of landowners in the catchment and with the many government agencies that have a stake in land management in the area was vital, both in developing and implementing the plan.
3. Ultimately, conservation of many threatened species and, especially, threatened ecological communities, depends upon retaining beneficial ecological processes across the landscape. All landowners and other agencies influencing land management need to be involved in the recovery process.

CONCLUSIONS

Recovery plans have evolved rapidly in Australia during the past few years and many have proved to be valuable guides for the recovery of threatened species. However, there is room for improvement in the content and clarity of plans, and more taxa and ecological communities, especially those that are critically endangered, need to be the subject of recovery plans. Based on my experience in writing and implementing recovery plans in Western Australia, I have reached the following conclusions, summarized from the above discussion.

1. Recovery plans should be the sole type of document driving the recovery process in Australia. Research is but one recovery action and should be integrated into recovery plans.
2. Where there are insufficient data available to clearly define all necessary recovery actions, or urgency dictates that recovery actions can not await the production of a full recovery plan, the preparation of interim recovery plans that prescribe work over a three year period, including any research needed to more clearly define recovery actions, should be undertaken.
3. Recovery plans should be prepared by or in consultation with the recovery team that will be responsible for co-ordinating implementation.
4. Recovery plans must be outcome oriented, practical, easily understood documents containing recovery actions that are achievable with a reasonable input of resources.
5. Background sections of recovery plans should contain only sufficient information to enable the reader to understand the scientific basis of the objectives and actions that are prescribed.
6. The most important parts of a plan are the recovery objectives, criteria for success and failure, and recovery actions.
7. Costings should be clear and accurate, and detailed records should be kept of how costs were developed.
8. Recovery plans should be officially adopted by relevant Ministers and conservation agencies.
9. Recovery plans should be regularly reviewed and flexible, and be amended as knowledge improves or conditions change.
10. Recovery plans require commitment of staff and resources for their implementation. Thus recovery plans need to be endorsed by the relevant State and Territory Ministers and conservation agencies, as well as by the Federal Minister for the Environment under the Commonwealth ESP Act.

ACKNOWLEDGEMENTS

I thank John Blyth, Andrew Brown, Keiran McNamara and Tony Start who generously provided advice and/or reviewed parts or all of the paper.

REFERENCES

Backhouse, G. N. and Clark, T. W., 1995. Case studies and policy initiatives in endangered species recovery in Australia: Introduction. Pp. 3–14 *in* People and nature conservation. Perspectives on private land use and endangered species recovery ed by A. Bennett, G. Backhouse and T. Clark. Transactions of the Royal Zoological Society of New South Wales. Surrey Beatty & Sons: Chipping Norton.

Bowman Bishaw Gorham, Jim Davies and associates, and Rural Planning, 1992. Recovery Plan for Lake Toolibin and Surrounding Reserves. Unpublished report prepared for the Department of Conservation and Land Management, Western Australia, under the Australian National Parks and Wildlife Service Endangered Species Program, 1991/92. Department of Conservation and Land Management: Western Australia.

Burbidge, A. A., Kuchling, G., Fuller, P. J., Graham, G. and Miller, D., 1990. The Western Swamp Tortoise. Western Australian Wildlife Management Program No. 6. Department of Conservation and Land Management: Perth.

Burbidge, A. A. and Kuchling, G., 1994. Western Swamp Tortoise Recovery Plan. Western Australian Wildlife Management Program No. 11. Department of Conservation and Land Management: Como, Western Australia.

Danks, A., Burbidge, A. A., Burbidge, A. H. and Smith, G. T., 1996. Noisy Scrub-bird Recovery Plan. Wildlife Management Program No. 12. Department of Conservation and Land Management: Perth.

Hall, G. P., Nelson, L., Storr, R. and Robinson, A. C., 1991. Recovery Plan for the Brush-tailed Bettong or Woylie *(Bettongia penicillata)*. Report submitted to Australian National Parks and Wildlife Service, Endangered Species Program Project No. 149. Department of Conservation and Land Management: Perth.

IUCN, 1994. International Union for the Conservation of Nature Red List Categories. Prepared by the IUCN Species Survival Commission. IUCN: Gland.

Nelson, L. S., Storr, R. F. and Robinson, A. C., 1992. Plan of management for the Brush-tailed Bettong, *Bettongia penicillata* Gray, 1837 (Marsupalia, Potoroidae) in South Australia. National Parks and Wildlife Service, Department of Environment and Planning: Adelaide.

Start, T., Burbidge, A. and Armstrong, D., 1995. Woylie Recovery Plan. Western Australian Wildlife Management Program No. 16. Department of Conservation and Land Management: Como, Western Australia.

Tear, T. H., Scott, J. M., Hayward, P. H. and Griffiths, B., 1995. Recovery Plans and the Endangered Species Act: are criticisms supported by data? *Conserv. Biol.* 9(4): 182–95.

Incorporating science into recovery planning for threatened species

Christopher R. Dickman[1]

A species may be considered threatened if it (1) is represented by one or a few stable but small populations, or (2) was formerly widespread or abundant but is now in decline. In the first case, it may not be appropriate to attempt recovery of the species if it has not declined; monitoring and perhaps captive breeding should be the only management needed. In the second case, research is important to identify the processes causing population decline and to ensure that management for recovery is informed and effective. A rigorous scientific methodology is advocated to identify causes of declines and is applied in case studies of the Lord Howe Island Woodhen and a small dasyurid marsupial, the Mulgara. Four steps are involved: (1) establish the magnitude and rate of population decline, and list possible explanations for it (observations); (2) identify the most plausible threatening process or processes that may be causing the decline (model); (3) derive predictions from the models (hypothesis); and (4) test critically, using null hypotheses, whether the putative threatening processes identified in the second step are causing the observed population decline, by using mensurative or manipulative experiments. When one or more processes causing population decline have been identified, management should be initiated to alleviate them and subsequent population changes monitored. It is proposed further that scientific methodology be used to identify effective management techniques, evaluate contentious practices, and guide reintroduction programs. I discuss objections that science is sometimes irrelevant for recovery of threatened species, and propose that strengthened links between research and management would increase the effectiveness of recovery planning.

Key words: Science, recovery plan, threatened species, threatening process, observations, model, hypothesis, null hypothesis, hypothesis testing, Mulgara, Lord Howe Island Woodhen.

INTRODUCTION

THE central goals of conservation biology are to slow and halt the worldwide erosion of biological diversity and to ensure the continued functioning of ecological systems. Although overall rates of species losses are difficult to quantify (May *et al.* 1995), especially for small and poorly-known taxa (Roper 1993), many extinctions of conspicuous organisms have been well documented. In Australia, at least 23 species of vertebrates and 97 species of vascular plants are believed to have become extinct in just 200 years of European settlement (Briggs and Leigh 1988; Kennedy 1990). A further 119 species and subspecies of terrestrial vertebrates and 888 species and subspecies of plants are listed as endangered or vulnerable (ANCA 1994).

There are two general approaches to stemming losses of species. First, we can try to identify areas of high conservation value and preserve them (Humphries *et al.* 1995). Such areas may be local remnants of bushland, representative land systems, intact ecosystems, or even entire continents such as Antarctica. The advantage of this approach is that small, inconspicuous taxa should be preserved (in addition to many larger taxa), and ecological processes may be allowed to continue (but see Prendergast *et al.* 1993). The rationale and methodology of this approach has been well summarized by Pressey and Nicholls (1989a,b), Pressey *et al.* (1993, 1996) and Margules and Nicholls (1994); it is not considered further here. Second, we can list species (or subspecies, genetic groups or other entities; see Rojas 1992) that are of concern, and attempt recovery before extinction occurs. Recovery may entail population and genetic management, habitat or resource management, or captive breeding and subsequent reintroduction programs. This approach has been used widely by government and non-government agencies, such as the World Wide Fund for Nature, and has the advantage that conservation efforts can be directed efficiently at target species. However, the species recovery approach can also be fraught with problems. Some of these, such as limited funding, involvement of all major stakeholders, community concerns and the operation of recovery teams, are well known and are discussed by other contributors to this volume. Other problems,

[1]Institute of Wildlife Research and School of Biological Sciences, University of Sydney, New South Wales, Australia 2006.

Pages 63–73 *in* BACK FROM THE BRINK: REFINING THE THREATENED SPECIES RECOVERY PROCESS ed by S. Stephens and S. Maxwell. Surrey Beatty & Sons, 1996.

including which species to recover, how to go about it, and in particular the role of science in recovery planning, are fundamental but appear not to be widely acknowledged.

In the present paper, I describe briefly the process of recovery planning in Australia, and discuss which species should be selected for recovery. I advocate use of a more rigorous scientific approach to identifying causes of species declines than is often adopted, and examine the importance of science in advising management prescriptions. Finally, I discuss objections that science is not relevant in recovery planning.

WHICH SPECIES SHOULD BE SELECTED FOR RECOVERY PLANNING?

Species with small populations or restricted geographical distributions are most often considered to be of conservation concern, and hence are well represented on any listing of threatened species. However, not all need to be subject to recovery planning. Caughley (1994) drew a clear distinction between small populations that were stable and those that were declining, and noted that only the latter posed relevant problems for conservation. Thus, if a small population has long been small, perhaps because it occupies a restricted habitat, it is not appropriate to attempt recovery; monitoring and possibly captive breeding as a hedge against catastrophe may be all that is needed. Examples from ANCA (1994) would include the Pedra Branca Skink *Niveoscincus palfreymani,* Hamelin Ctenotus *Ctenotus zastictus* (see Cogger *et al.* 1993) and Barrow Island Euro *Macropus robustus isabellinus.* In contrast, if a once large or broadly distributed population is now small and continuing to decline, recovery planning is clearly warranted. Present population size and rate of decline should best prioritize species most in need of recovery action, but factors such as cuteness, biological "importance" and taxonomic uniqueness are often also invoked (Strahan 1989; Norton 1994).

PROCESS OF RECOVERY PLANNING

After identification of species (or ecological communities) in need of recovery action, a recovery team is formed and evaluates whether enough information is available to construct a national recovery plan (Fig. 1). If so, the plan is written by the team, circulated for comment and then forwarded by the Endangered Species Advisory Committee (ESAC) to the Minister for approval. The approved plan is then implemented. If insufficient information is available initially to write the recovery plan, research and survey actions may be carried out as specified under a research plan, and the results used to guide preparation and implementation of the recovery plan (Fig. 1). Recent discussion indicates that research plans should in future be subsumed within recovery plans (Workshop Reports, this volume); this would unify documentation and add urgency to the recovery process, but otherwise not result in substantive change to the actions depicted in Figure 1. The aims of recovery plans may be specific, such as to secure existing populations or establish new ones in parts of a species' former range, but also require an overall objective such as to down-list a species from endangered to vulnerable within a specified time frame (ANCA 1995). Recovery criteria, actions needed, and budgets must also be specified in a recovery plan.

Interestingly, the role of science in recovery planning has been assumed by ANCA (1995) but not stated. I have noted where scientific methodology should best inform the construction of recovery plans on Figure 1, and next provide relevant background and case studies.

SCIENTIFIC METHODOLOGY

Although there is some disagreement about the nature and prosecution of science (e.g., Nagel 1961; Kuhn 1962), especially conservation science (Murphy 1990), the refutationist logic of Popper (1968) provides a particularly rigorous structure to distinguish among competing explanations for any observation. Following observation of some pattern, such as a decline in population size or geographical distribution of a species, the next step is to identify a plausible explanation, or set of explanations, to account for it. Using the terminology of Underwood (1990, 1991), such explanations may be called models (Fig. 2).

Models explaining the decline of a target species might include over-hunting, depredation by introduced carnivores, loss of habitat or other environmental perturbations (Dickman *et al.* 1993a). When models have been listed, the next step is to derive predictions, or hypotheses, from them. For example, if a model explains a species' decline as due to loss of habitat, the hypothesis may be stated that the species will occur in greater numbers where its habitat remains intact but not where the habitat has been removed. Conversely, if a model invokes over-hunting, the hypothesis should be that the species will occur in greater numbers where hunting pressure is light or absent but not where hunting pressure is great. Where several plausible models exist, they can be distinguished only if contrasting hypotheses are derived from them (Underwood 1990). Evaluation of these hypotheses should then be carried out, ideally using experimental tests. Testing may use either mensurative experiments (as, for example, by measuring densities of a target species in intact and degraded habitats or in areas where it is subjected to different levels of hunting pressure) or manipulative experiments where treatments are imposed by the experimenter (Hurlbert 1984).

It is important to note here that the protocol for hypothesis testing is made simpler by seeking refutation rather than proof. For example, an hypothesis stating that species *a* does not occur in degraded habitat could be proven only if all degraded habitat was searched and the species demonstrated to be absent. Even then, proof for the hypothesis would depend on using reliable census methods, simultaneous search of all areas of degraded habitat to ensure that individuals of species *a* did not move evasively among census points, and certainty that all degraded habitat had been identified. Conversely, the hypothesis would be refuted unequivocally by the observation of a single individual of species *a* in degraded habitat at any time during the survey program.

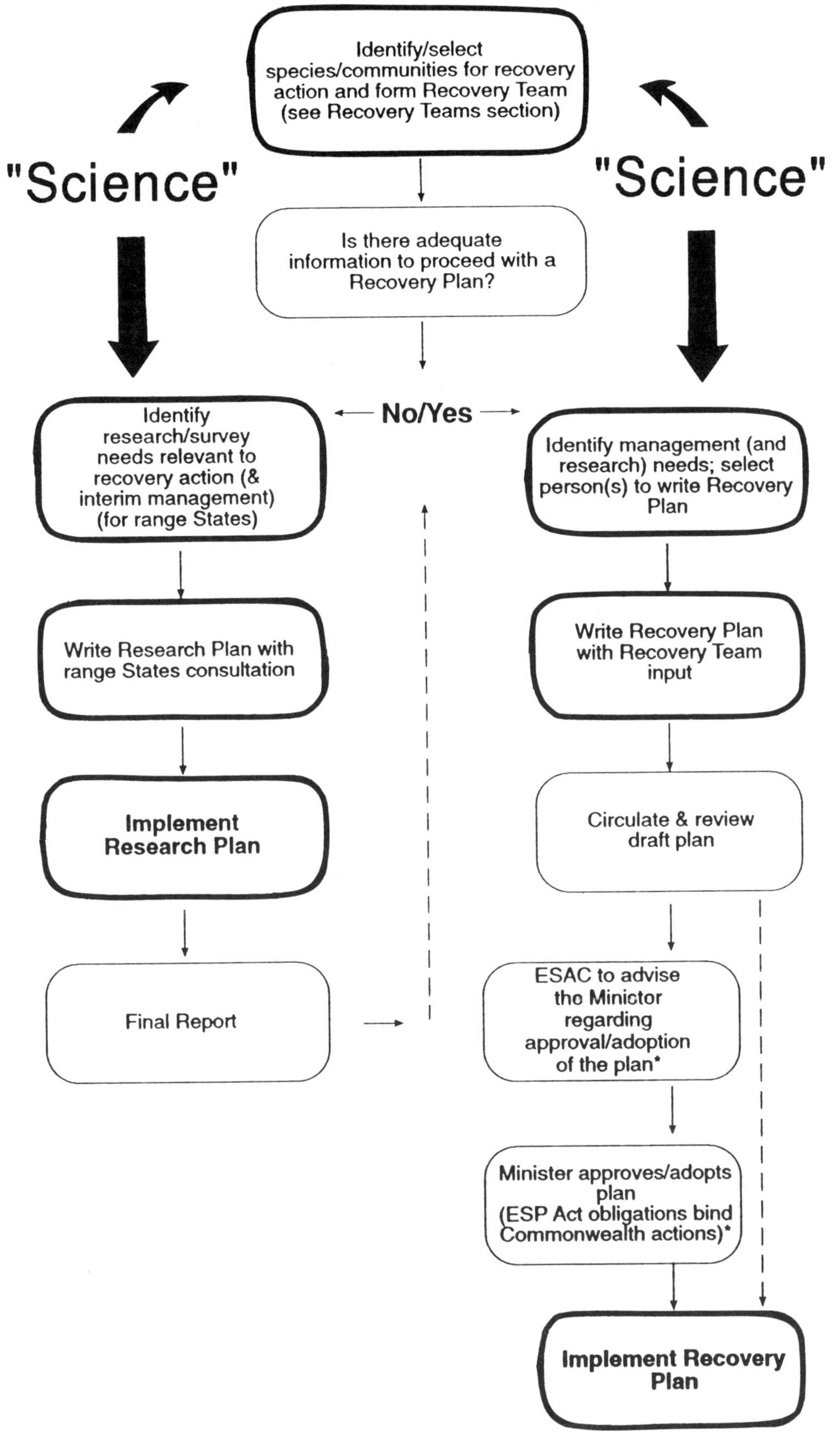

Fig. 1. Procedure for preparing a recovery plan, showing where science would be best incorporated. Boxes with thick lines show areas where scientific input can be most important, boxes with thin lines where it is least important or irrelevant. The steps identified with * are required if the plan is to be approved/adopted under the Commonwealth Endangered Species Protection Act. Modified from ANCA (1995).

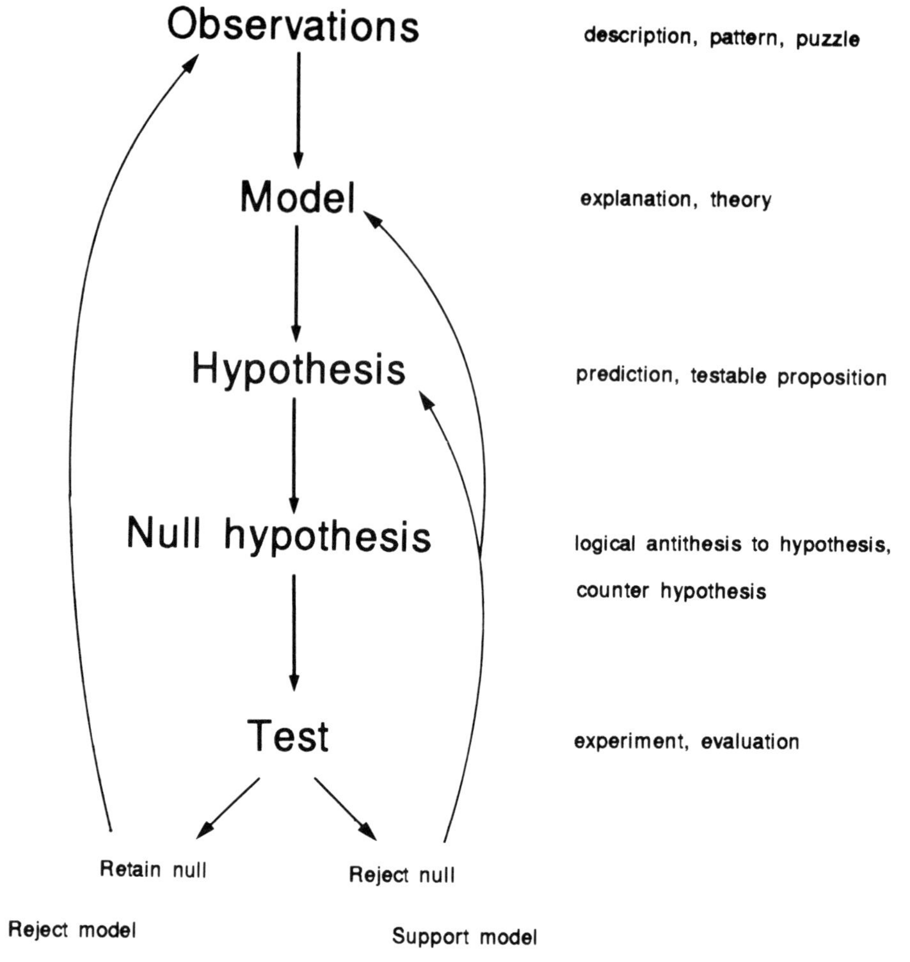

Fig. 2. Test procedure allowing experimental evaluation of explanations of any observed pattern or puzzle. Redrawn from Underwood (1991).

Because of the relative simplicity of obtaining disproof, and the certainty associated with finding it, hypothesis testing can be made tractable by seeking refutation of null hypotheses that express all logical alternatives to original hypotheses. In this context, null hypotheses include what are sometimes called counter hypotheses. Using the example above, the null hypothesis would be that there is no difference in occurrence of species *a* between intact and degraded habitats, or that species *a* occurs mostly or entirely in degraded habitat. This null hypothesis would be refuted if sampling showed species *a* to occur mostly or wholly in intact habitat. Using refutationist logic, rejection of a null hypothesis provides support (but not proof) for the original hypothesis and the model from which it was derived, whereas retention of a null hypothesis leads to rejection of the model (Fig. 2).

In most instances, hypothesis-testing can be achieved efficiently using appropriate experimental designs and statistical inference. Such designs, using controlled and replicated experiments, should be capable of yielding unambiguous results in the presence of potentially confounding factors, but must be developed with some care. There is little opportunity to describe all design prerequisites here, but consideration should at least be given to sample sizes, statistical power, ensuring independence of sampling units (replicate control and treatment areas), randomization or interspersion of treatments among replicates, and time needed before responses can be detected and verified. Detailed discussions of these topics are provided by Mead (1988), Underwood (1981, 1990, 1991) and Fletcher and Manly (1994) among others.

SCIENTIFIC METHODOLOGY IN RECOVERY PLANNING

The scientific method can be used to identify causes of species declines and to prescribe management requirements, as shown in Figure 1. These usages are considered separately below.

Identifying causes of decline

It should be obvious that a clear diagnosis is needed of the process or processes pushing a species toward extinction before effective steps can be taken to alleviate the decline (Simberloff 1988; Brussard 1991). This point has been made many times, most cogently and eloquently by Caughley (1994) and Caughley and Gunn (1996), yet management based on intuition and gut feeling about threatening processes remains prevalent. Following Figure 2, and also reformulating the recommendations of Caughley and Gunn (1996), the cause(s) of a species' decline can be identified by several steps:

1. OBSERVATIONS: Establish the magnitude and rate of a population's numerical or distributional decline. Trend analysis and historical/sub-fossil data can be used. List potentially threatening processes that may be causing the decline. Then use spatial and temporal correlation analyses and knowledge of the natural history of the target species to determine which of the potential threatening processes best explain the patterns of decline.
2. MODEL: Identify the threatening process or processes that are most likely to be causing the decline of the target species.
3. HYPOTHESIS: Derive a testable hypothesis from the model; if several models are plausible, derive contrasting hypotheses from them.
4. TEST: Critically test hypotheses, using mensurative or manipulative experiments. Testing is made possible by stating relevant *null hypotheses*.

General examples of how to detect trends in population or distributional declines are provided by Morton and Baynes (1985), Caughley and Sinclair (1994) and Caughley and Gunn (1996). Two case studies utilizing all steps are described briefly below.

A. Lord Howe Island Woodhen *Tricholimnas sylvestris*: Restricted to Lord Howe Island, the woodhen is a member of the Rallidae, a family that has suffered enormous loss throughout the Pacific region over the last few centuries (Steadman 1995). The woodhen once occurred over most of Lord Howe Island, but began to decline with the arrival of European visitors and settlers from the late 1700s and early 1800s. Detailed accounts of the decline and recent recovery of woodhens have been collated from several sources (Miller and Kingston 1980; Fullagar 1985; Miller and Mullette 1985; Harden 1990; Garnett 1992), and allow accurate diagnosis of the impact of one threatening process, the feral pig *Sus scrofa*. Steps were as follows:

1. OBSERVATIONS: By the 1970s, woodhen had declined to occupy two mountain summits on Lord Howe Island, Mt Gower and Mt Lidgbird. Only 20–25 birds (6–10 breeding pairs) remained from a former population of many hundreds. Decline was possibly caused by introduced predators (cats *Felis catus*, black rats *Rattus rattus*, goats *Capra hircus*, pigs, Tasmanian Masked Owls *Tyto novaehollandiae*), an endemic predator (Lord Howe Currawong *Strepera graculina crissalis*), lack of food or nest/roost sites, and disease, or by combinations of these factors. However, an inverse spatial relationship between pigs and woodhens implicated pigs as the most important threatening process. No spatial or temporal correlation occurred between woodhen and any other potential threatening process.
2. MODEL: Feral pigs caused the decline of the Lord Howe Island Woodhen.
3. HYPOTHESIS: If pigs were removed from areas formerly occupied by woodhens, the birds should increase in abundance in the newly-vacated areas.
4. NULL HYPOTHESIS: Removal of pigs will have no effect, or detrimental effects, on numbers of woodhen.

   TEST: Remove pigs, and measure the response of the birds. Virtually the whole pig population was removed over a two year period between 1979 and 1981. Woodhens moved immediately into pig-free areas and, supplemented by a captive breeding and release program, have since increased to 170–200 individuals and 50–60 breeding pairs, and occupy much of the island (Fig. 3). Reject the null hypothesis.

In this instance the experimental removal of pigs was neither controlled nor replicated, but the population response of woodhen was so great that the null hypothesis of no impact from pigs could be rejected unambiguously. The removal of pigs also provided little insight into how they impacted on woodhens; competition for food, disturbance to the habitat, and especially direct predation, could all be posited. However, while such insight might be of interest, it is of little relevance from a management point of view.

B. Mulgara *Dasycercus cristicauda*: Recent genetic analysis indicates that the "Mulgara" should be split into two species, the nominate form *Dasycercus cristicauda* and the Ampurta *D. hillieri* (M. Adams, pers. comm.). However, because of uncertainty about the distributional overlap of these species and their evident ecological similarity, for illustrative purposes I consider both together here. Once distributed broadly across most of inland Australia, the two species of *Dasycercus* had declined regionally from southern and eastern areas by 1950 (Fig. 4), and locally from northern areas in the 1970s (Woolley 1990). Declines in distribution and abundance continue, with a decrease at one known site in the Simpson Desert occurring precipitously between 1992 and 1993 (Fig. 5). At this latter site, small mammals have been studied in detail since 1990 (Dickman *et al.* 1993b, 1995; Predavec 1994); strong circumstantial evidence implicates depredation by feral cats and foxes as the cause of local decline of Mulgaras (Fig. 5). Experimental removals of these predators have been initiated recently to investigate the effects on Mulgaras and on native rodents (P. S. Mahon, pers. comm.; C. R. Dickman, unpubl.). Steps in the diagnostic procedure can be summarized:

1. OBSERVATIONS: Mulgaras have declined from the southern and eastern parts of their former range, and patchily elsewhere. Distribution and abundance continue to decline. Decline could possibly be caused by predators (foxes, cats), habitat change (fire, grazing, rabbits), and disease, or a combination of these factors. Spatial and temporal correlation analyses at one monitored site most strongly implicated foxes and feral cats as causes of decline. Dietary analyses of the predators and results of Mulgara radiotracking further incriminated the role of predation.
2. MODEL: Feral cats and/or foxes caused the local decline of Mulgaras.
3. HYPOTHESIS: If cats and foxes are removed from areas where Mulgaras have declined, Mulgara numbers should subsequently increase.

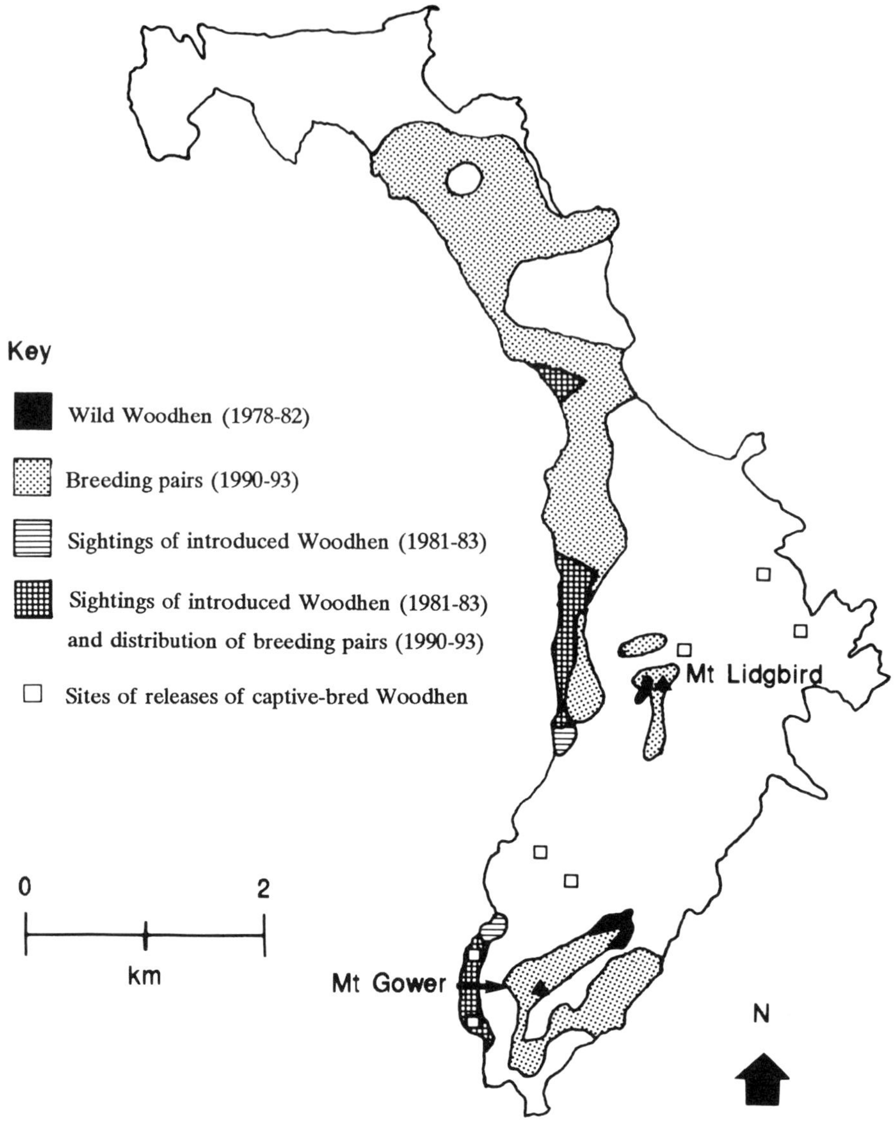

Fig. 3. Distribution of the Lord Howe Island Woodhen before and after recovery. Modified from Caughley and Gunn (1996).

4. NULL HYPOTHESIS: Removal of foxes and/or cats will have no effect, or detrimental effects, on numbers of Mulgaras.

   TEST: From replicated study areas, remove foxes alone, cats alone, foxes and cats together, and compare Mulgara numbers with those in replicated control areas where predators are left intact. Test for differences using 2-factor analysis of variance. Experimental manipulations and monitoring continue (P. S. Mahon, pers. comm., C. R. Dickman, unpubl.), hence acceptance or rejection of the null hypothesis cannot yet be specified.

Many further studies have successfully used scientific methodology to identify causes of declines (e.g., Savidge 1987; Kinnear *et al.* 1988; Van Rensburg and Bester 1988; review in Caughley and Gunn 1996); the work of Pearce *et al.* (1995) on the endangered Helmeted Honeyeater *Lichenostomus melanops cassidix* is particularly instructive. Other studies have eschewed scientific methodology, and their reliance on guesswork or diagnosis by gut feeling has resulted in costly and spectacular failure (Springer *et al.* 1978; Wiemeyer *et al.* 1988; Kiff 1989; Rabinowitz 1995; see also Irwin and Wigley 1993).

Guidance for management

Scientific methodology can assist in guiding management for species recovery in several ways.

First, following from the section above, it can identify processes causing species declines and thus ensure that management is directed at effective control. This may require culling of predators or competitors, modification or protection of habitat, provision of food or other resources, treatment of disease, or artificial boosting of a very small population to overcome threshold problems such as the Allee effect. Monitoring must also be carried out as part of the management process to ensure that control is effective and that recovery targets are achieved. When the decline of a species has two or more causes, it is also possible to prioritize possible management actions. In the simplest case, where the identified threatening processes do not interact, the level of control needed for each process can be gauged

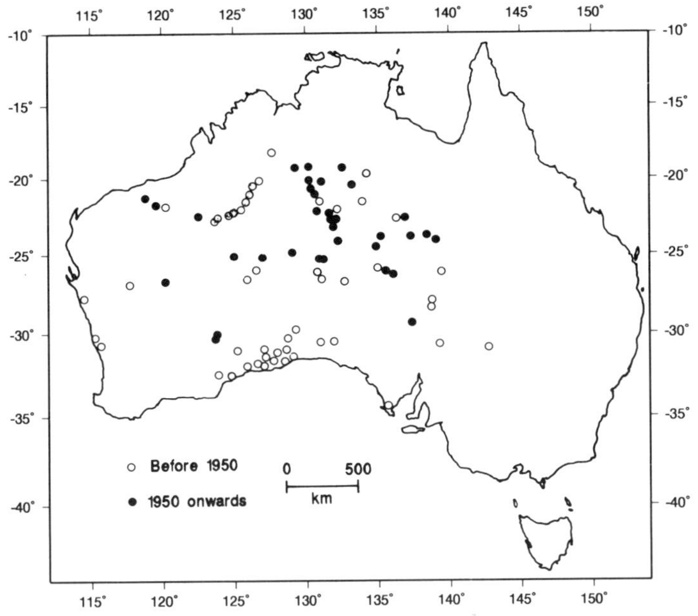

Fig. 4. Distribution of the Mulgara *Dasycercus cristicauda*, pre-1950 and from 1950 onwards. "Mulgaras" as depicted here may include two species, *D. cristicauda* and *D. hillieri*.

from its relative contribution to the species' decline. If there is interaction, control effort will need to be particularly responsive to monitoring and shift between the processes as needed. Examples of interpreting this more complex situation are given in Caughley and Gunn (1996).

Second, experimental evaluation may be used to determine the effectiveness of alternative control measures for any threatening process. For example, if an introduced species is identified as the cause of decline of a native species, the logical treatment is to control or remove it. If several potential culling methods are available and the hypothesis is that they differ in effectiveness, the null hypothesis may be stated simply that there is no difference in effectiveness. This can be tested by measuring the culling success (or cost-effectiveness) of each method; the null hypothesis is rejected if analysis reveals differences in culling success among the methods. The most effective method, or methods, of culling are then known and may inform management decisions. Experimental evaluations of baiting and culling methods have been made for several species of introduced mammals in Australia and New Zealand (e.g., rabbits, Foran *et al.* 1985; feral cats, Clapperton *et al.* 1994); an exemplary study is that of Williams and Moore (1995) which evaluated the separate and combined effects of four commonly-used methods for culling rabbits.

Third, experimental evaluation can be used to test the effectiveness of contentious management practices. Consider habitat corridors. These are often believed to allow movements of biota between remnant patches of habitat. On the one hand such movements may contribute to population persistence in remnant patches, as well as maintenance of genetic variation and regional diversity. On the other, they may allow spread of diseases or exotic pests (Simberloff and Cox 1987). Unfortunately, with one or two conspicuous exceptions (e.g., the "tunnel of love" for the Mountain Pygmy-possum *Burramys parvus*, Mansergh and Broome 1994), the relative merits of corridors remain largely speculative.

To evaluate critically the model that corridors enable movements of target biota into habitat remnants, an appropriate hypothesis is that remnants connected by corridors to source areas will be recolonized more quickly than remnants without corridors. The null hypothesis is that recolonization of remnants connected by corridors to source areas will be the same or slower than recolonization of unconnected remnants (Nicholls and Margules 1991). Testing the null hypothesis ideally requires creation of a number, n, of remnant patches located at similar distances from source habitat. Of these, $n/2$ should be connected to the source habitat and cleared of any target biota, and $n/2$ should be unconnected to the

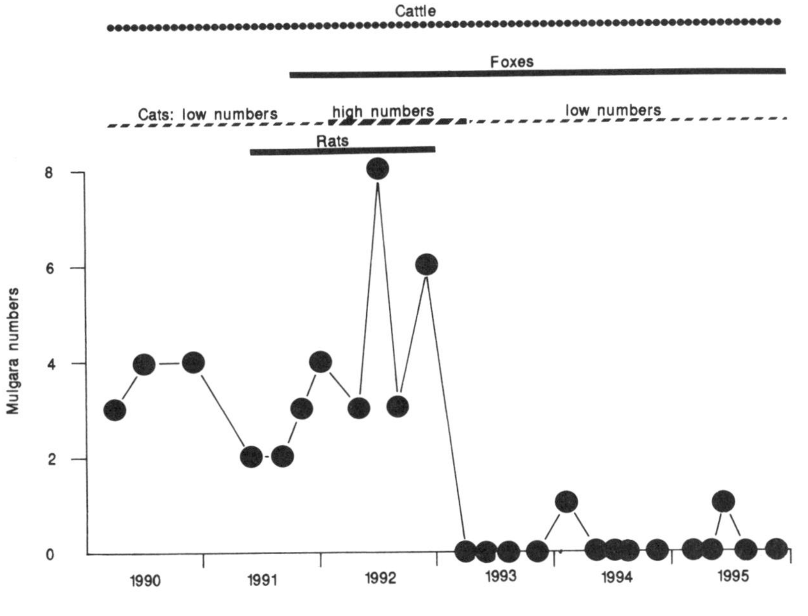

Fig. 5. Numbers of Mulgaras known to be alive from 1990 to 1995 at Ethabuka Station, Simpson Desert, Queensland, in relation to the presence of several possible threatening processes. C. R. Dickman and P. S. Mahon, unpubl.

source habitat and cleared of the target biota. If the null hypothesis is rejected, following sampling and analysis of numbers of the target biota in the remnant patches, corridors would be established as a reliable management option for allowing recolonization of remnant habitat. In this instance, clearing of the target biota from remnants prior to sampling ensures that only colonists are measured. Defaunation would not be necessary if colonists could be distinguished from residents by live-trapping and marking, or other means. However, movements of target biota into already occupied remnants could be slowed if territorial interactions occur or the habitat is saturated. More detailed consideration of alternative designs and statistical analyses of corridor experiments is provided by Nicholls and Margules (1991) and Inglis and Underwood (1992).

A final use of science in guiding management that is considered here is in reintroduction programs. If a species has declined throughout much of its geographical range and now exists in one or a small number of locations, its chance of persistence may be improved by establishing further populations. Such establishments should be preferably in locations where the species occurred previously, although this may not always be possible (IUCN 1987). The methodology for a successful reintroduction differs in one important respect from that discussed in the examples above. Thus, reintroductions take place into locations where no evaluation of threatening processes has been made, whereas diagnoses of population declines and evaluations of methods to reverse them are made where the problems occur. Sometimes the same threatening process may operate throughout the range of a species, so that neutralizing it in a reintroduction site should increase the chance of the reintroduction succeeding. However, threatening processes may also differ geographically. For example, Brush-tailed Bettongs *Bettongia penicillata* were eradicated on St Francis Island, South Australia, by a combination of habitat destruction and feral cats (Robinson and Smyth 1976), whereas extant populations in Western Australia were depressed by fox predation and appear to have increased following fox control (Christensen 1980; A. A. Burbidge, pers. comm.). To overcome the problem of geographical differences in threatening processes, pilot studies may be required. These should involve release of a small "probe group" of individuals (Caughley 1994) to confirm that putative threatening processes in the reintroduction site have been effectively controlled. A case study adopting this approach is outlined below.

Brush-tailed Phascogale *Phascogale tapoatafa*: This species occurs in coastal and sub-coastal forest in eastern, western and northern Australia and is considered to be uncommon to rare throughout its range (Soderquist 1995a). It has declined regionally in southern parts of its range, and populations are often localized. A reintroduction program was carried out by Soderquist (1995b) in Gippsland, Victoria, using the following steps:

1. OBSERVATIONS: Phascogales have declined over more than 40% of their original range in Victoria since European settlement. Extinctions occurred over most of Gippsland two decades ago. Animals appear to be less abundant in fragmented habitat, and the predominantly stringybark-silvertop forests of Gippsland appear also to provide less food than box-ironbark forests elsewhere in Victoria (Soderquist 1995b). In addition, predator densities (feral cats and foxes) are relatively high in Gippsland.
2. MODELS: (a) Predation from feral cats and foxes, (b) food limitation, and (c) habitat fragmentation, or combinations of these factors, have contributed to the observed declines.
3. HYPOTHESES: (a) Survival of phascogales will be greater if released into predator-controlled areas than into uncontrolled ones. (b) Foraging success, body weight, growth and condition will be less in phascogales released into stringybark-silvertop forest than into box-ironbark forest. (It is important to note that the predictions of this hypothesis refer only to the proximate effects of food limitation on phascogales. It is assumed that animals with poor foraging success and body condition ultimately have reduced survival or produce fewer young). (c) Phascogales released into fragmented habitat will survive more poorly than those released into continuous habitat.
4. NULL HYPOTHESIS (a): Survival of phascogales reintroduced to areas with and without foxes and feral cats will not differ, or will be greater in areas with predators.

   TEST: Remove predators from some areas but not others, and compare survival of probe groups of reintroduced phascogales. Results showed that control of foxes (cats could not be controlled) increased the survival of phascogales, thus allowing rejection of the null hypothesis. An additional protocol, to increase predator awareness among translocated juveniles, improved survival even more dramatically, further confirming the importance of predation.

   NULL HYPOTHESIS (b): Foraging success and condition of phascogales reintroduced to stringybark-silvertop forest and box-ironbark forest will not differ between forest types, or will be greater in the first forest type.

   TEST: Release probe groups of phascogales into the two forest types and monitor subsequent foraging success and condition. Results showed that, while diets of phascogales were similar between the forest types, foraging success and individual condition were less in the stringybark-silvertop forest than in the box-ironbark. The null hypothesis is rejected.

   NULL HYPOTHESIS (c): Survival of phascogales released into fragmented habitat will be the same or greater than those released into continuous habitat. This null hypothesis was not tested explicitly, although further specific hypotheses relating to male dispersal were.

Due to logistical constraints, Soderquist (1995b) was unable to adequately control or replicate his experiments, or to evaluate all models and combinations of models. However, rejection of the two null hypotheses tested was important and convincing due to the large amount of information obtained on experimental animals from direct observation and radiotracking, and critical evaluation of findings at all stages of monitoring. The study therefore clarified managerial requirements for a broader-scale reintroduction of *Phascogale tapoatafa* into the depleted Gippsland forests.

Several other studies that have successfully used scientific methodology to reintroduce taxa include Suckling and MacFarlane 1983 (see also Suckling and Goldstraw 1989), Walters *et al.* (1992) and Serena (1995), while further discussion of protocols for reintroductions is given by Dickman (1992) and Armstrong *et al.* (1995). Management based on intuition and gut feeling sometimes also produces successful reintroductions, but the list of failures is long (see Short *et al.* 1992 for a summary) and testament that intuitive management alone does not suffice.

Objections to the scientific approach to recovery planning

Despite the rigour of scientific methodology, it is sometimes perceived to be irrelevant in recovery planning. This perception is especially prevalent with respect to identifying causes of species' declines, and is based on several arguments. First, populations of threatened species are often so small or scattered that model construction and hypothesis testing are rendered impossible. It is true that small numbers impose difficulties. However, it should still be possible to construct a plausible model or models accounting for a species' decline even if very small or remnant populations remain. The Eastern Barred Bandicoot *Perameles gunnii* and Leadbeater's Possum *Gymnobelideus leadbeateri* are good examples of this (Maguire *et al.* 1990; Lindenmayer and Possingham 1994). It should be possible further to predict the consequences of imposing a management treatment on a small population. Although seldom stated explicitly, this is equivalent to erecting an hypothesis. If a null hypothesis is constructed that describes all logical alternatives to the hypothesis, it can then be tested explicitly by imposition of the management treatment. This approach has been used successfully on several very small populations, including the Lord Howe Island Woodhen, North Island Kokako *Callaeas cinerea wilsoni* (Innes 1992–95; Innes and Hay 1996) and Helmeted Honeyeater (Pearce *et al.* 1995).

Although convincing results may be obtained from experimental management of single populations, stronger inferences can be derived if two or more populations of threatened species are available and control-treatment comparisons are made. A variety of experimental designs is available, including before-after-control-impact (BACI) and asymmetrical designs, as described in Potvin and Roff (1993), Fletcher and Manly (1994) and Caughley and Gunn (1996).

A second argument for dismissing science in recovery planning is that it often just confirms what we already suspect. For example, if we suspect that a predator is depressing the numbers of some target species and then confirm this by experiment, it may be tempting to view the scientific method as a waste of time or at best as an expensive luxury. In this instance, however, predator management would have a sound basis, and appropriate levels of predator control might be deduced from the experimental results. Conversely, if our initial suspicion was wrong and predators have no impact on the target species, the scientific method could prevent wasteful allocation of management resources. A relevant example is provided by the management of the Aleutian Canada Goose *Branta canadensis leucopareia* on islands in the Bering Sea. A suspicion that bird numbers were declining due to predation from the Arctic Fox *Alopex lagopus* led to a costly program of fox eradication on the islands between 1949 and 1976. The program did nothing to stem the decline, hence confounding the original suspicion. Goose numbers increased eventually when human hunting was banned at wintering grounds distant from the Aleutian Islands (Springer *et al.* 1978).

Finally, rates of decline of some species populations may be so great that there appears to be insufficient time for testing hypotheses about threatening processes. In this situation, it may be argued that if there is time for a management treatment to be imposed at all, there should be time also to evaluate its effects. If evaluation is set up as a test of an appropriate null hypothesis (the model here is the putative threatening process, the hypothesis may be that removal of the threatening process will produce an increase in population size of the target species), ideally by comparing treatment with control populations, scientific methodology will have been used. Successful use of the scientific approach has been made on rapidly declining populations of Kokako in New Zealand (Innes 1992–95) and on several further species summarized in Caughley and Gunn (1996).

CONCLUSIONS

The scientific method can be a powerful tool in recovery planning, firstly in identifying causes of species declines, and secondly in evaluating recovery techniques. It is important to emphasize, however, that science is only one of several elements in the recovery process, and serves best to advise management decisions. Several authors have noted the need for increasing the level of communication among all concerned with recovering threatened species, especially between researchers and managers (e.g., Murphy and Noon 1992; Schemske *et al.* 1994). A particularly successful linkage has been developed for North Island Kokako using a "research-by-management" approach (Innes 1992-95), and a framework integrating ecological research and management decisions has been proposed recently by Underwood (1995). Strengthened and explicit linkages between the scientific and managerial processes should increase the effectiveness of the recovery process.

ACKNOWLEDGEMENTS

I thank Sally Stephens for the invitation to prepare this contribution, numerous colleagues at the Back from the Brink conference for discussion of ideas, Alan Saunders for insight into the research-by-management approach being pioneered in New Zealand, and Gee Chapman, Paul Mahon, Tony Underwood and two anonymous referees for constructive comments on the manuscript.

REFERENCES

Armstrong, D. P., Soderquist, T. and Southgate, R., 1995. Designing experimental reintroductions as experiments. Pp. 27–29 *in* Reintroduction Biology of Australian and New Zealand Fauna ed by M. Serena. Surrey Beatty & Sons: Chipping Norton.

Australian Nature Conservation Agency (ANCA), 1994. *Endangered Species Protection Act 1992*. Schedules 1, 2 and 3 — July 1994. Australian Nature Conservation Agency: Canberra.

Australian Nature Conservation Agency (ANCA), 1995. Recovery Plan Guidelines for Endangered and Vulnerable Species and Endangered Ecological Communities. Australian Nature Conservation Agency: Canberra.

Briggs, J. D. and Leigh, J. H., 1988. Rare or Threatened Australian Plants. Australian National Parks and Wildlife Service: Canberra.

Brussard, P. F., 1991. The role of ecology in biological conservation. *Ecol. Appl.* **1**: 6–12.

Caughley, G., 1994. Directions in conservation biology. *J. Anim. Ecol.* **63**: 215–44.

Caughley, G. and Gunn, A., 1996. Conservation Biology in Theory and Practice. Blackwell Science: Oxford.

Caughley, G. and Sinclair, A. R. E., 1994. Wildlife Ecology and Management. Blackwell Scientific Publications: Oxford.

Christensen, P. E. S., 1980. A sad day for native fauna. *Forest Focus* **23**: 3–12.

Clapperton, B. K., Eason, C. T., Weston, R. J., Woolhouse, A. D. and Morgan, D. R., 1994. Development and testing of attractants for feral cats, *Felis catus* L. *Wildl. Res.* **21**: 389–99.

Cogger, H. G., Cameron, E. E., Sadlier, R. A. and Eggler, P., 1993. The Action Plan for Australian Reptiles. Australian Nature Conservation Agency: Canberra.

Dickman, C. R., 1992. Predation and habitat shift in the House Mouse, *Mus domesticus*. *Ecology* **73**: 313–22.

Dickman, C. R., 1993. The Biology and Management of Native Rodents of the Arid Zone in New South Wales. New South Wales National Parks and Wildlife Service: Sydney.

Dickman, C. R., Downey, F. J. and Predavec, M., 1993b. The Hairy-footed Dunnart *Sminthopsis hirtipes* (Marsupialia: Dasyuridae) in Queensland. *Aust. Mammal.* **16**: 69–72.

Dickman, C. R., Predavec, M. and Downey, F. J., 1995. Long-range movements of small mammals in arid Australia: Implications for land management. *J. Arid Env.* **31**: 441–52.

Dickman, C. R., Pressey, R. L., Lim, L. and Parnaby, H. E., 1993a. Mammals of particular conservation concern in the Western Division of New South Wales. *Biol. Conserv.* **65**: 219–48.

Fletcher, D. J. and Manly, B. F. J. (eds), 1994. Statistics in Ecology and Environmental Monitoring. University of Otago Press: Dunedin.

Foran, B. D., Low, W. A. and Strong, B. W., 1985. The response of rabbit populations and vegetation to rabbit control on a calcareous shrubby grassland in central Australia. *Aust. Wildl. Res.* **12**: 237–47.

Fullagar, P. J., 1985. The Woodhens of Lord Howe Island. *Avicult. Mag.* **91**: 15–30.

Garnett, S. (ed.), 1992. Threatened and Extinct Birds of Australia. Royal Australasian Ornithologists' Union, Melbourne, and Australian National Parks and Wildlife Service: Canberra.

Harden, R. H., 1990. Interim report on the Lord Howe Island Woodhen census. Unpubl. Rep., New South Wales National Parks and Wildlife Service: Sydney.

Humphries, C. J., Williams, P. H. and Vane-Wright, R. I., 1995. Measuring biodiversity value for conservation. *Ann. Rev. Ecol. Syst.* **26**: 93–111.

Hurlbert, S. H., 1984. Pseudoreplication and the design of ecological field experiments. *Ecol. Monogr.* **54**: 187–211.

Inglis, G. and Underwood, A. J., 1992. Comments on some designs proposed for experiments on the biological importance of corridors. *Conserv. Biol.* **6**: 581–86.

Innes, J., 1992–95. 1992/93 and 1994/95 Kokako research-by-management reports. Unpubl. reps, Dept. of Conservation: Wellington.

Innes, J. and Hay, R., 1996. A direct test of a population limitation hypothesis on the New Zealand mainland. *Aust. Wildl. Manage. Soc. Newsl.* **9**: 9.

Irwin, L. L. and Wigley, T. B., 1993. Toward an experimental basis for protecting forest wildlife. *Ecol. Appl.* **3**: 213–17.

IUCN, 1987. The IUCN Position Statement on Translocation of Living Organisms. Species Survival Commission: Gland.

Kennedy, M. (ed.), 1990. Australia's Endangered Species. Simon and Schuster: Sydney.

Kiff, L. F., 1989. DDE and the California Condor *Gymnogyps californianus*: The end of a story? Pp. 477–80 *in* Raptors in the Modern World ed by B. U. Meyburg and R. D. Chancellor. World Working Group on Birds of Prey and Owls: Berlin.

Kinnear, J. E., Onus, M. L. and Bromilow, R. N., 1988. Fox control and rock-wallaby population dynamics. *Aust. Wildl. Res.* **15**: 435–50.

Kuhn, T. S., 1962. The Structure of Scientific Revolutions. University of Chicago Press: Chicago.

Lindenmayer, D. B. and Possingham, H. P., 1994. The Risk of Extinction. CRES, Australian National University: Canberra.

Maguire, L. A., Lacy, R. C., Begg, R. J. and Clark, T. W., 1990. An analysis of alternative strategies for recovering the Eastern Barred Bandicoot in Victoria. Pp. 147–64 *in* Management and Conservation of Small Populations ed by T. W. Clark and J. H. Seebeck. Chicago Zoological Society: Chicago.

Mansergh, I. and Broome, L., 1994. The Mountain Pygmy-possum of the Australian Alps. New South Wales University Press: Sydney.

Margules, C. R. and Nicholls, A. O., 1994. Where should nature reserves be located? Pp. 339–46 *in* Conservation Biology in Australia and Oceania ed by C. Moritz and J. Kikkawa. Surrey Beatty & Sons: Chipping Norton.

May, R. M., Lawton, J. H. and Stork, N. E., 1995. Assessing extinction rates. Pp. 1–24 *in* Extinction Rates ed by J. H. Lawton and R. M. May. Oxford University Press: Oxford.

Mead, R., 1988. The Design of Experiments. Statistical Principles for Practical Application. Cambridge University Press: Cambridge.

Miller, B. and Kingston, T., 1980. Lord Howe Island Woodhen *Tricholimnas sylvestris*. Pp. 17–26 *in* Endangered Animals of New South Wales ed by C. Haigh. New South Wales National Parks and Wildlife Service: Sydney.

Miller, B. and Mullette, K. J., 1985. Rehabilitation of an endangered Australian bird: the Lord Howe Island Woodhen *Tricholimnas sylvestris* (Sclater). *Biol. Conserv.* **34**: 55–95.

Morton, S. R. and Baynes, A., 1985. Small mammal assemblages in arid Australia: a reappraisal. *Aust. Mammal.* **8**: 159–69.

Murphy, D. D., 1990. Conservation biology and scientific method. *Conserv. Biol.* **4**: 203–04.

Murphy, D. D. and Noon, B. R., 1992. Integrating scientific methods with habitat conservation planning: reserve design for Northern Spotted Owls. *Ecol. Appl.* **2**: 3–17.

Nagel, E., 1961. The Structure of Science. Harcourt, Brace, World: London.

Nicholls, A. O. and Margules, C. R., 1991. The design of studies to demonstrate the biological importance of corridors. Pp. 49–61 *in* Nature Conservation 2: The Role of Corridors ed by D. A. Saunders and R. J. Hobbs. Surrey Beatty & Sons: Chipping Norton.

Norton, B. G., 1994. On what we should save: the role of culture in determining conservation targets. Pp. 23–39 *in* Systematics and Conservation Evaluation ed by P. L. Forey, C. J. Humphries and R. I. Vane-Wright. Clarendon: Oxford.

Pearce, J., Menkhorst, P. and Burgman, M. A., 1995. Niche overlap and competition for habitat between the Helmeted Honeyeater and the Bell Miner. *Wildl. Res.* **22**: 633–46.

Popper, K. R., 1968. The Logic of Scientific Discovery. Hutchinson: London.

Potvin, C. and Roff, D. A., 1993. Distribution-free and robust statistical methods: viable alternatives to parametric statistics? *Ecology* **74**: 1617–628.

Predavec, M., 1994. Population dynamics and environmental changes during natural irruptions of Australian desert rodents. *Wildl. Res.* **21**: 569–82.

Prendergast, J. R., Quinn, R. M., Lawton, J. H., Eversham, B. C. and Gibbons, D. W., 1993. Rare species, the coincidence of diversity hot spots, and conservation strategies. *Nature* **365**: 335–37.

Pressey, R. L., Humphries, C. J., Margules, C. R., Vane-Wright, R. I. and Williams, P. H., 1993. Beyond opportunism: key principles for systematic reserve selection. *Trends Ecol. Evol.* **8**: 124–28.

Pressey, R. L. and Nicholls, A. O., 1989a. Efficiency in conservation evaluation: scoring versus iterative approaches. *Biol. Conserv.* **50**: 199–218.

Pressey, R. L. and Nicholls, A. O., 1989b. Application of a numerical algorithm to the selection of reserves in semi-arid New South Wales. *Biol. Conserv.* **50**: 263–78.

Pressey, R. L., Possingham, H. P. and Margules, C. R., 1996. Optimality in reserve selection algorithms: when does it matter and how much? *Biol. Conserv.* **76**: 259–67.

Rabinowitz, A., 1995. Helping a species go extinct: the Sumatran Rhino in Borneo. *Conserv. Biol.* **9**: 482–88.

Robinson, A. C. and Smyth, M. E. B., 1976. The vertebrate fauna of Nuyts Archipelago, South Australia. *Trans. Roy. Soc. S. Aust.* **100**: 171–76.

Rojas, M., 1992. The species problem and conservation: what are we protecting? *Conserv. Biol.* **6**: 170–78.

Roper, M. M., 1993. Biological diversity of micro-organisms: an Australian perspective. *Pac. Conserv. Biol.* **1**: 21–28.

Savidge, J. A., 1987. Extinction of an island forest avifauna by an introduced snake. *Ecology* **68**: 660–68.

Schemske, D. W., Husband, B. C., Ruckelshaus, M. H., Goodwillie, C., Parker, I. M. and Bishop, J. G., 1994. Evaluating approaches to the conservation of rare and endangered plants. *Ecology* **75**: 584–606.

Serena, M. (ed.), 1995. Reintroduction Biology of Australian and New Zealand Fauna. Surrey Beatty & Sons: Chipping Norton.

Short, J., Bradshaw, S. D., Giles, J., Prince, R. I. T. and Wilson, G. R., 1992. Reintroduction of macropods (Marsupialia: Macropodoidea) in Australia — a review. *Biol. Conserv.* **62**: 189–204.

Simberloff, D., 1988. The contribution of population and community biology to conservation science. *Ann. Rev. Ecol. Syst.* **19**: 473–511.

Simberloff, D. and Cox, J., 1987. Consequences and costs of conservation corridors. *Conserv. Biol.* **1**: 63–71.

Soderquist, T. R., 1995a. Brush-tailed Phascogale *Phascogale tapoatafa*. Pp. 104–06 *in* The Mammals of Australia ed by R. Strahan. Reed Books: Sydney.

Soderquist, T. R., 1995b. The importance of hypothesis testing in reintroduction biology: examples from the reintroduction of the carnivorous marsupial *Phascogale tapoatafa*. Pp. 159–64 *in* Reintroduction Biology of Australian and New Zealand Fauna ed by M. Serena. Surrey Beatty & Sons: Chipping Norton.

Springer, P. F., Byrd, G. V. and Woolington, D. W., 1978. Reestablishing Aleutian Canada Geese. Pp. 331–38 *in* Endangered Birds. Management Techniques for Preserving Threatened Species ed by S. A. Temple. University of Wisconsin Press: Madison.

Steadman, D. W., 1995. Prehistoric extinctions of Pacific island birds: biodiversity meets zooarchaeology. *Science* **267**: 1123–131.

Strahan, R., 1989. Why save endangered species and which ones do we save? Pp. 2–17 *in* Endangered Species: Social, Scientific, Economic and Legal Aspects in Australia and the South Pacific ed by S. Burgin. Total Environment Centre: Sydney.

Suckling, G. C. and Goldstraw, P., 1989. Progress of Sugar Glider, *Petaurus breviceps,* establishment at the Tower Hill State Game Reserve, Victoria. *Vic. Nat.* **106**: 179–83.

Suckling, G. C. and Macfarlane, M. A., 1983. Introduction of the Sugar Glider, *Petaurus breviceps,* into re-established forest of the Tower Hill State Game Reserve, Victoria. *Aust. Wildl. Res.* **10**: 249–58.

Underwood, A. J., 1981. Techniques of analysis of variance in experimental marine biology and ecology. *Oceanogr. Mar. Biol. Ann. Rev.* **19**: 513–605.

Underwood, A. J., 1990. Experiments in ecology and management: their logics, functions and interpretations. *Aust. J. Ecol.* **15**: 365–89.

Underwood, A. J., 1991. The logic of ecological experiments: a case history from studies of the distribution of macro-algae on rocky intertidal shores. *J. Mar. Biol. Assoc. U.K.* **71**: 841–66.

Underwood, A. J., 1995. Ecological research and (and research into) environmental management. *Ecol. Appl.* **5**: 232–47.

Van Rensburg, P. J. and Bester, M. N., 1988. The effect of cat *Felis catus* predation on three breeding Procellariidae species on Marion Island. *S. Afr. J. Zool.* **23**: 301–05.

Walters, J. R., Copeyon, C. K. and Carter, J. H. III, 1992. Test of ecological basis of co-operative breeding in Red-cockaded Woodpeckers. *Auk.* **109**: 90–97.

Wiemeyer, S. N., Scott, J. M., Anderson, M. P., Bloom, P. H. and Stafford, C. J., 1988. Environmental contaminants in California Condors. *J. Wildl. Manage.* **52**: 238–47.

Williams, C. K. and Moore, R. J., 1995. Effectiveness and cost-efficiency of control of the wild Rabbit, *Oryctolagus cuniculus* (L.), by combinations of poisoning, ripping, fumigation and maintenance fumigation. *Wildl. Res.* **22**: 253–69.

Woolley, P. A., 1990. Mulgaras, *Dasycercus cristicauda* (Marsupialia: Dasyuridae); their burrows, and records of attempts to collect live animals between 1966 and 1979. *Aust. Mammal.* **13**: 61–64.

Collaboration of science and management in endangered species recovery

Peter J. Jarman[1] and Margaret A. Brock[2]

This paper clarifies the context for science in recovery management, aiming to help managers and scientists to appreciate their distinct roles in working for the recovery of endangered species. Saving endangered species is not itself a scientific aim; it is a management aim. Science uses well established methods to generate and verify facts, prove relationships, make predictions, and demonstrate the effectiveness of management actions. It does so by answering carefully framed questions. Management, on the other hand, sets aims and makes decisions about actions to achieve its aims. Science alone will not make decisions for management, but can assist decision-making by answering some kinds of questions. Managers will get the most out of scientists by framing the right questions, answerable by scientific methods. Those questions may include ones that test the effectiveness of management actions. There should be a formal place in recovery planning for collaboration between scientists and managers.

INTRODUCTION

WE sometimes see tensions between managers of endangered species and scientists. Those tensions are worth exploring because they may get in the way of full and fruitful co-operation in endangered species recovery management. Those tensions sometimes arise from managers and scientists not understanding either each other's roles or what science can and cannot contribute to species recovery. Rather than give specific examples of such tensions, we invite readers to consider recovery programs with which they are familiar and consider where tensions might have arisen, and how they were, or could have been, avoided.

Management and science have different broad cultures, goals and timetables. Science is concerned fundamentally with answering questions that advance knowledge. It is an exploratory, creative activity, that never comes to an end; it will always generate new questions to be answered or pose new explanations to be tested. Management, on the other hand, is concerned with making decisions and initiating actions to achieve the ends that people want. The open-endedness of science may frustrate a manager who, when dealing with an endangered species, needs to work quickly towards specific short-term objectives. Both scientists and managers need to recognize this difference in their disciplines' basic thinking. It does not mean that managers cannot benefit from having a scientist work on a management problem, nor that management problems are not fruitful fields of enquiry for scientists. Above all, it does not mean that the scientific *method* is irrelevant to answering management questions.

Both managers and scientists also need to see that scientists cannot answer all management questions, because some management questions are simply not scientific (i.e., answerable by the scientific method alone); and managers should also see that scientists may want to pursue some inquiries that do not seem relevant to management in the short term. A cost-benefit analysis, for example, may be a necessary part of a species recovery plan, but falls outside the expertise of most scientists. And a molecular evaluation of the phylogeny of an endangered species may not be the manager's first choice as an important step towards saving it, however interesting it may be to a scientist.

There is no harm in managers and scientists having different interests and priorities provided that the researcher is fully committed to the aims of the recovery plan and open to guidance by the recovery process. Researchers fully funded from within the recovery program invariably are committed to it. However, scientists funded from outside the recovery program need to be drawn into the recovery process, planning their research to be compatible with the objectives of the recovery plan whenever possible.

Science and management may have different attitudes towards facts and ideas. Scientists feel an ownership of ideas and data, and want to claim ownership by publishing their findings (when they are ready to). Managers take pride in making good decisions, but generally do not feel they "own" the facts or ideas on which the decisions depended. But for sound management reasons they sometimes need to keep facts secret. For example, a manager may not want the sites where an endangered species occurs to be

[1]Department of Ecosystem Management, University of New England, Armidale, New South Wales, Australia 2351.
[2]Department of Botany, University of New England, Armidale, New South Wales, Australia 2351.

Pages 74–78 *in* BACK FROM THE BRINK: REFINING THE THREATENED SPECIES RECOVERY PROCESS ed by S. Stephens and S. Maxwell. Surrey Beatty & Sons, 1996.

widely known. These cultural differences need not be a problem if they are recognized and accommodated by both sides.

This paper clarifies the context for science in recovery management, exploring the kinds of contributions that science can (and cannot) make. Other publications have had similar aims, and the recent book by Caughley and Gunn (1996) addresses conservation biology for scientists. We address both managers and scientists (whether working for a conservation agency or as "academic" scientists), in the hope that this paper will help them to value each other's roles and to see their own more clearly.

THE SCIENTIFIC METHOD

1. Gathering facts

Science has developed methods that are good for gathering verifiable facts, demonstrating relationships in processes and assigning causes; and (above all) saying how probable it is that what has been deduced is real. Those methods include framing the right questions, picking the best measurements to make, designing experiments or observations that will best answer the questions, and calculating how many repeated situations or measurements are needed to give a reliable answer. It has also developed a vast array of techniques for making the measurements, calculations and comparisons.

Natural history, by contrast, relies upon opportunistic or guided observations, without specific questions in mind, and those will produce descriptions only. Yet such naturalistic observations are often a critical starting point for scientific inquiry. For example, studies of the behaviour of animals often begin by observing, listing and describing the distinct kinds of actions (behaviours) that they perform. Then, knowing what can be measured, the scientist can answer specific questions such as: "Do peacocks that often display their tails produce more offspring than those displaying less often?" A full natural-history knowledge of a species is an invaluable start to any scientific investigation, but some points may need to be verified by planned data-gathering.

Science works by taking sets of facts or measurements and comparing them with each other or with other known facts, or finding relationships amongst them. It has developed various ways (usually called "statistics") of telling us how sure we can be that the differences or relationships are real, not just the product of chance selection of the measurements we have looked at. Science tends to use purposely collected facts (data) so that we can be sure how the facts were obtained. It would be misleading to compare two sets of facts collected in quite different ways and then to say that there were real differences between the sets. Major features of the scientific method are being discriminating about which facts are useable, and planning the ways in which to collect data to assure the most reliable interpretations.

2. Posing the right questions

Since 1960 several taxa of mammals that were thought to be extinct in Australia have been "rediscovered" by chance: Mountain Pygmy Possum *Burramys parvus* in 1966 (Warneke 1967); Bridled Nailtail Wallaby *Onychogalea fraenata* in 1973 (Gordon and Lawrie 1980); the mainland Rufous Hare-wallaby or Mala *Lagorchestes hirsutus hirsutus* in 1973 (Parker 1973); and Gilbert's Potoroo *Potorous gilbertii* in 1995.

Each opportunistic rediscovery produced a single, exciting fact: this species is still alive at this site. But that is all! To know more about the population, and especially to know enough to manage it, we need to gather further information, not by chance, but in a planned fashion. Our first questions would be: how widespread and how abundant is the population; what habitats and resources are being used; and what limits the population? Just by posing these questions we see what facts we need to answer them and can work out how to get those facts. Those original opportunistic sightings were invaluable, and many recovery programs try to increase the chances that they will happen by alerting the public to our need to hear of sightings of rare species, (e.g., through publicity campaigns for Brush-tailed Rock-wallaby *Petrogale penicillata* in New South Wales, and Proserpine Rock-wallaby *Petrogale persephone,* B. Nolan, pers. comm.). But to be really useful they must lead on to planned, scientific collection of relevant data.

Science gathers and uses facts most effectively when it has a tightly specified question to answer. Such questions often arise from a stated hypothesis, a simple statement of what might be true, which has its "null hypothesis" converse. For example, if we were interested in the size of trees in which koalas sleep, we might start with the question: Do koalas prefer to sleep in tall trees? We would then state as our hypothesis: Koalas sleep more often in tall trees than expected by chance; and the null hypothesis would be: Koalas sleep in tall trees at a random rate (i.e., as frequently as tall trees occur in relation to any other trees). We can then gather facts on sleeping koalas, the trees in which they sleep, and all the trees available to them, and test whether their rate of use of tall trees is greater than would be expected by chance; and we can describe statistically how reliable our answer is. In such a case, several questions could be asked and answered using the same data set (what species of trees do they prefer to use; do males and females prefer the same trees; does choice vary seasonally) provided we think about them in advance and collect the right data.

People may ask one question thinking they are asking another. The question "What trees do koalas use?" is different from, but sometimes used to mean, the question "What trees do koalas prefer to use?" The first is answered with a list of trees in which koalas have been seen, perhaps with the frequencies of their being seen in each kind; the second needs to be answered by comparing the frequencies of koalas being seen in each species of tree with the frequencies of those trees being available for koalas. Koalas might prefer a very rare tree, yet, just because the tree is so rare, be seen in specimens of it very rarely. Loose thinking about questions bedevils science. Scientists wanting to know whether a predator is having an impact on a population of a rare animal species have sometimes set out to study the diet of the predator (seeing what they can find in its scats). That only tells us whether the predator does or does not eat that rare animal, and perhaps how often relative to other prey. While of natural history interest, that does not tell us whether, by eating it, the predator has an important impact on the rare species' population. That is another question and needs a different approach.

Part of the scientific skill, then, is to frame questions, that can generate hypotheses, that can be expressed as null hypotheses, that can be tested by collectable data. "How can we save this species?" is not a scientific question because it generates no testable hypothesis; it is a management question because it presupposes decisions that must be made about actions. One of those possible management actions might be food supplementation (for example). "Will doubling spring food availability increase the fledging rate of this species' nestlings?" is a question that science can readily answer, and the answer can be used by managers in choosing the best actions. To get the most out of collaboration with scientists, managers need to learn how to choose and frame questions so that they can be answered scientifically.

In doing this managers can usefully use intuition (gut feel). Almost anything that you have a gut feel about can be framed as a testable scientific question. Not all the questions may be worthwhile, and some gut feelings may need to be broken down into a series of answerable questions; but no one should be ashamed of using intuition as a starting point for designing a scientific investigation, as long as they use the intuition to frame the right questions.

Soderquist (1994) and Southgate (1994) have recently emphasized how important it is that scientists should generate and test hypotheses in the course of attempted reintroductions or other recovery actions. Appropriately chosen, these will contribute greatly to the programs in the long term, even when there appear to be short-term failures. Southgate (1994) cited attempts to reintroduce Greater Bilbies *Macrotis lagotis* in which the animals failed to survive; yet much was learned because well framed hypotheses were being tested.

3. Interpreting facts

Science by itself does not notice things. The human mind is supremely good at noticing patterns, changes and differences. Unprompted, scientific methodology does not do that, but intuitively skilled scientists can use scientific techniques for analysis of data to look for patterns and changes, and to verify them. Verification is essential if we are to be as sure as reasonably possible that the relationships or patterns we find are real rather than a product of chance or a figment of the fertile human imagination.

Much science works by answering narrowly specific questions. Interpretation is then limited to saying whether or not we can be reasonably sure of the answer. Almost everything that we measure in biology is more or less naturally variable. That means we are always likely to get a spread of values for any measured attribute. If we are comparing or relating two or more attributes, there is almost always a chance that the data that is being analysed may look importantly different (or importantly related) simply because we happen to have sampled the upper extreme of measurements from one set and compared it, by ill luck, to samples from the lower extreme of the other set. Such bad luck might make us say falsely that the two sets differed. Science uses statistical tests simply to tell us what the chance or probability is that we have, by bad luck, misinterpreted differences or relationships. The common convention is to say that we will accept as real (or "significant") any answer for which there is no more than a 1-in-20 chance of its being produced by bad luck in the sampled measurements.

However, this is only a convention. Working to a significance level of 1-in-20 may require larger samples of measurements, or measurements from more repeated situations, than we can get when working with rare species. Moreover, management may want *guidance* from science, rather than absolute surety. A manager of a rare species, under pressure to help the population and anxious to do what is probably the best for it, may risk a 1-in-10 chance rather than a 1-in-20 chance that the difference or relationship might not be real after all. Each case has to be judged on its merits, but a manager may be able to accept this greater risk of being wrong, knowing that actions based upon the supposed answer either will quickly test its correctness or can be initiated without great harm being done while more data are gathered.

Managers can lose faith in science when they hear scientists arguing about the statistical validity of rival interpretations of data, and especially when those managers have paid for the scientific investigation and they see its conclusions being questioned by other scientists. Most biologists have not been taught statistics very well; it is a technology whose theory lies outside biology itself. (Most biologists used to think they did not need to understand statistics any more than they needed to understand how a diesel engine worked; yet they would use each in their research. This attitude has changed.) The availability of software has made application of statistics easier, but has not necessarily improved scientists' understanding.

The solution may be for managers and scientists on a recovery team to ensure that statistical procedures are planned in consultation with a trusted statistician. However, conventional statistics and statisticians use fairly large samples; endangered species management is often restricted to very small numbers of individuals, and there may be too few populations or sites to replicate sampling adequately (although even one replicate is better than none). If those numbers are too small or replication is impossible, conventional statistics will not work or conclusions cannot be safely generalized. The scientist and the manager may have to admit that a situation simply does not allow statistical verification; it is then the manager's job to make a decision. Both conservation biologists and managers would benefit, however, by consulting a statistician who has specific experience of devising statistical tests for the small-sample situations so typical of endangered species management.

4. Prediction

Science can be used to make predictions. Science extrapolates from particular observations and measurements, to make generalized statements about what should be found again in the same circumstances. This basic type of prediction simply uses scientifically demonstrated facts to say what is likely to be found: if research has shown that kangaroo females, on average, begin breeding when two years old, then, all else being equal, at an unstudied site you should expect kangaroo females to begin breeding when two years old. This use of facts gives you best guesses (or rather better than guesses) as to what to expect of species, communities and circumstances.

An extension of this is to use well collected data to make models of the relationships between measurable things; for example, measurements of changes in population size through time can be used to model population growth. Such a model can then be used to make predictions; it can answer the questions: If we leave this population for another five years, how big (or small) will it be? Moreover, if the data on which the model was based were suitable, we can say how precise that modelled estimate is.

Some models are useful not because they tell us a concrete value (like how big a population will be in five years time), but because they tell us how likely it is that something will happen. For example, there are models that, with the right data, can tell us how likely it is that a small population of a species will go extinct within a specified length of time, or how likely it is to be rescued from extinction if a management action has a particular outcome. Many of these models are of little interest in "pure" science, as they may not deepen understanding of a species; they are an excellent example of the scientific method being used for management ends.

Estimating the probability of things happening is often important in making management decisions, especially when establishing priorities for the use of scarce resources.

5. Monitoring outcomes of management

The scientific method is our best way of gaining reliable information. A major use to be made of it is in confirming whether management actions are having their desired effect. Management of endangered species is usually costly and sometimes risky, and we want to be sure, as soon as we can, that we are doing the right thing. The best way of doing that is to design management actions as though they were experimental treatments, i.e., to plan and design the actions to maximize our chance of detecting their effects. This will usually mean leaving some sites or populations or individuals of the target species untreated, to act as a control against which to demonstrate the effects of the treatment. Only managers, not scientists, can decide whether they can afford not to treat all the endangered species; but they should listen carefully to scientific advice as to whether or not they will be able to demonstrate the effectiveness of their actions afterwards. The demonstration that poisoning of Red Foxes *Vulpes vulpes* allowed populations of Black-footed Rock-wallabies *Petrogale lateralis* to grow (Kinnear *et al.* 1988) would merely have been an unsupported and therefore arguable interpretation of observations if there had been no control sites where foxes remained unpoisoned.

SCIENCE IN THE MANAGEMENT OF ENDANGERED SPECIES

Endangered or rare species are almost inevitably poorly known. They may have small populations, or be restricted to small areas, and may be difficult to detect and study. Where they now live may be only a fraction of their former range, and may not reflect their optimum habitat (being merely where whatever has reduced them has had least impact or has operated for the shortest time). For all these reasons they are scientifically difficult to study and to make predictions about. Management must therefore try to be patient. There are sure to be new scientific approaches and techniques that will emerge in response to the peculiar challenges of these species. An excellent example of scientific technology rising to a management challenge is the use of DNA profiling to identify individual Northern Hairy-nosed Wombats *Lasiorhinus krefftii* from remotely gathered hair samples (W. B. Sherwin and A. C. Taylor, pers. comm.).

The science of conservation biology is rather new, as is the management of endangered species. Both scientists and managers have much to experience yet. Both can learn in co-operation, by each developing their understanding of the problems facing the other and the skills brought to bear upon those problems. This will help define the questions that can be answered only by management or only by science, and the area where their skills overlap and they can collaborate in formulating and answering questions. We have tried to emphasize in this paper how important it is that managers and scientists mutually understand how to ask the right (which means answerable) questions.

Managers of endangered species should also consider encouraging scientists to undertake some curiosity-driven research on the species or to interpret their findings as a contribution to fundamental science, rather than confining their investigations to immediately obvious management-related questions. There are several reasons why management might value such an investment. Firstly, the information it produces may be useful later. Secondly, the scientist may then seek extra funds from outside the recovery programme for further research that will complement the objectives of the recovery plan. Thirdly, and perhaps most importantly, it fosters a commitment by a scientist to the endangered species and its recovery and justifies their taking part in the recovery planning process, not as substitute managers, but as informed commentators who are deeply knowledgeable in some areas. This has already happened with most recovery teams, and we suggest that funds should be committed to it. Menkhorst (pers. comm.) has suggested that each recovery team should include a scientist charged specifically with the task of ensuring scientific rigour in all deliberations, in the setting of priorities for funding, and in the framing of questions for investigation. This scientific role would extend from the planning phase into the implementation and monitoring phases of the recovery program.

In this paper we have tried to suggest that scientists can bring a particular set of skills to the recovery process. The managers that understand those skills will get most out of scientists. In particular, it is important that managers learn how scientists frame questions, so that managers and scientists can pose and answer questions together. The scientific method is not a magical cure for all management problems, but for many management questions it is the best guarantee that the answers will be reliable. For this reason, it is helpful to have a formally identified place and process in the recovery plan for consultation between managers and scientists on the validity of the existing factual information about the species or community, the planned obtaining of more information, predictions to be made from the data about management and its outcomes, and ways to monitor the effectiveness of management.

ACKNOWLEDGEMENTS

We are particularly grateful to Chris Johnson and Peter Menkhorst for their thoughtful comments on a draft of this paper, and to those colleagues whose experiences in endangered species recovery planning, research and management (in Australia and elsewhere) we have drawn upon.

REFERENCES

Caughley, G. and Gunn, A., 1996. Conservation Biology in Theory and Practice. Blackwell: Cambridge, Mass. USA. Pp. xii + 459.

Gordon, G. and Lawrie, B. C., 1980. The rediscovery of the Bridled Nailtail Wallaby, *Onychogalea fraenata* (Gould) (Marsupialia: Macropodidae) in Queensland. *Aust. Wildl. Res.* 7: 339–45.

Kinnear, J. E., Onus, M. L. and Bromilow, R. N., 1988. Fox control and rock-wallaby population dynamics. *Aust. Wildl. Res.* **15**: 435–50.

Parker, S. A., 1973. An annotated checklist of the native land mammals of the Northern Territory. *Rec. S.A. Mus.* **11**: 1–57.

Soderquist, T. R., 1994. The importance of hypothesis testing in reintroduction biology: examples from the reintroduction of the carnivorous marsupial *Phascogale tapoatafa*. Pp. 159–64 *in* Reintroduction Biology of Australian and New Zealand Fauna ed by M. Serena. Surrey Beatty & Sons: Chipping Norton.

Southgate, R., 1994. Why reintroduce the Bilby? Pp. 165–70 *in* Reintroduction Biology of Australian and New Zealand Fauna ed by M. Serena. Surrey Beatty & Sons: Chipping Norton.

Warneke, R. M., 1967. Discovery of a living *Burramys*. *Bull. Aust. Mammal Soc.* **2**: 94–95.

Involving farmers in recovery programs

Ian C. McClintock[1]

There is a limit to the number and appropriateness of national parks and wilderness areas that can, in practical terms, be utilized for endangered species recovery programs. Many land owners have a deep interest in nature conservation and many more would develop one if the incentives to do so were right. Predators, particularly ferals such as foxes, cats and pigs etc., jeopardize the release of many species. The option of a network of reserves including feral-free reserves covering a much wider and more diversified range of environments, managed by farmers, could offer an order of magnitude difference in the success rate of recovery programs. Landcare and the Total Catchment Management or Integrated Catchment Management processes are the most exciting development in reversing environmental problems and stimulating interest in nature conservation. They nourish a comprehensive, holistic outlook and attitude to land management taking into account a balance between the economic, social and environmental factors. There needs to be trust by both government and landholders for this to succeed. The unfortunate example set by the New South Wales Government's environment policy introduction strategy, illustrates just how fragile the standing of this confidence and trust can be. With an enlightened approach and the right information, assistance and incentives, farmers would be willing to play their part.

Key words: Viability, voluntary, trust, understanding, ownership, positive incentives, balance.

INTRODUCTION

THIS paper is designed to address from the farmer's point of view, the broad question of how best to sustainably conserve our biological heritage, given the reality of an exponentially increasing human population and the subsequent, inevitable change and threat that we humans impose on our ecosystems.

I will argue that, to have any chance of success at all in this quest, it is essential that we enlist and attain the *enthusiastic* support of the farming community.

I would like to quote from Professor Andrew Beattie, (1992) Macquarie University, Sydney, who states in part;

> ". . . a very wide spectrum of Australians . . . now appreciate, that we are organisms that depend on millions of others for our well-being and that it is the farming community that we rely on to establish sound and productive relationships with them."

The concern of the wide spectrum of Australians referred to by Beattie, which includes the great majority of farmers, is warranted, for the very future of human life will become increasingly dependent on the retention of the widest assemblage possible of our biological heritage.

To achieve this it is most important to understand as accurately as possible the attitudes, opinions, aspirations, pressures and requirements of farmers in relation to this matter.

I am passionate about farming and I have a fascination with nature. Farmers work with nature to make their living and the vast majority are seriously interested in doing so on a sustainable basis.

Farming is a business and it must be viable to ensure sustainability, including environmental and species sustainability.

It is important to realize that regulations and legislation that have the effect of reducing farm viability, threaten our ability to pro-actively manage and protect the environment. There is a very clear message here.

CHALLENGE

I would like to issue a challenge, which is to ask *exactly what are we trying to achieve with recovery and conservation programs outside nature reserves?*

We need to know this if we are to set in place an *appropriate* structure that is *acceptable*, and *capable* of achieving our objectives.

I would define the *goal* then, as: "to protect and sustain as much of the existing biological diversity as is reasonably possible".

The difficult question then becomes *how* can we best accomplish this?

I suggest this turns out to be a much more difficult question to answer than most assume, and this is where the psychological approach employed (knowingly or

[1]"Milford Park", Cootamundra, New South Wales, Australia 2590.

Pages 79–82 *in* BACK FROM THE BRINK: REFINING THE THREATENED SPECIES RECOVERY PROCESS ed by S. Stephens and S. Maxwell. Surrey Beatty & Sons, 1996.

unknowingly), as perceived by farmers, is critical to the successful adoption or otherwise of on-farm conservation measures.

To be able to answer this question with some insight we should *first*, briefly look at the conditions prevailing when the current spectrum of indigenous biota developed and changed.

There are three major periods to consider, white settlement over the last 200 years, Aboriginal settlement over some 60 000 years and the pre-human habitation period over many millions of years.

The pre-Aboriginal condition

Prior to Aboriginal settlement a very diverse spectrum of flora and fauna evolved and developed, different in both extent, composition and proportion to that of their time.

Eucalypts were rare and sclerophyll forest was very limited according to pollen samples obtained from cores taken from the continental shelf off Cairns and Townsville.

There was a range of some 60 large Australian vertebrate herbivores, including interestingly a hoofed kangaroo. They were typified by weighing only about one third as much as their overseas counterparts, and there were only two or three mammalian carnivores.

The herbivores extensively grazed the grasslands and woodlands, creating disturbance and limiting the intensity of wildfire due to fuel build-up, and an equilibrium was established.

Aboriginal settlement conditions

With the arrival of the Aboriginals some 60 000 years ago, it is postulated that the rapid extinction of the megafauna that subsequently occurred, some 94%, by hunting, the subsequent fuel build-up and the effects of fire, caused a rapid change in the composition and distribution of the flora and fauna as well as other significant environmental repercussions.

A new equilibrium was eventually established dominated by fire tolerant species, as depicted by many descriptions of the prevailing conditions in eastern Australia when the early Europeans arrived.

Dr Tim Flannery (1994) in his book *The Future Eaters*, writes;

> "As a result of the extensive Aboriginal use of fire, the plant communities seen and recorded by Banks and other explorers were very different from those that exist at the same locality today.
>
> "The lack of recognition of this change has critical implications for the management of national parks and wilderness areas today."

I would like to quote three of these early reports (cited in Ryan *et al.* 1995) to illustrate what those conditions were like.

Joseph Banks noted in his journal while in Botany Bay:

> ". . . very barren place without wood . . . very few tree species, but every place was covered with vast quantities of grass . . . the trees were not very large and stood separate from each other without the least underwood."

Charles Darwin wrote in 1836:

> "The extreme uniformity of the vegetation is the most remarkable feature in the landscape of the greater part of New South Wales. Everywhere we have open woodland; the ground being partially covered with a very thin pasture."

And:

> "In the whole country I scarcely saw a place without a trace of fire; whether these had been more or less recent — whether the stumps were more or less black, was the greatest change which varied the uniformality, so wearisome to the traveller's eye."

And finally, on the regrowth that occurred with the cessation of burning, Mitchell in 1848 wrote:

> "The omission of the annual periodical burning by natives, of the grass and young saplings, has already produced in the open forest lands nearest to Sydney, thick forests of young trees, where formerly, a man might gallop without impediment, and see whole miles before him. Kangaroos are no longer seen there, the grass is choked by underwood; neither are there natives to burn the grass, nor is fire longer desirable there amongst the fences of the settler."

It is clear that without regular burning, the open woodlands rapidly developed into the heavily timbered forests we are familiar with today, creating quite a different habitat and by definition, a new and different range of species.

White settlement conditions

With the advent of white (initially) settlement in Australia an entirely new regime has developed.

European farming methods were introduced including clearing and ploughing, as were livestock, crops, grasses, carnivores such as the fox and cat, ferals such as the rabbit, goat, pig and carp and many species of weeds and exotic plants, over 2 000 of which have become naturalized and are not going to go away.

As we all know, this has cumulatively put tremendous pressure on our native biota, which we are now attempting to redress.

The major underlying problem that is not being addressed however, is the increasing population of humans which demand food, clothing, shelter, transport and the range of goods and services that we all aspire to, and that over time this pressure will inexorably increase.

We have in the past concentrated on shutting up certain areas, creating national parks where farming and other operations are forbidden, anticipating that this will provide the necessary habitat, space and conditions for our biota to survive and hopefully prosper.

Let us look critically again at this supposition.

THE NATIONAL PARK CONDITION

We can see from the early records that the national parks and wilderness areas we have so carefully set aside generally do not accurately represent conditions that existed in Aboriginal stewardship times and certainly do not represent the period prior to that. They are in effect an experiment under a new and different set of conditions.

While species that enjoy these conditions may prosper, others that depend on the previously open woodland and grassland fire stick mosaic of Aboriginal times will not.

This is a high risk strategy, as the build up of combustible fuel will lead almost inevitably to high intensity wildfire events, such as the recent Sydney fires. These can devastate the carefully protected areas leaving little or no food for any surviving fauna and may destroy the most sensitive flora species.

This is especially true on discrete farm-sized fenced off areas and so the question of how these areas should be sustainably managed becomes critically important.

RESERVE MANAGEMENT PRINCIPLES

- Firstly, it is not sufficient to simply fence off an area, de-stock it and expect it to achieve the desired result without any management input.
- It is clear that either a regular, low intensity mosaic fire regime and/or sensitive strategic grazing of these areas is going to be required to maximize the biological diversity able to be sustained in them.

  As Attiwill (1996) correctly states:

  > "Disturbance is common to all ecosystems. It drives the cyclic renewal of community composition, and it releases resources and promotes nutrient recycling . . . It is a basic tenet that biodiversity is maximum at some intermediate level of disturbance."

- The discriminating use of cattle and sheep, when necessary, is a practical on-farm option to achieve the desired results in private reserve areas.
- Weed encroachment, feral animal control or (if possible) eradication, are important ongoing management tasks, as are:
  - — Repair and maintenance of fencing, creek crossings, removal of straying stock, etc.
  - — Fire prevention and/or control.

FARMERS

Farmers manage some 60% of the Australian land area (most of the rest is desert or mountain ranges) and at least 77% of the mainland eastern States, so their participation is critical.

Many land owners have a deep interest in nature conservation and many more would develop one if the incentives to do so were right.

Predators, particularly ferals such as foxes, cats and dogs, jeopardize the survival of many species, and those such as rabbits, goats and carp degrade the habitat.

The option of a network or web of reserves including feral-free reserves covering a much wider and more diversified range of environments, managed by farmers, could offer an order of magnitude difference in the success rate of biodiversity protection.

The key question then becomes: *what is the best way to involve farmers?*

There are two main techniques:

Command and control

1. The regulate, command and control, "big stick" approach

This has a number of advantages for government:

- It is quick and easy to implement (stroke of the pen).
- It can cause practices that are perceived to be bad for the environment to be stopped immediately.
- It is relatively cheap, although administration, monitoring and policing can become very expensive, but income from fines can offset some of these costs.
- It suits the short-term time horizon of most politicians.
- It is usually the first response concerned citizens advocate, and so can have political popularity bonuses for government.
- And it can work. There are many successful examples.

It is entirely appropriate for simple straightforward problems such as ensuring everyone drives on the left-hand side of the road, and it receives strong community support.

However, *as you move away from simple solutions, the command and control approach becomes less relevant, less effective and less acceptable.*

For example, in the USA if an endangered species moves onto your farm, it may be classified as "critical habitat" and a range of restrictions on use will apply, without any redress or compensation.

The powerful incentive then is to *"shoot, shovel and shut up"* before anyone finds out.

This is exactly the opposite result to that envisaged by the legislators and *is the inevitable, perverse consequence* of this type of negative incentive approach.

There needs to be *trust* by both Government and landholders for environmental policy *to succeed.*

The unfortunate example set by the New South Wales Government's environment policy introduction strategy, again illustrates just how fragile the standing of this confidence and trust can be.

The environment is at the extreme end of the simple: complex continuum and it is vital to recognize this and develop a more appropriate, effective, practical and acceptable strategy.

Bottom up

2. The education, bottom up, distributed intelligence, assistance, trust, co-operation, ownership, holistic approach is by far the most potent method

This is typified by the Landcare movement. Landcare and the Total Catchment Management or Integrated Catchment Management processes are the most exciting development in reversing environmental problems and stimulating interest in nature conservation.

They remove the fear and nourish a comprehensive, holistic outlook and attitude to land management taking into account the necessary balance between the economic, social and environmental factors.

The National Farmers' Federation strongly supports this proven second option.

Incentives

There are many forms positive incentives can take including:

- Grants for fencing and ongoing management or stewardship of reserved areas.
- Compensation for production and income foregone, land value reduction (where that occurs), restrictions or obligations imposed, etc.
- Possibility of bidding for elements in species recovery plans.
- Bounty payments for the breeding up of endangered species.
- Commercialization of sanctuaries for certain native animals or plants can ensure their survival.
- Provision of long-term low interest loans.
- Rate and taxation concessions and rebates.

CONCLUSIONS

If we are serious about retaining the extant biological diversity we have in Australia, then we have no option but to enlist the co-operation of farmers.

To do this successfully, there is only one option, which is the bottom up, distributed intelligence, Landcare type model.

If we choose this approach and ensure that the incentives are right, farmers will respond in an enthusiastic manner, lending their local knowledge, wisdom, and financial and physical support.

This will maximize our success rate in this daunting but important task.

REFERENCES

Attiwill, P. M., 1996. Two sides to complementary management: production forests and forest reserves. A paper presented at Conservation outside Nature Reserves: a conference to look at nature conservation in landscapes managed primarily for other uses. Centre for Conservation Biology, University of Queensland, 5–8 February 1996.

Beattie, A., 1992. Letter to the *New South Wales Farmers' Federation News* (June).

Flannery, T. F., 1994. The future eaters. Reed Books, Chatswood.

Ryan, D. G., Ryan, J. and Starr, B. J., 1995. Australian landscape — observations of explorers and early settlers. Prepared for the Murrumbidgee Catchment Management Committee.

Species recovery planning: some New Zealand case studies

Alan Saunders[1]

Recovery plans have been used as a basis for managing a growing number of New Zealand's threatened species since 1989. The process of plan development and refinement has been important in facilitating better integrated and effective actions, while plans are being used to guide decision making and to increasingly promote wider understanding of conservation issues.

Some case studies are presented to illustrate not only specific benefits of the process as applied in New Zealand, but also some of the difficulties which have been encountered.

Issues for discussion include acknowledging different perspectives, marrying expectations and achievements, integrating research and management and single species recovery as a precursor to ecosystem-based restoration.

INTRODUCTION

WHILE some important successes had been achieved in averting extinctions and recovering threatened species in New Zealand, by the late 1980s it was recognized by the Department of Conservation that a comprehensive and standardized recovery process was needed. Key considerations were the growing number and array of plant and animal taxa threatened with extinction for which recovery management was being undertaken or proposed, the increasing complexity of management techniques, and the growing number of practitioners and "stake-holders" involved in species recovery.

A process of species recovery planning was initiated by the Department in 1989 with recovery plans prepared for Kakapo *Strigops habroptilus* and Blue Duck *Hymenolaimus malacorhynchos*.

In response to the need for more effective national co-ordination and technical support for recovery programs the Department of Conservation also established in 1989 a six-person Threatened Species Unit (TSU) within its Protected Species Policy Division. The main roles of the unit are to co-ordinate threatened species programs nationally, and to provide technical advice and practical support to project operators and others as required. Priority was given by the TSU to the refinement and application of the recovery planning process. Recovery plans are now the key tool by which goals, objectives, tasks and priorities for management are identified and are an important basis for departmental planning. Recovery groups are fulfilling roles not only in developing and overseeing recovery plans, but also as technical advisers to the Department.

Since 1989 over 40 recovery plans have been initiated. Of these, 22 have been formally approved (Appendix 1). The remainder are at earlier stages in the consultation and refinement process.

With over 600 New Zealand taxa so far identified as threatened with extinction (Molloy and Davis 1994) it is clear that recovery plans cannot be prepared for each. A trend is emerging towards the development of recovery plans for multiple taxa where there are common management issues. For example, while a range of taxa are involved, the Giant Land Snail Recovery Plan (Parrish *et al.* 1995) focuses on common threats and management issues relevant to all Kauri and Flax Snails in the far north. Similarly, a plan is being drafted for a dozen or so species of threatened Weta found at various sites throughout New Zealand. "Omnibus" plans are also being drafted which focus on Chatham Islands plants and forest birds. While such a trend makes good sense it is important that plans do not become so generic that their values as operational plans which identify specific objectives and tasks are lost.

REVIEW OF RECOVERY PLANNING GUIDELINES

Seven years later the Department is undertaking a review of its recovery planning guidelines. This is an acknowledgement not only of the important strengths of the process, but also of some weaknesses which need to be addressed. Some general observations can be made in relation to these strengths and weaknesses.

Strengths of the recovery planning process

A key attribute of the process is that it provides a formal mechanism for communication between recovery practitioners and other stakeholders. As the

[1]New Zealand Department of Conservation, P.O. Box 10–420, Wellington, New Zealand.
Pages 83–86 *in* BACK FROM THE BRINK: REFINING THE THREATENED SPECIES RECOVERY PROCESS ed by S. Stephens and S. Maxwell. Surrey Beatty & Sons, 1996.

needs of a growing range of stakeholders become clearer, effective communication will be vital. Demands for recovery plans continue to grow from a range of users indicating the values of these plans as information sources and as the basis for consultation. The identification of recovery goals, objectives, tasks and priorities in recovery plans has allowed for more comprehensive support of recovery programs — both by the Department and by other agencies. The process of consultation leading to agreement on objectives and priorities has also led to improved integration of research and management activities, and between various recovery activities and other ecological management programs.

Compared to the relatively "*ad hoc*/in-house" approaches of the past these features constitute important and timely advances.

Weaknesses of the recovery planning process

As the array of species recovery issues being addressed in recovery plans expands, it is perhaps not surprising that some weaknesses in the process have emerged.

In particular, as the range of participants in recovery programs and the number of plan users grow, there is some confusion about the purpose of recovery plans and the roles of recovery groups. While initially developed as operational plans, recovery plans are increasingly being seen as strategic documents which contain policy statements in addition to containing detailed background information on the species and management issues.

Feedback from a few recovery groups suggests that input costs for group members associated with plan development and oversight, are seen as being unacceptably high in relation to conservation outcomes. In some cases the demands on recovery group leaders, in particular, in consulting and resolving very complex technical and logistical issues have had a high cost on individual recovery group members. "Burn out" is now seen as a serious threat in a few groups.

While these observed weaknesses are significant, all are manageable provided appropriate changes are made.

Four New Zealand case studies are examined to consider their particular features in relation to the recovery planning process.

Case Study 1: Tuatara recovery program

The Tuatara *Sphenodon punctatus* recovery program is one of the longest running and successful programs co-ordinated by the Department. Features include a high level of public interest and support for Tuatara recovery, and a relatively good understanding of Tuatara ecology upon which management initiatives can be based. Because Tuatara are now confined to about 20 relatively small islands, threat management (predators, habitat disturbance) is by and large achievable using modern techniques. Furthermore, neither species of Tuatara is critically endangered, and the pressure on managers to "get it right first time" is less of a consideration than in programs where more critically endangered species are involved. A further important feature of the Tuatara program is the long-term involvement of a cadre of knowledgeable and enthusiastic scientists and practitioners from several agencies.

Important advances which have been made in recovering Tuatara populations which may be related directly or indirectly to the recovery planning process have included an increased involvement by various "stakeholders", including Iwi Maori (tribes and families) in recovery activities, and ongoing support for strategic research into Tuatara ecology. As a result of the support being given to the Tuatara program, and the results being achieved, the Tuatara has emerged as a "flagship" species for island restoration programs more generally.

Case Study 2: Kakapo recovery program

The Kakapo *Strigops habroptilus* recovery program is one of the largest and most comprehensive currently undertaken by the Department. In addition to being critically endangered, there is a high level of public awareness of the plight of the Kakapo, and strong support exists for its recovery. With the transfer of the entire population to "safer" islands, risks and potential benefits of management are significant. Expectations of success in the program have become, in some quarters, unreasonably high. The involvement of a major commercial sponsor, while constituting an important advantage for the program, has also served to raise tensions amongst project managers and practitioners. Despite some intensive research, understanding of key features of Kakapo ecology which may lead to successful new management approaches is still relatively limited.

At the end of the first plan period a review was recently undertaken which evaluated the general directions of the program, progress in relation to planned objectives, and any necessary changes. As a result of the review significant changes have been instituted in recovery projects, and new research initiated. New program management mechanisms and procedures have also been introduced including the establishment of a separately-managed National Kakapo Team and a Kakapo Scientific and Technical Advisory Committee replacing the Kakapo Recovery Group. While it is clear that separate teams and procedures are neither desirable nor practical for all recovery programs, it is likely that important lessons will emerge from the Kakapo program which may have relevance to the co-ordination and management of other recovery programs.

Case Study 3: Kokako recovery program

With several thousand individuals still persisting in several populations throughout the upper North Island, the Kokako *Callaeas cinerea wilsoni* recovery program is a good example of a program where opportunities exist to employ an experimental management approach without jeopardizing the taxon as a whole.

With its hauntingly melodious call, relatively large size and distinctive face mask and wattles, the Kokako enjoys a very high public profile. Intensive research has shown the effects of forest habitat destruction and modification on Kokako populations, as well as the direct impacts of competition and predation by introduced mammals such as Brush-tailed Possums *Trichosurus vulpecula* and Ship Rats *Rattus rattus*. The Kokako has emerged as a symbol for forest protection in the North Island, and as a flagship for ecological restoration activities.

As a result of a relatively good understanding of Kokako ecology, and of threats to their survival, coupled with strong public support for their recovery, significant advances have been made following the approval of the first Kokako Recovery Plan (Rasch 1992) in 1993. In particular the application of a "Research-by-Management" (Innes, unpubl. report) program whereby managed and non-managed Kokako populations were compared, and treatments reversed to determine responses to specific management regimes has been notable. Dramatic increases in Kokako numbers have been observed at managed sites where pest animals have been effectively controlled (Bradfield and Flux 1996). While progress in other ecological management activities has also been important, successes in achieving planned Kokako recovery objectives have probably been the single greatest stimulus for New Zealand conservation managers to attempt more ambitious species recovery and habitat management programs. In particular, intensive control of suites of alien pest animals at forested "mainland" sites to benefit specified resident native species has emerged as the "new wave" of conservation management in New Zealand. The Kokako program continues to be very important in breaking new ground as "mainland island" programs are developed.

Due in part to the important successes achieved, the Kokako Recovery Group is a good example of a functional group, despite it being one of the largest, including virtually every one of the Department's eight conservancies which cover the North Island.

Case Study 4: Brown Teal recovery program

The Brown Teal *Anas chlorotis* program features a very successful captive breeding program undertaken by Ducks Unlimited (NZ) members. Unfortunately, population and habitat management has not advanced apace and Teal populations have virtually disappeared from the main islands. As a wetland species the Brown Teal provides particular management challenges which have yet to be addressed comprehensively in New Zealand.

After several years of discussions there is now a large measure of agreement between the Department, Ducks Unlimited and other "stakeholders" for recovery objectives. The recently approved Brown Teal (Pateke) Recovery Plan (Williams and Dumbell, 1996) sets out an experimental approach to habitat management at key sites, and serves to better-integrate *in situ* and *ex situ* activities. It can be anticipated that significant potential benefits to wetlands more generally, as well as to Brown Teal in particular, will accrue as this program proceeds.

These case studies indicate that the recovery planning process has played an important role in stimulating effective dialogue between practitioners, scientists and other "stakeholders" in species recovery, and that important advances have been made as a result of agreement on recovery goals, objectives and priorities. A flexible approach must be retained so that the process may be adapted as the needs of practitioners and "stakeholders" evolve.

CONCLUSIONS

The initiation of a recovery planning process in 1989 constituted an important advance in conservation management in New Zealand. In identifying recovery goals, objectives, tasks and priorities, recovery plans have allowed the Department of Conservation, as well as other agencies, to determine priorities for the allocation of limited resources. In addition the establishment of recovery groups consisting of representatives of key "stakeholder" groups to oversee the refinement and implementation of recovery plans has also had significant benefits. After seven years a trend is emerging towards multi-taxa plans, although there will continue to be a need for single-species plans in some cases.

In conclusion, the availability of recovery plans which focus on important conservation management issues, and recovery groups where consultation and agreement between stakeholders is promoted, will provide an important base from which further advances in conserving New Zealand's biodiversity will be made.

REFERENCES

Bradfield, P. and Flux, I., 1996. The Mapara Kokako Project: 1989–1996. A Summary Report. Department of Conservation: Private Bag 3072, Hamilton, New Zealand.

Innes, J. G., 1995. Kokako Research-by-Management: 1994–95. Unpublished report. Manaaki Whenua Landcare Research: Hamilton, New Zealand.

Molloy, J. and Davis, A., 1994. Setting priorities for the conservation of New Zealand's threatened plants and animals. Department of Conservation: P.O. Box 10–420, Wellington, New Zealand.

Parrish, R., Sherley, G. and Aviss, M., 1995. Giant Land Snail Recovery Plan. Threatened Species Recovery Plan Series. Department of Conservation: P.O. Box 10–420, Wellington, New Zealand.

Rasch, G., 1992. Recovery Plan for the North Island Kokako. Threatened Species Recovery Plan Series. Department of Conservation: P.O. Box 10–420, Wellington, New Zealand.

Williams, M. J. and Dumbell, G., 1996. Brown Teal (Pateke) *Anas chlorotis* Recovery Plan. Threatened Species Recovery Plan Series. Department of Conservation: P.O. Box 10–420, Wellington, New Zealand.

APPENDIX 1

NZ Department of Conservation species recovery plans

Approved Plans

Taxon	Plan Period
Kakapo *Strigops habroptilus*	1989–1994
Blue Duck *Hymenolaimus malachorhynchos*	1989–1994
North Island Kokako *Callaeas cinerea wilsoni*	1991–1996
Yellow-eyed Penguin *Megadyptes antipodes*	1991–1996
Kiwi *Apteryx* spp.	1991–1996
Chevron Skink *Leiolopisma homalonotum*	1992–1997
Whitaker's and Robust Skink *Cyclodina whitakeri* and *C. alani*	1992–1997
Black Stilt *Himantopus novaeseelandiae*	1993–1998
Kakabeak *Clianthus puniceus*	1993–1998
Mohua *Mohoua ochrocephala*	1993–1998
New Zealand Dotterel *Charadrius obscurus*	1993–1998
Tuatara *Sphenodon punctatus*	1993–1998
Sub-Antarctic Teal *Anas aucklandica, A. a. nesiotis*	1993–1998
Takahe *Porphyrio mantelli*	1994–1999
South Island Saddleback *Philesturnus carunculatus carunculatus*	1994–1999
Otago and Grand Skink *Leiolopisma otagense* and *L. grande*	1995–2000
Giant Land Snails *Placostylus* and *Paryphanta* spp.	1995–2000
Bat *Mystacina* spp., *Chalinolobus tuberculatus*	1995–2000
Stitchbird *Notiomystis cincta*	1995–2000
Hooker's Sealion *Phocarctas hookeri*	1995–2005
Dactylanthus taylorii	1995–2000
Brown Teal *Anas chlorotis*	1995–2005

Draft plans (as at April 1996)

Taxon
Black Robin *Petrioica traversi*
Weka *Gallirallus australis australis, G. a. hectori, G. a. greyi, G. a. scotti*
Chatham Island Oystercatcher *Haematopus chathamensis*
Chatham Island Petrel *Pterodroma axillaris*
Chatham Island Taiko *Pterodroma magentae*
Chatham Island Pigeon *Hemiphaga novaeseelandiae chathamensis*
Fairy Tern *Sterna nereis*
New Zealand Shore Plover *Thinornis novaeseelandiae*
Hector's Dolphin *Cephalorhynchus hectori*
Dwarf Inanga *Galaxias gracilis*
Canterbury Mudfish *Neochanna burrowsius*
Native Frog *Leiopelma hamiltoni, L. hochstetteri, L. archeyi*
Powelliphantid Snail *Powelliphanta* spp.
Threatened Weta *Dienacrida* spp.
Carmichaelia kirkii
Tecomanthe speciosa
Pennantia baylisiana
Muehlenbeckia astonii
Leptinella nana
Lepidium spp.
Helichrysum dimorphum
Duvallia "Puketi"

Note: Published recovery plans are available from the Science and Research Publications Section, Department of Conservation, P.O. Box 10–420, Wellington, New Zealand.

Community involvement in the species recovery process: insights into successful partnerships

Stephanie Williams[1,2]

This paper contains the personal views of the author, which are not necessarily the policy of the Chicago Zoological Society or the South Australian Department of Environment and Natural Resources.

Involving the general public in the recovery of threatened species and ecological communities provides discrete short-term benefits for conservation programmes as well as long-term gains in developing social responsibility for Australia's natural heritage. Guidelines for successful engagement of the community in the species recovery process, based on personal experience, are outlined. It is suggested that government agencies provide community endeavours with honesty, support, expertise and a sensitivity to the community's concerns for conservation. This will help to develop effective partnerships in species recovery initiatives.

INTRODUCTION

AGENCIES responsible for managing recovery programs for threatened species and ecological communities are becoming increasingly aware of the need to involve members of the local, national and international community. Traditionally these groups have been excluded from the process of developing and implementing species recovery programs (Bennett *et al.* 1995; Saunders *et al.* 1996). Nonetheless few recovery programs in Australia have carried this philosophy of community involvement to its full potential.

The aim of this paper is to outline ways in which community involvement can be encouraged; what has and has not worked; and what gains can be made in species recovery by taking such an approach. I will not delve into the sociology of individual, group or institutional commitment to preserving global biological diversity. These sociological principles have been reviewed elsewhere (Ostrom 1990; Blyth *et al.* 1996; Moore 1996). Instead I will consider the practical side of these principles based on my experience working with the community on conservation initiatives and on the experience of colleagues who are responsible for implementing recovery programs and managing natural areas.

I define the community as groups of people or individuals who are not involved in conservation initiatives in a professional capacity. Community, therefore, includes private landholders, traditional owners, local government, members of non-government conservation agencies, industry, research students from academic institutions, as well as interest groups concerned with recreation, hunting and natural history. Community members can be local, regional, national or international. All have a role to play in the species recovery process.

CASE STUDY 1: MALLEEFOWL RECOVERY PROGRAMS IN SOUTH AUSTRALIA

This project illustrates how links between government agencies and the community can be developed to carry out conservation programs that neither member of the partnership could accomplish alone.

Community contributions to species recovery programs can be very important in regions where native vegetation is inadequately protected

My involvement with the Malleefowl recovery program commenced in 1991, prior to the writing of the research phase for the Malleefowl Recovery Plan. I was contracted, along with Rick Barratt and Craig Nixon, by the South Australian Department of Environment and Natural Resources (SADENR) to develop management protocols for Heritage Agreement areas in the Murray Mallee region of the State. The State initiated the Heritage Agreement scheme in 1980 to encourage the retention and conservation of native vegetation on private land. The government recognized that much of the State's remnant vegetation is owned and managed by private landholders. The program has evolved to its current status under the *Native Vegetation Act 1991* with varying degrees of success (Prescott 1996). At June 1995, 21% of South Australia

[1]Conservation Associate, Chicago Zoological Society, P.O. Box 1304, Berri, South Australia, Australia 5343.
[2]Current address: National Parks Service, Department of Natural Resources and Environment, P.O. Box 41, East Melbourne, Victoria, Australia 3002.
Pages 87–96 *in* BACK FROM THE BRINK: REFINING THE THREATENED SPECIES RECOVERY PROCESS ed by S. Stephens and S. Maxwell.
Surrey Beatty & Sons, 1996.

was protected in the State reserve system. Nine-hundred-and-ninety-seven Heritage Agreement areas contributed a further 0.52% to the State's protected areas (Dendy 1996).

The team's role was to lead the participants in the Heritage Agreement scheme from the complacency of "legislative conservation" toward a personal responsibility for and ownership of management programs to protect and enhance the biological diversity of their Heritage Agreement areas. We were contracted to produce documents outlining management protocols for conservation of mallee remnants on farms.

SADENR wanted the team to address threats to species diversity and to maximize the protection of biological diversity within the Murray Mallee region of South Australia. The region provided a useful trial of the effectiveness of off-reserve conservation because nearly half of the remnant native vegetation is held on private land and most State reserves are small and highly fragmented.

Managing conservation issues locally led to the development of community-based conservation programs

Meeting SADENR's goals for conservation regionally required the team to develop a co-ordinated strategy for conservation programs amongst private and public landholders. We developed a model for co-ordinated management programs within the context of the individual holding, the district and throughout the South Australian Murray Mallee region as a whole. The program was trialed in Mantung and Maggea, a small farming district in the northern Murray Mallee of South Australia.

The Mantung/Maggea Malleefowl project has been described in detail elsewhere (Williams 1996). The Mantung/Maggea community incorporated conservation management programs into their traditional farming practices with the aim of enhancing the quality of Malleefowl habitat on their farms and increasing the bird's breeding densities. They fenced remnant vegetation to exclude livestock and commenced an on-going fox and rabbit control program.

The development of a personal working relationship with the landholders in the Mantung/Maggea district led to my subsequent involvement in community stewardship of the Malleefowl recovery program. The Malleefowl project evolved during several years of interrelated research and management initiatives. Students, nature conservation organizations and other interest groups became involved in controlled experiments and long-term monitoring programs designed to document the effectiveness of these management initiatives for Malleefowl recovery. The evolutionary process of increasing community awareness and building a commitment to species recovery programs has inspired this contribution to these proceedings.

By adopting principles for successful extension programs later outlined by Prescott (1996) we increased the Mantung/Maggea community's acceptance of our management protocols and encouraged the development of conservation initiatives. We accomplished this by providing information to increase the landholders' knowledge of the habitats they are managing and outlining the ecological concepts and ecological consequences of particular management actions (Barratt *et al.* 1991 a,b,c). This information was provided not only via publications and field days but also by spending time in the paddock with the landholders. We made an effort to answer landholders' questions about their land and to demonstrate how each landholder could conserve the plants and animals on their farm.

The success of the Mantung/Maggea Malleefowl project suggests that the achievements of many other recovery programs can be enhanced by increasing the general community's understanding of conservation issues and by demonstrating how these issues relate to them personally. This personal identification with conservation problems helps change value systems toward a more sustainable conservation ethic. By helping change personal value systems, conservation biologists can encourage ownership of conservation initiatives by the community.

CASE STUDY 2: THE BLACK-EARED MINER RECOVERY PROGRAM

The Black-eared Miner recovery program is one in which the general community was initially unable to contribute to most research and recovery actions. This project also illustrates how members of the community eventually provided conservation biologists with information and expertise crucial to the recovery of this endangered species.

Community contributions to Black-eared Miner recovery actions were not initially necessary for the bird's protection

Black-eared Miners *Manorina melanotis* have declined at an alarming rate during the past 10 to 20 years (Starks 1987; McLaughlin 1994). They are one of the most endangered vertebrates in Australia (ANZECC 1991; Garnett 1992). Debate over taxonomic status of the species (Sibley and Monroe 1990; Christidis and Boles 1994) initially slowed the progress of recovery initiatives. Nonetheless rigorous scientific studies into its status and threatening processes in Victoria have enabled the development of priorities for recovery actions (Starks 1987; McLaughlin 1990, 1992, 1993, 1994; Backhouse *et al.* 1995).

In contrast to Malleefowl in South Australia, the Victorian Government placed a high priority on the recovery of this species and funded much of the early research and management of the species accordingly. Adequate staff and resources decreased the immediate need for community and volunteer support for the research phase of the Black-eared Miner recovery program.

Surveys during the 1980s in the Murray Mallee of New South Wales, Victoria and South Australia suggested that Victoria was the species' current stronghold (Blakers *et al.* 1984; Joseph 1986; Land Conservation Council 1987). Subsequent to recommendations by the Victorian Land Conservation Council (Land Conservation Council 1989) to protect what remained of the Black-eared Miner's remnant habitat in Victoria, large areas of the Victorian Sunset Country were incorporated into the State reserve system.

Virtually none of the Black-eared Miner's critical habitat in Victoria occurs on privately owned or privately managed land. This made community consultation during the research phase of the recovery plan less important to the recovery of the Black-eared Miner than consultation has been for the recovery of Malleefowl. Total government control also meant that critical habitat assessment and the development of priorities for managing threats, such as controlled burns, to protect colonies of birds could proceed quickly without extensive negotiation with land-owners.

Constructive community contributions to the Black-eared Miner recovery program can be difficult to obtain

Unlike the Malleefowl, it is difficult for the general community to make major contributions to our understanding of the status or biology of Black-eared Miners. Therefore, community involvement in the Black-eared Miner recovery program initially focused on volunteer participation in searches co-ordinated by the Royal Australasian Ornithologist's Union, a community-based conservation organization (McLaughlin 1994). Even these contributions provided challenges.

The Black-eared Miner is particularly difficult to study

The species is extremely uncommon and rare. Many people wish to contribute to survey efforts simply to see one of Australia's rarest species. Once they have achieved this goal, they may not be interested in further participation. This makes training and consistency amongst survey efforts a challenge. Volunteers can also be discouraged and disappointed when they spend long hours with little or no reward from seeing a bird.

The species is very difficult to identify from its congener, the Yellow-throated Miner *Manorina flavigula* (Joseph 1986; Starks 1987; McLaughlin 1990). This difficulty makes sightings reported by volunteer surveyors difficult to vet and also discourages community participation. Volunteers become frustrated when they can not distinguish the Black-eared Miner from its common congener or from a hybrid bird.

Involving large groups of people in surveys of this rare species may compromise the protection of birds from collectors

Egg and skin specimens of very rare species are valuable on the black market. This value makes protecting Black-eared Miner colonies a high conservation priority that is difficult to implement in remote areas. Both of these consequences of rarity have discouraged agency staff and conservation biologists from involving the community in research and survey programs.

Most of the areas under scrutiny were in remote areas of Victoria

Health and safety considerations associated with large teams of volunteers conducting surveys in remote and physically challenging conditions can be substantial. These issues are also applicable to Malleefowl population surveys.

The observations of amateur ornithologists eventually changed the scope of the Black-eared Miner recovery program

Despite the reasons, outlined above, for not having the general community involved in the Victorian recovery programs, the outlook for conserving Black-eared Miners has improved largely due to the dedication of birdwatchers in South Australia.

Joseph (1986) summarized his research into the status of Black-eared Miners in South Australia by stating that they may have been close to extinction in that State. He suggested that local extinction associated with clearing mallee and unsuccessful searches for birds in large areas of remnant mallee, did not bode well for Black-eared Miners in South Australia. Most subsequent survey and research efforts focused on the Victorian mallee, which was thought to be their last stronghold in Australia.

Despite efforts to conserve habitat and manage threats associated with fire, the population in Victoria continued to decline. Of eight individuals known in 1987 (Starks 1987) none now exist (McLaughlin 1993). It appeared probable that the Victorian populations would soon be extinct if the perceived rate of decline continued (Backhouse *et al.* 1995). These findings shifted the attention of the recovery team back to potential populations reported by Bob Goodfellow in South Australia.

For the past 20 years Mr Bob Goodfellow, a talented birdwatcher from the Riverland region of South Australia, has been observing mallee birds. Goodfellow commonly saw Black-eared Miner-like birds on his bird watching trips in the Riverland during the 1960s. At that time he was unaware of the growing concern for the survival of Black-eared Miners. To his knowledge they were still called Dusky Miners. When the president of the South Australian Ornithological Association inquired about sightings and explained to Goodfellow how rare the species had become, his keen interest in these obscure miners ensued.

Despite the consensus formed in the 1980s that the Black-eared Miner was probably extinct in South Australia, Goodfellow was convinced he had evidence of their continued existence. He questioned the validity of applying the Victorian scientific findings to the South Australian mallee systems he knew so well. Upon pursuit of Goodfellow's sightings, the recovery team has confirmed that the large areas of mallee north-west of the River Murray in South Australia do sustain a large number of Black-eared Miner-like birds. The open-minded nature of Goodfellow's observations of miners in the Riverland region has substantially improved the recovery team's understanding of the status and habitat requirements of Black-eared Miners.

The Black-eared Miner Recovery Team has always acknowledged the importance of community ownership of conservation programs. In South Australia local residents and private land managers will have more of a role to play in the recovery of Black-eared Miners than they have had in Victoria. Bob Goodfellow is now a member of the recovery team. Along with other keen natural historians, he has assumed the major responsibility for determining the location of Black-eared Miners, Yellow-throated Miners and hybrid birds in the Riverland region of South Australia. Biologists capable of accurate identification of the birds will use Goodfellow's sightings to confirm the abundance and distribution of Black-eared Miners. Land management decisions in South Australia, associated with the occurrence of Black-eared Miners on public and private land, will be based on Goodfellow's findings. A significant number of colonies now appear to occur on pastoral leases where threats associated with fire management are an issue. The community will play a strong role in the resolution of fire management issues.

The Black-eared Miner partnership between scientists and dedicated birdwatchers supports the conclusions of Young *et al.* (1996): such partnerships are the most efficient and cost effective path to species recovery. The research priorities for recovering Black-eared Miners will be carried out by scientists who have specialist expertise in collaboration with landholders and natural historians who have the knowledge of the local land and its management.

COMMUNITY OWNERSHIP OF SPECIES RECOVERY PROGRAMS

I will outline a number of the Mantung/Maggea project's attributes and what I have learned from involving the community in recovery initiatives, in the hope that these may provide conservation managers some insight into the successful development of similar community-driven species recovery programs. Blyth *et al.* (1996) provided guidelines for the establishment of partnerships between government-driven conservation programs and the community. These guidelines have helped shape the following discussion.

Some attributes of communities are conducive for the species recovery process

A number of the successes of the Malleefowl project can be linked to the social structure of the Mantung/Maggea community. Many of the attributes of the Mantung/Maggea community highlight the potential for developing other community-based conservation initiatives.

The population is small, relatively isolated and has a long-term stability within the region

Most of the farming families we worked with had been farming in the region for generations. Their sense of identification with and responsibility for the land and its inherent natural values was high. Some of the larger communities I have worked with, such as the townships that comprise the Riverland region, possess less total cohesiveness than the Mantung/Maggea area. Local parochialism can lead to conflicting views on conservation issues, which can make engaging the community in species recovery programs more of a challenge.

The Mantung/Maggea farmers shared a common concern for the conservation of Malleefowl

Although the composition of the Mantung/Maggea community was relatively diverse and the background of individual farmers within it varied greatly, they had a similar fascination for a rare bird. A shared concern for a rare and valued plant or animal species can instil a conservation ethic even amongst the most anti-conservationist elements of a community. I believe that this common cause is essential for establishing a constructive commitment to a conservation program in which the cost to benefit ratio for participation may vary greatly amongst participants. Indeed, despite the general commitment to protecting Malleefowl, not all members of the Mantung/Maggea community maintained their farm management programs. This inequality amongst farmers in contributions to the community's co-ordinated vermin control programs can undermine the continuing success of the Malleefowl program.

In contrast to the fascination instilled by Malleefowl, the Black-eared Miner is small, difficult to see and difficult for most mallee farmers to distinguish from other bush birds. They are not as unique as are Malleefowl amongst Australia's birds. Such traits make it difficult to convince land holders of the personal value of saving Black-eared Miners. This lack of personal value presents a challenge to gaining compliance with recovery plans by managers.

For example, the protection of critical habitat from fire is crucial for the recovery of Black-eared Miners. Sheep farmers on marginal mallee lands have traditionally used fire to increase the carrying capacity of their grazing leases. Because leaseholders do not place a high personal value on maintaining miner populations on their leases, financial incentives may be more effective in the short-term for recovering Black-eared Miners. The development of a personal value for the birds is likely to take time.

The utility of providing financial incentives to encourage species conservation has been demonstrated in other recovery programs. Very small financial subsidies were successful in protecting Cape Barren Goose populations on Victorian farms. The farmers were essentially paid to agist geese during the breeding season (Department of Conservation and Natural Resources 1994). Compensation for livestock fodder that the geese ate was a simple, effective solution to an on-going conflict of interests (Temby 1995).

Some attributes of species recovery programs are conducive to community involvement

Attributes of the Mantung/Maggea Malleefowl recovery program increased its success. Acceptance of landholders' values and insights, accessibility, continuity of individuals involved in facilitation, honesty and the subsequent development of mutual trust amongst the stakeholders led to agency staff being accepted as part of the Mantung/Maggea community working along with landholders to achieve a common goal.

The project commenced as a community initiative

The community had a growing concern about the plight of the Malleefowl populations on their farms. They were, however, uncertain of the causes for the birds' decline. The group solicited SADENR's advice on how to conserve Malleefowl populations in their district. The group encouraged me to undertake research on the birds in their farming properties. They needed recommendations for managing threatening processes, such as fox control, for which protocols had not previously been developed in rural South Australia.

The community had a clear definition of the problem

They observed a declining Malleefowl population within their farming district and wanted to reverse this trend. Their clarity of focus opened the way for agency staff to suggest changes in their traditional farming practice to resolve the problem. Short and long-term goals that are achievable and whose success can be monitored are key features of effective co-operative conservation initiatives.

The community had a common aim to conserve the Malleefowl on their farms and in their local reserves

This common goal encouraged members who had conflicting vested interests to co-operate with initiatives that were jointly agreed upon by the majority of the group. Some individuals were initially sceptical of the intentions of government agency staff. Eventually they accepted and supported management initiatives because staff were able to provide them with supportive evidence that justified the recommended conservation programs. Other members of the community who already understood the need for such initiatives increased the program's credibility amongst the uncommitted members. Acceptance of and participation in conservation management initiatives within this farming community soon spread.

The community had continuity in their interactions with individuals representing the government agencies who are legislatively responsible for conservation in the district

The South Australian National Parks and Wildlife Service's ranger-in-charge of State reserves and wildlife management in the district was invited to join a formalized conservation management group, the Mantung/Maggea Land Management Group. The group co-ordinated management initiatives and obtained funding for their initiatives. The South Australian National Parks and Wildlife Service ranger attended meetings and participated in all stages of the development and implementation of programs. The group invited me to become their official scientific adviser. In that capacity, I assisted in the development of their initiatives by bringing other agency input and providing scientific and conservation management information for discussion at the group's regular meetings. This continuity in agency involvement with the project instilled trust amongst members of the community, both for the staff personally as well as for the agencies we represented.

Conservation biologists and managers were accessible and responsive to the needs and concerns of community members

I believe that the long-term stability of a community conservation initiative is correlated directly with the ownership of programs by the community. Nonetheless, government agencies must assert their legislative responsibility for protecting natural resources. They should also use their ecological expertise to guide conservation programs and to resolve conflicts and raise awareness within the community. SADENR's and my abilities to be open with the Mantung/Maggea community, to listen to and understand their needs and concerns were critical to their successful engagement. Staff gained credibility and respect by taking their concerns seriously and developing programs that addressed conservation issues without compromising their farm management requirements. When conservation programs can be incorporated into landholders' goals for economic return, then adoption of a conservation ethic by the community is more easily achievable and is longer lasting.

The community's motivation was maintained through regular feedback on their achievements

The progress of many species recovery processes can be slow. A community's interest in programs that are expensive in terms of time and conflicting economic priorities will wane unless the effectiveness of their efforts is clearly understood. The Mantung/Maggea Malleefowl project was a challenge in this regard. Controlling foxes and cats in fragmented mallee habitats is a very difficult task. Foxes immigrate quickly into such landscapes even after a very rigorous baiting program. Although agency staff made it clear from the onset of the program that predator control would be an on-going commitment, the farmers were often discouraged and sometimes ready to abandon their baiting program when the presence of foxes appeared to be little changed.

An increase in the Malleefowl population in response to the fox baiting would have helped to maintain their commitment to the program. Because Malleefowl are long-lived birds that probably always had a low recruitment rate, increases in Malleefowl density can take many years to achieve and are difficult to demonstrate. Until it could be shown that fox baiting was leading to recruitment into the Malleefowl breeding population, it was necessary to regularly review the immediate benefits of the program to both the farmers and the wildlife on their farms. It was necessary to remind the group that continued high numbers of bait takes by foxes might be partly explained by neighbouring farmers who were not baiting.

Positive results, such as increased lambing rates and increased observations of young Malleefowl, encouraged new predator management initiatives and increased community participation. Regular surveys of breeding Malleefowl density and clear interpretation of the results will remain critical to maintaining the continuity of the community's conservation initiatives. Positive results are necessary motivators.

Community conservation initiatives can sometimes slow the progress of species recovery programs

Not all aspects of community involvement in conservation programs can be positive. The benefits of community involvement have outweighed the costs, in my experience, but the costs and benefits must be weighed individually for each recovery program.

Loss of control over how and when management programs are initiated can decrease the effectiveness of recovery actions

Agency staff conducting recovery programs on State or national reserves usually have total control over the development and initiation of recovery actions. Priorities for action can then be based on current scientific knowledge without the need for compromises to address the concerns and interests of the general community. When significant areas of threatened species' critical habitat occur on privately managed land, changes to land management practices that benefit endangered species must be negotiated. Major alterations to traditional management practices must be clearly justified.

Instilling in private land managers an understanding of the need to conserve threatened species and the potential for negative impacts from traditional land use can take time and may require training that many conservation biologists do not have. Critically endangered species, such as the Black-eared Miner, cannot always afford the luxury of time.

The development of protocols for species recovery sometimes requires technical expertise that is difficult to maintain in a community-based program

When new conservation initiatives are being developed, it is often necessary to maintain tight scientific control over the methods employed in order to interpret the results of a trial. For example, SADENR undertook a program to re-establish a population of Brush-tailed Bettongs *Bettongia penicillata* in Venus Bay Conservation Park. This was to be one of the first attempts at re-establishing a mainland population of this species. SADENR contracted a biologist to undertake a rigorously documented series of management programs to decrease the threats to the bettongs. Cat, fox and rabbit control were a major component of the management program. The effectiveness of controlling non-indigenous species was tested by monitoring feral pest populations and by monitoring the survival of radio-collared bettongs after their release.

Interpreting the results of these trials and effectively refining the program would have been difficult unless the pest control programs and the monitoring programs were undertaken in a consistently rigorous manner (D. Armstrong, pers. comm.). Community participation in this process may have compromised the success of this phase of the program.

Now that the protocol for translocating bettongs has been established, the community has a very important role to play in the Venus Bay project. SADENR is involving local community groups as much as possible to educate and instil ownership of the management initiatives. Landcare groups and schools are contributing to programs, such as revegetating corridors and maintaining labour-intensive feral pest control, that SADENR would otherwise have difficulty resourcing.

HOW TO FACILITATE COMMUNITY OWNERSHIP OF SPECIES RECOVERY PROGRAMS

In my experience with State, Commonwealth and international conservation agencies and a range of species recovery and conservation initiatives, I have found that some approaches are generally effective while others are less likely to be successful.

What has worked to engage the community in conservation programs

Have a focus for conservation initiatives that the community identifies with personally

Biologists and managers should understand which aspects of their environment are important to local communities and employ this understanding when developing community initiatives. Effective foci vary amongst communities as illustrated by examples of successful initiatives:

- Malleefowl as a flagship has instilled an interest in conserving mallee ecosystems in general.
- Many people in the Riverland region of South Australia are concerned about riverine ecosystems because they enjoy fishing, camping, hunting, canoeing and bird watching. Their personal values encourage them to alter unsustainable traditional land and wetland uses, such as modifying fishing techniques to protect declining populations of native fishes, decreasing livestock grazing pressure on degraded floodplain habitats to control erosion, and closing traditional campgrounds to protect waterbird rookeries.
- A general fascination many people have with Southern Hairy-nosed Wombats *Lasiorhinus latifrons* increased the South Australian National Parks and Wildlife Services' capacity to promote the formation of a Friends of Brookfield Conservation Park group. Cultivating the Friends' interest in wombats soon led to a willing adoption of education and conservation programs for arid rangeland ecosystems.

Encourage community involvement from the earliest developmental stages of a conservation or species recovery initiative

This involvement will instil a sense of ownership for the program and increase the longevity of commitment. The Mantung/Maggea Malleefowl program illustrates the power of this strategy. Temby (1993) stresses the importance of involving people in practical demonstrations rather than just providing well-substantiated information that can be easily ignored. Walters (1986) suggests most people's basic attitudes can be changed only if they construct the arguments for change themselves. Few respond to arguments of other people. Involvement leads to understanding. Understanding leads to attitudinal change.

Develop programs that are beneficial for the community as well as conservation

Make it clear to all stakeholders that everyone has something to gain, directly and indirectly, from species recovery programs. Be specific about the gains to dispel fears of the personal costs of conservation initiatives. The Mantung/Maggea farmers understand that their Malleefowl conservation initiatives also increase the productivity of their farms. This can be an important incentive to some members of the Mantung/Maggea community.

Listen to the community's concerns

Understand and appreciate their values. Make a concerted effort to incorporate their priorities for action into a program that effectively achieves everyone's long-term conservation goals and personal visions for our future heritage. In the case of the Mantung/Maggea farmers, I could have jeopardized co-operation if I insisted that fox baiting be carried out on all farms every two months. This recommendation could have been justified from research undertaken in Western Australia to conserve Black-footed Rock Wallabies *Petrogale lateralis* (Kinnear *et al.* 1988; Kinnear 1990). Nonetheless such a protocol would have been unreasonable for wheat and sheep farmers. A compromise to bait twice a year when foxes and Malleefowl were breeding and sheep were lambing was easily negotiated. This strategy has been effective for Malleefowl conservation and realistic for farm management.

Gain the community's trust

Trust can lead to reciprocity amongst stakeholders and a long-term commitment to acting on behalf of the common good. Trust can be instilled by continuity and consistency in personal interactions between biologists, managers and the community.

Provide the community with the appropriate information at the appropriate level and at the appropriate time

Acknowledge the community's understanding of conservation issues and respond to their perceptions. Major changes in basic attitudes cannot be expected to occur overnight. Regular review of programs, feedback on results, and provision of additional information at the level of complexity the community wants can lead to positive changes in attitude toward conservation.

What does not work

Failing to recognize the community's understanding of ecological concepts can create resentments

Ideally, conservation managers should build on the community's current level of understanding to assist their progression toward ownership of conservation programs. The pace of this progress must be determined by the community. The Mantung/Maggea farmers were not interested in the ecology of non-indigenous predators until they had accumulated experience with control programs and began to question their results. It would not have been appropriate for me to provide them findings of scientific studies on introduced predators before the interest arose.

Failing to appreciate what the community hopes to gain from participating in a recovery program can dampen their enthusiasm to contribute

The community demands respect for their personal viewpoints. Even dedicated conservationists, such as Bob Goodfellow, can resent being held to the same standards as trained ecologists. He is willing to contribute a 100% effort within his own capacity and on his own terms. He has already proved the value of his contributions.

Failing to provide appropriate support after community-based programs have been initiated can threaten the community's continued commitment to the project

The need for support and the type of support will vary amongst recovery programs and composition of the community. The Mantung/Maggea farmers need technical support to reassess the effectiveness of their management programs. Community groups participating in Malleefowl surveys need feedback on the results of their surveys and on the importance of their contributions to the recovery of Malleefowl.

Government personnel often have conflicting professional responsibilities. This makes it hard for them to either remain receptive to this need to support community-based conservation programs or to cater to this need. Few community groups survive without the requisite support from agencies. Subsequent animosity toward the institution can be difficult to repair.

Support can be as simple as being available to answer questions about Malleefowl biology or facilitating school monitoring programs by simplifying research permit procedures. It can be as complex as knowing when it is appropriate to use discretion in applying administrative protocol. Sensitivity for corporate responsibilities can be maintained while meeting expectations of participants in community endeavours.

Agency staff can provide incentives, such as not charging camping fees, to groups who are actively involved in park programs. This shows that the park agency values the contributions made by volunteers. Malleefowl surveys involve 15 to 20 people searching remote mallee habitats, often in hot weather, for up to three days. Most parks in South Australia provide free accommodation to these volunteer groups. Districts with strict business plans sometimes find it difficult to relinquish charges. Volunteers who are providing the time, energy and skills to undertake conservation programs do not always appreciate why entrance fees or camping fees cannot be subsidized by the agency. If financial and staff resources continue to dwindle in many State conservation agencies, such partnerships will become increasingly important to nurture.

When agency staff behave as if wildlife and park management belong only to them, the public are discouraged from developing a personal responsibility for conservation

In carrying out their legislative responsibilities for protecting wildlife and wildlife habitats, managers can inadvertently develop a sense of "us and them" ownership that is conveyed in their interactions with the public. Managers must consciously break down such barriers and instil the sentiment that Australia's natural heritage belongs to all Australians. Ownership barriers discourage the community from assuming a more personal responsibility for protecting that heritage.

Agency staff were insensitive to the community's personal identification with the land prior to the dedication of Coffin Bay National Park in South Australia. This led to deep resentment and mistrust of South Australian National Parks and Wildlife Service's philosophy for land management. The community considered the reserve to be part of their commons when it was privately owned. They associated the feral populations of ponies that had been living in the reserve as a part of the cultural heritage that was very important to the identity of the local people. The ponies had been "theirs" for between 90 and 100 years.

When South Australian National Parks and Wildlife Service tried to enforce their legislative responsibility to protect the landscape by eradicating non-indigenous animals, they failed to consult with the community. The insistence that all ponies were to be shot without negotiation or compromise led to six years of turmoil. South Australian National Parks and Wildlife Service was perceived as working outside the community's best interest. The Coffin Bay Pony Preservation Society began an energetic campaign to stop government staff from killing the ponies.

Eventually South Australian National Parks and Wildlife Service staff, who were new to the district, propounded a broader interpretation of their legislative charter. A compromise was successfully negotiated: two pony herds would be exterminated and the remaining herd would be culled and maintained in an already degraded area of the Park. The Coffin Bay Pony Preservation Society agreed to assume the responsibility for culling and managing the ponies with the assistance of South Australian National Parks and Wildlife Service staff. It took three years in which Park staff worked openly and honestly with the community to achieve the agreed goals for pony management before the community's mistrust waned. The Preservation Society has maintained their responsibility for managing the pony population and the South Australian National Parks and Wildlife Service has regained credibility amongst the local community.

WHAT GAINS CAN BE MADE TO SPECIES RECOVERY BY ENGAGING THE PARTICIPATION OF THE COMMUNITY?

It is well known that long-term sustainability of ecological communities and small populations of species is compromised by fragmentation, isolation and degradation of remnant native vegetation (Saunders *et al.* 1987; Saunders and Hobbs 1991; Saunders *et al.* 1993). Most State reserve systems were not developed strategically to protect representatives of all ecosystems. Many threatened Australian animals, such as Superb Parrots, Regent Parrots and Brush-tailed Bettongs, require large tracts of suitable habitat to feed, breed and successfully recruit young. Their need to disperse does not stop at park boundaries. Although South Australia's protected areas incorporate 21% of the State, 26% of the State's 176 plant communities are inadequately protected in the State's reserve system and an additional 19% is only moderately conserved (Neagle 1995).

The preservation of many of Australia's threatened species depends upon the ability to maintain and enhance populations and communities on private land. Federal and State programs such as Save the Bush, the South Australian Heritage Agreement Scheme and the Victorian Land for Wildlife Scheme were initiated in recognition of the importance of wildlife conservation on private land.

Engaging people in conservation initiatives has additional benefits that are, I believe, equally or substantially more important than the recovery of individual populations. Long-term maintenance of Australia's biodiversity will entail more than programs to recover threatened species and ecosystems. It also requires societal acceptance of responsibility for conserving biodiversity, an awareness of the plight of our continent and the adoption of a conservation ethic that incorporates every aspect of society and every aspect of an individual's livelihood. Societal change generally takes place gradually and in stages. It appears to take place more rapidly when individuals have hands-on involvement with issues.

Because the interactions within the total ecosystem are so complex and poorly understood, concentration on one threatened species out of many can focus the community's attention and produce tangible outcomes. The challenges associated with recreating areas of mallee that have low levels of human-induced disturbances to protect Malleefowl from foxes, cats, rabbits, sheep and fire has increased the Mantung/Maggea farmers' regard for their remnant mallee vegetation. In many cases, simplification aids comprehension.

Such an involvement with species recovery programs can, I believe, enhance society's progress toward developing a strong conservation ethic. Ownership of conservation initiatives, stewardship for species recovery and the acceptance that conservation management is everyone's responsibility is an attitudinal change that must take place in society if one is to hold any hope for maintaining what remains of Australia's natural heritage.

CONCLUSIONS

Conservation biologists and land managers must be flexible in their approaches to species and ecosystem recovery programs if they are to achieve cultural change in society's attitudes toward conservation. Each element of society and each community will bring different interests and opinions to the need for conserving local, regional and global biodiversity. Conservation biologists and managers can increase their effectiveness through creative approaches to developing strategies for species recovery. Managers must guarantee that conservation initiatives requiring major changes in locally accepted land use practices evolve at a pace that facilitates the community's understanding and acceptance of the need for change. Biologists and managers also have to be receptive and allow the community to mould *their* understanding of how to develop a more sustainable landscape in Australia.

ACKNOWLEDGEMENTS

I am grateful for the co-operation and inspiration I have received from the Mantung/Maggea farmers. The projects described have taken place over a period of five years. I acknowledge the contributions of the large number of staff and volunteers who made these programs a success and thank you all. P. Hussey and M. Moore provided helpful comments on an early draft. J. McLaughlin, I. Sluiter, D. Armstrong, B. Goodfellow and SA National Parks and Wildlife Service Murraylands regional staff made useful suggestions for improving this manuscript.

REFERENCES

ANZECC, 1991. List of endangered vertebrate fauna in Australia. Australia and New Zealand Environment and Conservation Council. Australian National Parks and Wildlife Service: Canberra.

Backhouse, G., Bennett, S. and McLaughlin, J., 1995. Recovery Plan for the Black-eared Miner, *Manorina melanotis*. Department of Conservation and Natural Resources, Royal Australasian Ornithologists Union, Museum of Victoria, La Trobe University, Zoological Board of Victoria, Australian Nature Conservation Agency: Melbourne.

Barratt, R., Williams, S. and Nixon, C., 1991a. How to Manage Native Vegetation in the Murray Mallee: a Conservation Handbook. Department of Environment and Planning Government of South Australia: Adelaide.

Barratt, R., Williams, S. and Nixon, C., 1991b. How to Manage Native Vegetation in the Murray Mallee: a District Guide. Department of Environment and Planning. Government of South Australia: Adelaide.

Barratt, R., Williams, S. and Nixon, C., 1991c. How to Manage Native Vegetation in the Murray Mallee: Property Plans. Department of Environment and Planning. Government of South Australia: Adelaide.

Bennett, A., Backhouse, G. and Clark, T. (eds), 1995. People and Nature Conservation. Perspectives on Private Land Use and Endangered Species Recovery. Transactions of the Royal Zoological Society of New South Wales. Surrey Beatty & Sons: Chipping Norton.

Blakers, M., Davies, S. J. J. F. and Reilly, P., 1984. The Atlas of Australian Birds. Royal Australasian Ornithologists Union and Melbourne University Press: Melbourne.

Blyth, J. D., Burbidge, A. A. and Brown, A. P., 1996. Achieving co-operation between government agencies and the community for nature conservation, with examples from the recovery of threatened species and ecological communities. Pp. 343–67 *in* Nature Conservation 4: The Role of Networks ed by D. A. Saunders, J. L. Craig and E. M. Mattiske. Surrey Beatty & Sons: Chipping Norton.

Christidis, L. and Boles, W. E., 1994. The Taxonomy and Species of Birds in Australia and it is Territories. R.A.O.U. Monograph No. 2. Royal Australasian Ornithologists Union: Melbourne.

Dendy, T., 1996. Protected Area Management in South Australia. Report for IUCN Commission on National Parks and Protected Areas Meeting, 8–10 June 1996.

Department of Conservation and Natural Resources, Victoria., 1994. File No. 85-5004 Cape Barren Geese — Problems with.

Garnett, S., 1992. Action Plan for Australian Birds. Australian National Parks and Wildlife Service: Canberra.

Joseph, L., 1986. The decline and present status of the Black-eared Miner in South Australia. *SA Ornith.* **30**: 5–13.

Kinnear, J. E., Onus, M. L. and Bromilow, R. N., 1988. Fox control and rock wallaby population dynamics. *Aust. Wildl. Res.* **15**: 435–450.

Kinnear, J., 1990. Trapping of success. *Landscope* **5**(3): 35–40.

Land Conservation Council, 1987. Mallee Area Review. Land Conservation Council: Melbourne.

Land Conservation Council, 1989. Mallee Area Review. Final Recommendations. Land Conservation Council: Melbourne.

McLaughlin, J., 1990. Surveys and observations of the Black-eared Miner *Manorina melanotis* in Victoria. Royal Australasian Ornithologists Union Report No. 71.

McLaughlin, J., 1992. The floristic and structural features of Black-eared Miner *Manorina melanotis* habitat. Royal Australasian Ornithologists Union Report No. 84.

McLaughlin, J., 1993. Recovery Plan Research Phase for the Black-eared Miner *Manorina melanotis* (Wilson 1911). Royal Australasian Ornithologists Union: Melbourne.

McLaughlin, J., 1994. Searches for the Black-eared Miner *Manorina melanotis* in the Victorian Murray Mallee, 1992/1993. Royal Australasian Ornithologists Union Report No. 93.

Moore, S. A., 1996. The role of trust in social networks: formation, function and fragility. Pp. 148–54 *in* Nature Conservation 4. The Role of Networks ed by D. A. Saunders, J. L. Craig and E. M. Mattiske. Surrey Beatty & Sons: Chipping Norton.

Neagle, N., 1995. An Update of the Conservation Status of the Major Plant Associations of South Australia. Department of Environment and Natural Resources: Adelaide.

Ostrom, E., 1990. Governing the Commons: the Evolution of Institutions for Collective Action. Cambridge University Press: Cambridge.

Prescott, A. M., 1996. Successful extension programmes for nature conservation in South Australia's rural areas. Pp. 601–08 *in* Nature Conservation 4: The Role of Networks ed by D. A. Saunders, J. L. Craig and E. M. Mattiske. Surrey Beatty & Sons: Chipping Norton.

Saunders, D. A., Arnold, J. A., Burbidge, A. A. and Hopkins, A. J. M. (eds), 1987. Nature Conservation: the Role of Remnants of Native Vegetation. Surrey Beatty & Sons: Chipping Norton.

Saunders, D. A. and Hobbs, R. J. (eds), 1991. Nature Conservation 2: The Role of Corridors. Surrey Beatty and Sons: Chipping Norton.

Saunders, D. A., Hobbs, R. J. and Ehrlich, P. R. (eds), 1993. Nature Conservation 3: The Reconstruction of Fragmented Ecosystems. Surrey Beatty & Sons: Chipping Norton.

Saunders, D. A., Craig, J. L. and Mattiske, E. M. (eds), 1996. Nature Conservation 4: The Role of Networks. Surrey Beatty & Sons: Chipping Norton.

Sibley, C. G. and Monroe, B. L., 1990. Distribution and Taxonomy of Birds of the World. Yale University Press: New Haven and London.

Starks, J., 1987. The status and distribution of the Black-eared Miner *(Manorina melanotis)* in Victoria. Arthur Rylah Institute Technical Report No. 49.

Temby, I., 1995. Perceptions of wildlife as pests: you can teach an old dogma new tricks. Pp. 174–80 *in* People and Nature Conservation. Perspectives on Private Land Use and Endangered Species Recovery ed by A. Bennett, G. Backhouse and T. Clark. Transactions of the Royal Zoological Society of New South Wales. Surrey Beatty & Sons: Chipping Norton.

Walters, C., 1986. Adaptive Management of Renewable Resources. MacMillan: New York.

Williams, S. L., 1996. Malleefowl as a flagship for conservation on farms in the Murray Mallee of South Australia. Pp. 316–20 in Nature Conservation 4: The Role of Networks ed by D. A. Saunders, J. L. Craig and E. M. Mattiske. Surrey Beatty & Sons: Chipping Norton.

Young, M. D., Gunningham, N., Elix, J., Lambert, J., Howard, B., Grabosky, P. and McCrone, E., 1996. Reimbursing the Future. An Evaluation of Motivational, Voluntary, Price-based, Property-right, and Regulatory Incentives for the Conservation of Biodiversity. Biodiversity Series, Paper No. 9. Department of Environment, Sport and Territories: Canberra.

Conservation of the endangered plant *Grevillea caleyi* (Proteaceae) in urban fire-prone habitats

Tony D. Auld[1] and Judith A. Scott[1]

The endangered plant *Grevillea caleyi* (Proteaceae) occurs in bushland that is adjacent to urban areas in the Sydney region. Within these areas, repeated and frequent fire threatens not only the endangered flora but life and property as well. These threats were well illustrated by the impact of the fires that occurred in January 1994 in Sydney.

Management of urban fire-prone areas needs to identify those fire regimes likely to drive the endangered flora to extinction, as well as identifying if any populations of endangered flora occur in locations that pose a fire hazard for the protection of life and property. Research into the population dynamics of *G. caleyi*, as part of the development of a recovery plan for the species, indicates that a regime of frequent fire will lead to local population declines and extinction. Consequently, burning on a frequent basis for hazard reduction to protect property assets in the vicinity of *G. caleyi* is inappropriate for the conservation of this plant. Instead, a minimum fire-free interval of 8–12 years is recommended for the conservation of *G. caleyi*. Additionally, areas unburnt for 20–25 years should be monitored for adult plant survival and seedling recruitment. If all or most adults have died and there is no seedling recruitment then consideration should be given to burning such sites.

Key words: Fire, fire frequency, fire regime, urban bushland.

INTRODUCTION

GREVILLEA CALEYI R.Br. is a tall open shrub, restricted to roughly a 6 km × 6 km area centred on the northern Sydney suburb of Terrey Hills (Fig. 1). Within this area, *G. caleyi* occurs on laterite soils in open forests or low open forests and three major areas of suitable habitat exist. At present, the major habitat areas are highly fragmented among natural, urban, semi-rural and commercial areas, resulting in the loss of around 85% of the former habitat of *G. caleyi*. This has left remnant fragments of former populations of *G. caleyi* in some 17 patches and two additional sites, one of dubious origin and one in a rehabilitated area. Only two of the remaining sites are relatively undisturbed and only one of these two has any long-term protection as it is in a national park.

The remaining large patches of *G. caleyi* are highly disturbed, with many of the small fragments of plants subject to very high edge-related disturbance effects. The remnant patches of *G. caleyi* are likely to be important components of the genetic diversity within the species and are urgently in need of adequate protection from further disturbance.

The major threats to the survival of *G. caleyi* were identified in the recovery plan for this species (Scott *et al.* 1995) and arise from human interference in the environment. These threats range from loss of habitat through changes in land use, urbanization and associated infrastructure, an inappropriate fire regime, to localized disturbance at sites including invasion of weed species and the dumping of rubbish. At all sites it is the interaction of a range of threats that currently impacts on *G. caleyi*.

Considerable research into the population dynamics of *G. caleyi* has been undertaken (see Table 1 and Scott *et al.* 1995), allowing an assessment of the role fire plays in maintaining the remaining fragments of *G. caleyi*, that is , determining what fire regimes are compatible with the conservation of *G. caleyi*. These requirements form an essential component of the recovery plan for *G. caleyi* (Scott *et al.* 1995).

Many remnant patches occur on the urban/bushland interface (Fig. 2), with some 60% of the habitat of *G. caleyi* burnt during 1994. Such fragments are in danger of being burnt too frequently for fuel reduction purposes. Other fragments are surrounded by urban areas and are threatened by long-term fire exclusion. Here we present a summary of research findings for *G. caleyi* in relation to fire management, the major disturbance event in its habitat.

THE FIRE REGIME AND*GREVILLEA CALEYI*

In managing plants, animals and fires, it is important to remember that fires at a particular place can occur at different frequencies (fire frequency); will burn at different intensities (fire intensity); occur at different times of the year (fire season); and will burn different sized areas from small patches to huge expanses (fire spatial extent). These four components make up what is known as the fire regime. For the long-term survival

[1]Environmental Survey and Research Division, NSW National Parks and Wildlife Service, P.O. Box 1967, Hurstville, New South Wales, Australia 2220.
Pages 97–100 *in* BACK FROM THE BRINK: REFINING THE THREATENED SPECIES RECOVERY PROCESS ed by S. Stephens and S. Maxwell.
Surrey Beatty & Sons, 1996.

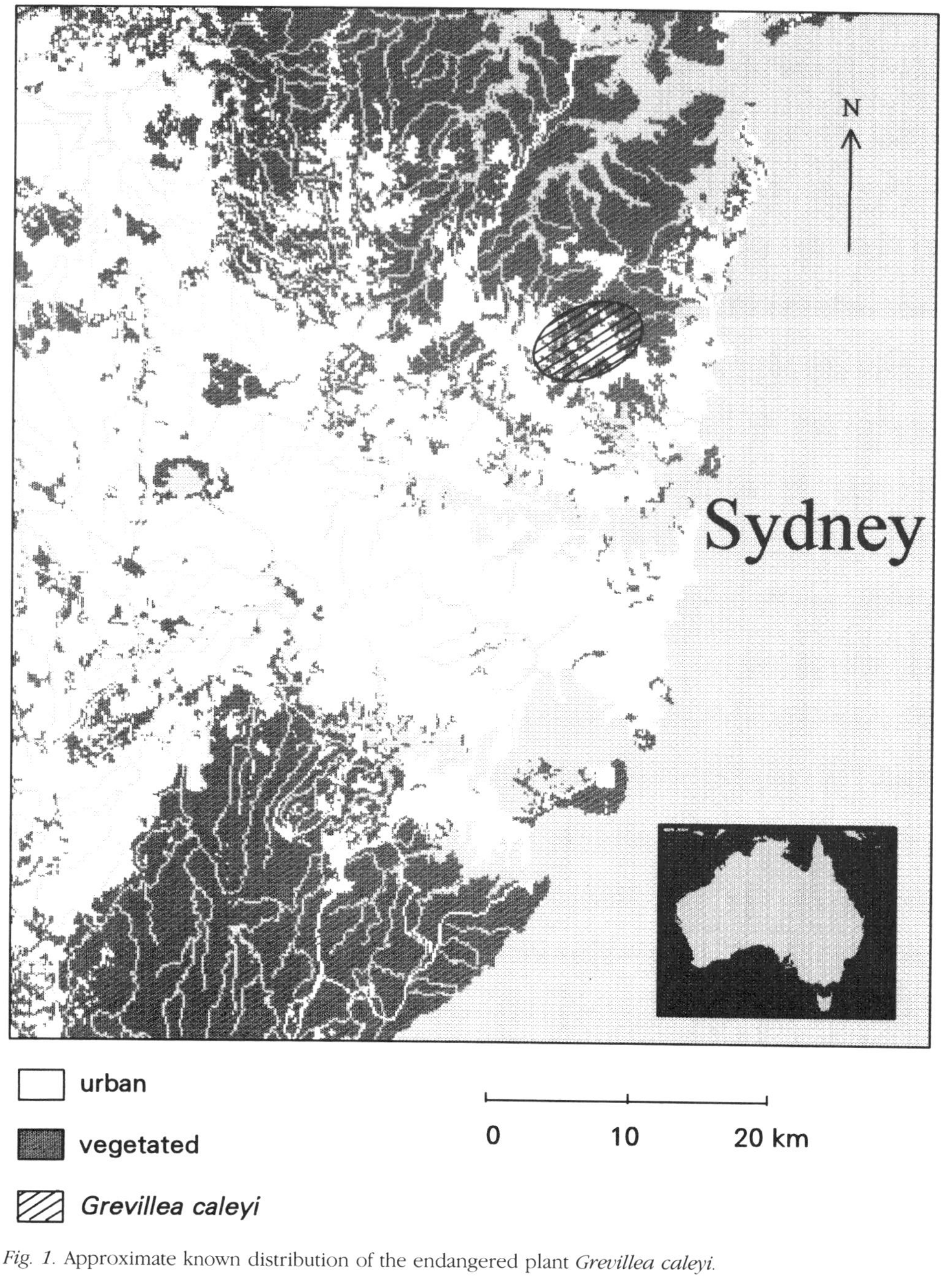

Fig. 1. Approximate known distribution of the endangered plant *Grevillea caleyi*.

of endangered plants such as *G. caleyi*, we need to know how they are affected by the fire regime. We now consider how each of these four components of the fire regime affect *G. caleyi*, remembering though, that all four operate together in nature.

Fire frequency

Grevillea caleyi is killed by fire (fire-sensitive) and relies on seed germination from a soil seedbank after fires to survive at a particular location. For such fire-sensitive plants, there must be sufficient time between successive fires for seedlings to mature and produce more seeds. For *G. caleyi*, it takes 2–5 years for seedlings to mature and few seeds are produced on young plants. It takes several more years for plants to grow big enough to produce many seeds (Table 1). As well, seeds of *G. caleyi* and other *Grevillea* species are consumed by native mammals (*Rattus fuscipes* and *Wallabia bicolor*) after release. Levels of seed losses to mammals are very high (Table 1) and will severely retard the replenishment of the soil seedbank after fire. At the same time, much of the available seedbank is exhausted after any one fire (55–83%, Auld 1994). These data suggest that a minimum fire-free interval of 8–12 years is needed to allow the soil seedbank to be sufficiently replenished. If fires occur more frequently than this, population declines are likely. Should two fires occur close enough together that the seedlings from the first fire are killed before they mature, then dramatic declines are expected. Auld (1994) has shown that under such a scenario there was a 55% reduction in seedling abundance after the second fire and that no residual seeds remained in

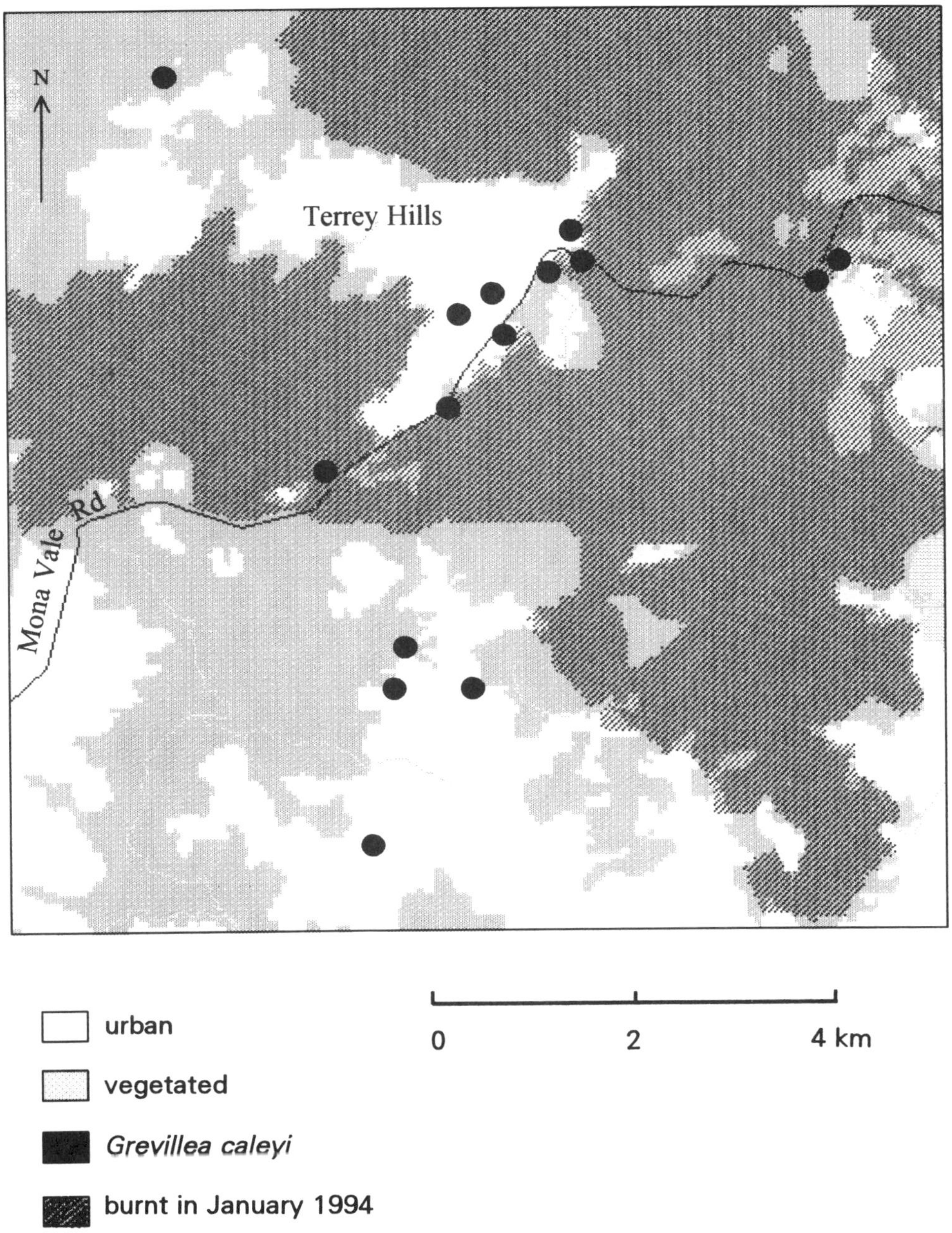

Fig. 2. The urban/bushland interface showing approximate locations of fragments of *Grevillea caleyi* sites.

the soil. Consequently, three fires in short succession may be sufficient to cause local extinction in *G. caleyi*.

At sites now surrounded by urban development (Fig. 2), fire may be excluded for long periods of time. As senescence of adult plants is high in sites over 12–15 years unburnt (Table 1) and the soil seedbank will decay, there will be a limit to how long *G. caleyi* can persist at a site if the site does not get burnt. A maximum fire-free interval to maintain plants of *G. caleyi* at a site is in the order of 20–25 years. If areas are left unburnt longer than this, the species may be lost. Sites unburnt for between 15–20 years should be monitored for the level of adult survival and the level of seedling recruitment that is occurring. If no seedlings can be found and there is high adult mortality, consideration should be given to burning the site.

Fire intensity

There is a pulse of germination after fire in *G. caleyi* (Table 1), like in several other *Grevillea* species and other fire-sensitive plants such as wattles (Auld and Tozer 1995). However, at present the mechanism controlling the breaking of seed dormancy in the soil and subsequent germination is not understood. We also need to understand how such a germination cue may be influenced by fire, and in particular, fire intensity. Auld and Tozer (1995) showed that soil heating during the passage of a fire did not break seed dormancy in *Grevillea*, but other fire cues such as smoke, charred wood or nitrates may be important (cf. Keeley 1991; Brown 1993; Dixon *et al.* 1995; Thanos and Rundel 1995). How such factors may be influenced by fire intensity is currently unknown, but

Table 1. Demographic features of the endangered plant *Grevillea caleyi.*

Characteristic	*Grevillea caleyi*
Primary juvenile period	2.5–5 years
Flowering	annually
Fecundity post-fire	low initially, with increasing fecundity with plant size/age
Mortality rate[1]	0.22–0.5
Predispersal seed predation (% seeds lost)	20–31
Postdispersal seed predation (% seeds lost)	82–100
Dispersal agent	none known
Seed viability (%)	69–95
Seed dormancy at release (%)	7.6 years
Seed longevity in the soil (half-life)[2]	97–98
Dormancy broken by heat	no
Post-fire seedling pulse	yes

[1]for adult plants greater than 10 years old and assuming log linear decay.
[2]based on exponential decay rates of seeds in the soil.

will influence the use of fire as a management tool in urban bushland.

Fire season

Given the prevalence of arson and widespread hazard reduction burning for the protection of life and property, fires can now be expected to occur at just about any time of the year in the Sydney region. The question is, for the survival of *G. caleyi* and other co-habiting plants and animals, does it matter when the vegetation is burnt? The impact of season of burn on germination and survival of seedlings of *G. caleyi* is currently unknown and at present no management directions with regard to fire season can be made.

Fire spatial extent

Managers should avoid burning small patches in or around *G. caleyi* populations, as such small burns are prone to high mammal grazing which may be detrimental to seedlings. Such areas also have a significantly increased probability of being burnt by a second fire soon after the first which increases the risk of local extinction in *G. caleyi* through high fire frequency.

In addition, there is the possibility that the impact of mammalian seed predators (Table 1, see above) may be reduced if large surrounding areas are also burnt when *G. caleyi* plants are burnt. This idea relies on the fire reducing the population size of the mammals and hence reducing their impact in the first few fruiting seasons after fire and allowing a more rapid replenishment of the soil seedbank post-fire. Recent data (Auld and Denham, unpubl. data) suggested that losses of seeds to mammals approached 100% after an extensive wildfire and that there was no beneficial effect of burning a large surrounding area. Presumably, the urban/bushland interface provides sufficient refuges for the survival of mammal species.

CONCLUSIONS

Fire plays an important role in the maintenance of the remaining remnant patches of the endangered plant *G. caleyi.* While aspects of the impact of components of the fire regime, notably fire intensity and fire season, need to be further investigated, consideration of fire frequency is essential to the successful management of this species. For the long-term survival of *G. caleyi* under a regime of repeated fires, extreme care should be taken to accurately record the fire histories of all population fragments. This provides an information base for planning and hopefully avoids *ad hoc* decisions to burn areas. It is critical that the interval between successive fires is not less than 8–12 years at a site or population declines will occur. Repeated fires at intervals of less than eight years will cause the local extinction of *G. caleyi* from a site. Care should also be taken, where possible, to ensure that sites of *G. caleyi* isolated in urban areas do not remain unburnt for too long a period. This can be monitored by scoring the survival of adult plants and whether any recruitment of seedlings is occurring. Such management recommendations are broadly compatible with maintaining other plant species in the habitats within which *G. caleyi* occurs (Bradstock *et al.* 1995).

The management of fire in the vicinity of remnant fragments of *G. caleyi* will involve NSW National Parks and Wildlife Service input via the bushfire planning committees. This will help ensure that the conservation of *G. caleyi* is given a high priority.

ACKNOWLEDGEMENTS

Research into the population dynamics of *G. caleyi* was funded by NSW National Parks and Wildlife Service, World Wide Fund for Nature Australia and the Australian Nature Conservation Agency.

REFERENCES

Auld, T. D., 1994. The role of soil seedbanks in maintaining plant communities in fire-prone habitats. Proc. 2nd Int. Conf. Forest Fire Research, Vol II. Pp. 1069–078. Coimbra, Portugal.

Auld, T. D. and Tozer, M., 1995. Patterns in emergence of *Acacia* and *Grevillea* seedlings after fire. *Proc. Lin. Soc. New South Wales* **115**: 5–15.

Bradstock, R. A., Keith, D. A. and Auld, T. D., 1995. Fire and conservation: imperatives and constraints on managing for diversity. Pp. 323–33 *in* Conserving Biodiversity: Threats and Solutions ed by R. A. Bradstock, T. D. Auld, D. A. Keith, R. T. Kingsford, D. Lunney and D. P. Sivertsen. Surrey Beatty & Sons: Chipping Norton.

Brown, N. A. C., 1993. Promotion of germination of Fynbos seeds by plant-derived smoke. *New Phytologist* **123**: 575–83.

Dixon, K. W., Roche, S. and Pate, J. S., 1995. The promotive effect of smoke derived from burnt native vegetation on seed germination of Western Australian plants. *Oecologia* **101**: 185–92.

Keeley, J. E., 1991. Seed germination and life history syndromes in the California chaparral. *The Botanical Review* **57**: 81–116.

Scott, J., Marshall, A. and Auld, T. D., 1995. Conservation research statement and recovery plan for *Grevillea caleyi* R.Br. Australian Nature Conservation Agency, Endangered Species Program, Endangered Species Project No. 456.

Thanos, C. A. and Rundel, P. W., 1995. Fire-followers in chaparral: Nitrogenous compounds trigger seed germination. *J. Ecol.* **83**: 207–16.

Rediscovery program for the endangered plant *Haloragodendron lucasii*

Marita A. Sydes[1], Mark Williams[2], Rob Blackall[2] and Tony D. Auld[3]

The *Haloragodendron lucasii* Rediscovery Team was established to try and find new locations of this plant in the wild. Prior to the initiation of the team only three sites were known with a total of four genetically distinct individuals. Each of these individuals is effectively male sterile. Finding more locations of this endangered plant will lead to the protection of more individuals, the possibility of discovering male fertile plants, as well as assisting the planning of conservation measures. The rediscovery team involved joint co-ordinating efforts by NSW National Parks and Wildlife Service, The Australian National University and Ku-ring-gai Council. Community involvement was encouraged through the use of volunteer groups to search for *H. lucasii* in the field. Instruction to community groups involved an evening session, where the details of the recovery of *H. lucasii* and associated genetic research were discussed, through to field days where the public were shown what the plant looks like in the wild. The value of the involvement of the community groups for the rediscovery program is highlighted by the discovery of a new location for *H. lucasii* in late September 1995.

Key words: Genetic variation, community involvement, survey.

WHAT IS *HALORAGODENDRON LUCASII*?

HALORAGODENDRON LUCASII (Maiden and E. Betche) Orch. is an endangered plant known only from the Hornsby-Gordon area in the northern suburbs of Sydney (Fig. 1). The species was thought to be extinct until 1986 when it was rediscovered in urban bushland in the northern Sydney suburb of St Ives. Since its original rediscovery, it has been found in two more nearby locations. Biologically, the species is somewhat special. It is an erect bushy shrub that grows up to 1.5 m tall whilst spreading vegetatively from underground rhizomes. It is highly clonal, producing numerous aerial stems. The recovery plan for *H. lucasii* (Nash and Matthes 1994) recommended studies of the breeding system and genetic variation in this species as well as searches for possible new locations. Recent studies of genetic variation have found that only four genetically distinct individuals exist at the three studied locations (two locations each have a single genetically distinct individual and the third location has two genetically distinct individuals).

While *H. lucasii* produces lots of showy white flowers in spring, no successful seed set has been observed in the wild. The existing plants are effectively male sterile, with little dehiscence of the anthers, very low pollen counts and less than 5% pollen viability. This probably accounts for the lack of successful fruit set in the species, as ovules are produced in flowers. As well as this biological limitation there are a number of threats to the long-term survival of *H. lucasii* plants in the known locations. These threats include an inappropriate fire regime, habitat loss through urbanization, nutrient enrichment of sites and weed invasion. Management actions for these threats have been addressed in the recovery plan for the species (Nash and Matthes 1994).

THE REDISCOVERY TEAM

Given that only four genetically distinct individuals were known and that these were all effectively male sterile, there was a need to search for any undiscovered populations of *H. lucasii* so that they may be protected from loss through future urban development. Finding new locations of *H. lucasii* was seen as a priority in order to potentially increase the number of genetic individuals known and, it was hoped, to locate a male fertile individual so that it might be possible to re-initiate sexual reproduction in the species.

Research into the biology, breeding system and patterns of genetic variation of *H. lucasii* involved researchers from Australian National University (ANU) and the NSW National Parks and Wildlife Service. Management of the species in the wild was the responsibility of district staff from NSW National Parks and Wildlife Service and Ku-ring-gai Council. Consequently, these groups came together to form the nucleus of a co-ordination effort to search for any remaining locations of *H. lucasii*. Their responsibility was to assess the rationale and priorities for any searches, to administer and co-ordinate the rediscovery process and to unite a wide range of people with skills in research, management, conservation biology, botany

[1]Division of Botany and Zoology, The Australian National University, Canberra, Australian Capital Territory, Australia 0200.
[2]NSW National Parks and Wildlife Service, Garigal National Park, P.O. Box 134, Forestville, New South Wales, Australia 2087.
[3]Environmental Survey and Research Division, NSW National Parks and Wildlife Service, P.O. Box 1967, Hurstville, New South Wales, Australia 2220.
Please send correspondence to: Dr Tony Auld, NSW National Parks and Wildlife Service, P.O. Box 1967, Hurstville, New South Wales, Australia 2220.

Pages 101–04 *in* BACK FROM THE BRINK: REFINING THE THREATENED SPECIES RECOVERY PROCESS ed by S. Stephens and S. Maxwell. Surrey Beatty & Sons, 1996.

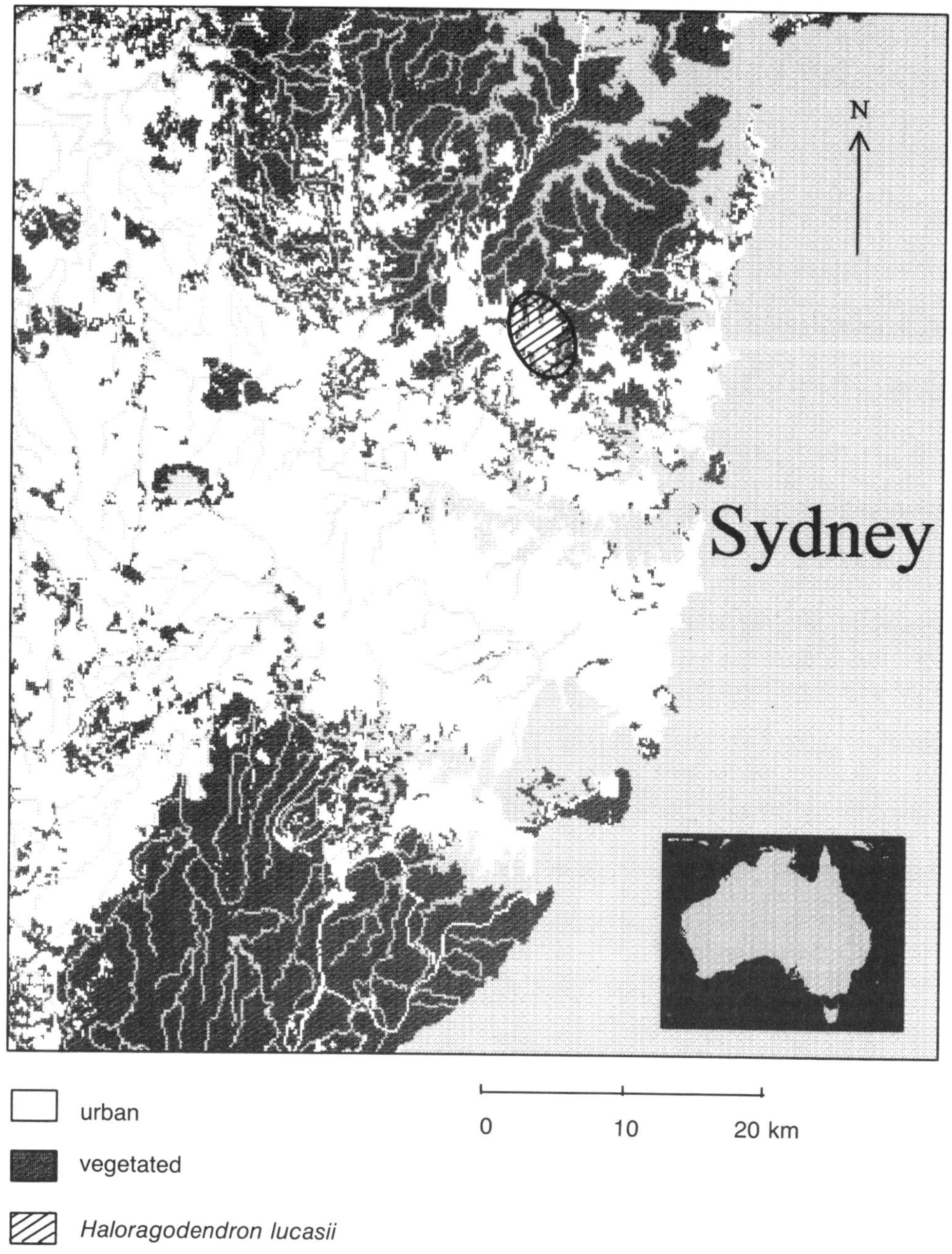

Fig. 1. Approximate location of *Haloragodendron lucasii* in the northern suburbs of Sydney.

or simply a keen interest in local bushland issues. The critical component of the rediscovery team were the many volunteers who generously donated their time and effort to helping the conservation of *H. lucasii.*

THE REDISCOVERY PROGRAM

Haloragodendron lucasii produces many showy white flowers in spring, making it far more observable than at other times of the year. Consequently, the rediscovery program was geared toward conducting field searches in spring and a program of recruitment of suitable volunteers and the education and orientation of these individuals was begun in winter 1995. The main sources of volunteers for the program were from a large pool of volunteer bush regenerators working for the NSW National Parks and Wildlife Service and Ku-ring-gai Council in the northern suburbs of Sydney. These groups were targeted as well as the local Society for Growing Australian Plants group and local bushwalking clubs. Organization of interested groups and individuals who were keen to become involved was co-ordinated by the Garigal NSW National Parks and Wildlife Service office, using organization techniques developed in the volunteer bush regeneration program. Volunteers were introduced to *H. lucasii* through several presentations giving information on the known distribution and biology of *H. lucasii,* the recovery plan process (and details of the recovery plan for *H. lucasii*), the rationale behind the need to find any remaining locations for this species and an outline of how the rediscovery process would operate. A handout containing information on *H. lucasii* was distributed (Fig. 2). Further

Haloragodendron lucasii

FAMILY: Haloragaceae

Conservation status of *Haloragodendron lucasii*

Haloragodendron lucasii is an endangered plant species that was presumed to be extinct until it's rediscovery in 1986 by St Ives residents. Prior to 1986 *H. lucasii* had remained undetected in the sandstone valleys of the northern suburbs of Sydney since 1926. *Haloragodendron lucasii* is now known from three different sites in the Hornsby - Mona Vale region, the third site only being discovered in late 1994.

Threats to *H. lucasii*

Unlike many endangered plant species, the known populations of *H. lucasii* do not appear to be in immediate danger. However, problems associated with the encroachment of suburbia, such as weed invasion, and run off may have an impact on the species in the future.

In contrast, two biological problems may be the most serious threats to the survival of the species. A detailed study currently being undertaken at the Australian National University has found that each population, though consisting of many stems, may in fact represent a single individual or clone. Thus, in the genetic sense, instead of apparently representing many individuals, these populations may in fact represent only a few individuals.

Tip of a flowering branch of *Haloragodendron lucasii* (approximately half actual size).

The second complication is that the known plants, while flowering prolifically, appear to be male sterile, so are unable to produce seed. Without reproduction by seed there is little chance of the species surviving in the long term.

It follows, that it is essential to find male fertile plants that can then be used in an *in situ* and *ex situ* breeding program. Until we find these plants, the species is merely a museum piece waiting for extinction; perhaps a fire, or even a subtle rise in temperature over the next few decades, may be all it takes to drive this species to extinction.

Action required

Present surveys for this endangered species are inadequate and there is the possibility that more populations exist in the areas near to existing populations that have not yet been found. It has become imperative that new populations of *H. lucasii* are located to help ensure the long term survival of the species. In particular the location of male fertile clones is critical so that sexual reproduction can occur.

Location of known populations

The known populations of *H. lucasii* occur in the catchment of Cowan Creek and in a side tributary of Rocky Creek which flow into the Middle Harbour Creek catchment. In these catchments *H. lucasii* grows on sheltered slopes, often at the base of rock faces and scrambling among rock boulders with a fern understorey.

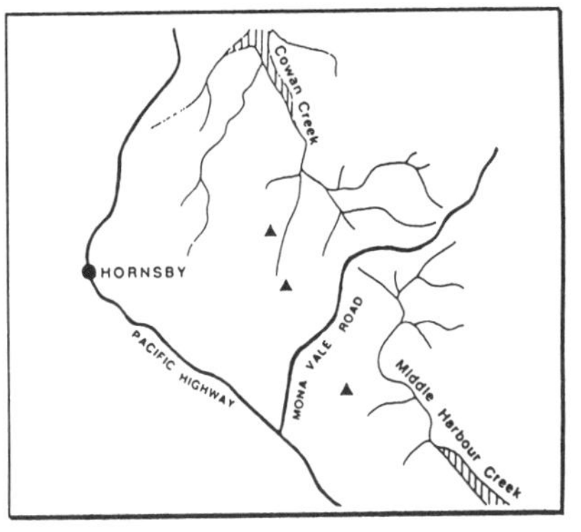

Location of the three known *H. lucasii* sites in the Hornsby area.

Identification of the species

The species is relatively easy to identify in the field. It is a green, multi stemmed herbaceous shrub to 1.8m tall and usually possesses fresh green growth on the ends of the branches. The stems are 4 angled and hairless. The leaves are opposite, narrow-elliptic, 25 to 50 mm long, and distinctly toothed on the margins. During the period from October to December, the tips of the branches are covered in white flowers each with 4 petals that are twisted in bud.

When not in flower single stemmed individuals of the species may be confused with other species in the area, particularly *Crowea saligna*, *Boronia* and *Eriostemon* species, but these species lack obvious teeth on the leaf margins and have distinct aromas when the leaves are crushed.

Reporting a possible new population

Before notifying anyone of a possible new population be sure to check the features of *H. lucasii* listed in the table below. A colour photo of this species can also be found in Fairley and Moore (1989) 'Native plants of the Sydney district: An identification guide', p 225.

	FEATURES
LEAVES	- toothed margins
	- not hairy or pubescent
	- no aroma when crushed
STEMS	- square and angular
FLOWERS	- white, 4 petals
	- cross shaped
	- twisted in bud

If you think that you have located a new population of *H. lucasii*, prepare an accurate sketch map of the population location, provide an estimate of the area covered by the population or the number of plants and contact one of the people listed on this brochure.

It will also be of great value for this search program if you can provide us with information on the areas you have searched for the species so that we can maximise the areas searched.

Low impact survey and NSW NPWS code of conduct in the bush

Remember, that when searching for this or any rare plant species it will be important to ensure that minimal damage is done to the habitat. Damage to the habitat can be minimised by following existing paths or natural breaks in the vegetation and with reduced group sizes so that trampling of the vegetation does not occur.

Persons to contact on a new sighting of the species

Mark Williams
NSW Bush Regeneration Technical Officer
Garigal National Park
PO Box 134, Forestville, NSW 2087
Phone: (02) 451 3479
Fax: (02) 451 7390

Marita Sydes or Rod Peakall
Division of Botany and Zoology
The Australian National University
Canberra, ACT 0200.
Phone: (06) 249 0022, 249 4364
Fax: (06) 249 5573

This is a pilot program involving people at the Australian National University, NSW National Parks and Wildlife Service, Ku-ring-gai Council and local volunteer groups in the Hornsby area.

Thankyou for your participation in the program.

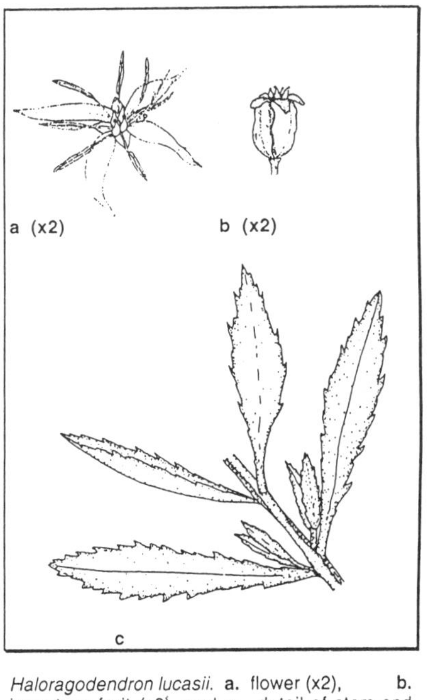

Haloragodendron lucasii. **a.** flower (x2), **b.** immature fruit (x2), and **c.** detail of stem and leaves (actual size).

Prepared by Marita Sydes, Botany and Zoology, The Australian National University, Canberra, ACT.

Fig. 2. A handout used to introduce volunteers to *Haloragodendron lucasii*.

issues relating to safety, minimal impact bushwalking, appropriate clothing, bush navigation, first aid, accident management and survey methodology were also addressed. Finally, site inspections of *H. lucasii* plants and associated habitat were made so that the volunteers were familiar with how the plant looked in the wild as well as gaining some appreciation of the plant species that frequently occurred with *H. lucasii.*

SUCCESS — A NEW *H. LUCASII* SITE FOUND

The spring flowering season of 1995 saw the first volunteer searches for *H. lucasii.* Despite the difficulties of searching in untracked bushland, many volunteers came forward — park neighbours, bush regeneration group members, botanists, bushwalkers, native plant propagators and others, giving a team of some 40 enthusiastic searchers.

Searches were conducted in habitats in and around the known catchments where *H. lucasii* occurs. Although initial searches proved fruitless, they did serve a role in eliminating possible locations and to allow the volunteers some conjecture as to just what was the preferred habitat of this endangered plant. Visits to all the known sites, collecting information on vegetation, slope, aspect, geology and disturbance, further focused the volunteers' understanding of the possible habitat preferences of *H. lucasii.*

Success at last came in late September 1995, when members of the Warringah Bushwalking Club chanced upon a new patch of *H. lucasii.* This site was some two kilometres from the nearest known locality and was the second biggest (in terms of numbers of aerial stems) found so far. As well, it was within Ku-ring-gai Chase National Park and so was protected from any development (all previous locations were just within, or near the boundaries of a national park, but also close to disturbance from adjacent urban areas).

Research has shown that this new location consists of two new genetically distinct individuals, raising the total number of known individuals to six. These new individuals are also thought to be male sterile, but further research in 1996 will examine this in more detail.

CONCLUSIONS

Further searches are planned for spring 1996. A *H. lucasii* newsletter has been developed and sent to all volunteers who joined in the searches in order to keep them informed of the success last year and of the results of current research on the species, such as the studies of genetic variation used to determine the number of individuals at the new site. The highlights of the program so far have been more than just the discovery of the new site for *H. lucasii.* The whole program is an excellent forum for increasing community awareness and involvement in conservation issues, especially in an area where resources are limited and need to be spread among many threatened species.

ACKNOWLEDGEMENTS

Thanks to Mark Coustin for his enthusiastic support for the program, Dr Rod Peakall for insights into plant genetics and to all the volunteers who made the program a success.

REFERENCES

Nash, S. and Matthes, M., 1994. Conservation Research Statement and Species Recovery Plan: *Haloragodendron lucasii.* Australian Nature Conservation Agency Endangered Species Program Project No. 302.

Threatened by discovery: research and management of the Wollemi Pine *Wollemia nobilis* Jones, Hill and Allen

John Benson[1]

The discovery of the Wollemi Pine *Wollemia nobilis* in 1994 not only brought to light a new genus in the Araucariaceae and a conifer with at least a 91 million year old Gondwanan history, it also increased the threat to the two known wild populations of 40 adults and about 130 seedlings. Although growing in an inaccessible, warm temperate rainforest-lined gorge in a large national park, the impacts of visitation, and indeed researchers, could prove costly to the species. The main threats from people are trampling of seedlings, compaction of the ground and introduction of pathogens. Another threat is wildfire which has the potential to destroy much of the population in one catastrophic event. A range of *in situ* ecological research and *ex situ* botanical and horticultural research is being conducted on the species to aid its conservation. A species recovery plan has also been prepared. In the short term, a key research program aims to discover the most efficient way to propagate and cultivate the species to meet market demand for garden plants. This would remove the pressure of illegal seed collection from the fragile wild populations. Since the Wollemi Pine is a relic species, "recovery" is not the question. Management should aim to maintain the current population and genetic variation. Translocation may arise as an issue in the long run, but there would need to be sound reasons for it to be undertaken.

Key words: Wollemi Pine, Species recovery plan, Endangered plant species, Araucariaceae.

INTRODUCTION

During a weekend recreational canyoning trip in August 1994, an employee of the NSW National Parks and Wildlife Service collected material from a strange looking tree growing in a deep gorge within Wollemi National Park. Initially, without information on the wilderness source of the material, botanists at the National Herbarium of New South Wales thought the juvenile foliage brought to them may have come from a garden escape of a Chinese conifer. However, once the adult foliage and cones were collected, it was obvious that a new species of Araucariaceae had been discovered. In fact, the taxon contained such distinct morphological features that it was described as a new genus *Wollemia* — a third genus in the Araucariaceae (Jones *et al.* 1995). The discovery made international headlines in late 1994 stimulating many scientific and general inquiries about the tree.

Ironically, the discovery of this ancient conifer and the subsequent publicity, has increased its chances of extinction mainly due to illegal visitation to the wild population by people wanting to see the tree. Visitation leads to trampling of seedlings, compaction of the soil in the vicinity of the trees and possible introduction of pathogens such as *Phytophthora*. The possibility of illegal seed collecting is also a management concern.

From the time of its discovery it was realized that the protection of the wild populations of the Wollemi Pine would be enhanced if rapid propagation and cultivation of the plant could satisfy market demand for the tree as a garden plant. Lessons had been learnt from the failure to prevent illegal seed collecting from the wild population of the threatened Foxtail Palm *Wodyetia bifurcata* at Cape Melville in north Queensland (Tucker 1988).

The Wollemi Pine is listed as an endangered species under the New South Wales *Threatened Species Conservation Act 1995* and is coded 2ECi in Briggs and Leigh (1996). It is a relic species once widespread in southern Australia, now close to "natural" extinction, and it is unlikely that its status has been significantly affected by recent human activities. One can only hypothesize why it survives in one gorge in the Blue Mountains.

NATURAL HISTORY

1. Origin and taxonomic position

Araucariaceae has fossil representation from the Triassic (200 million years ago) and reached its maximum diversity before the end of the Cretaceous (65 million years ago) with a world-wide distribution. A comparison of fossil pollen with that of the Wollemi Pine (Macphail *et al.* 1995) reveals that it is similar to the fossil pollen *Dilwynites* which is known from the late Cretaceous (*c.* 91 million years ago). Without definitive macro-fossil evidence, it is impossible to estimate when *Wollemia nobilis* evolved. It contains a number of advanced morphological features (K. Hill,

[1]National Herbarium Royal Botanic Gardens, Mrs Macquaries Road, Sydney, New South Wales, Australia 2000.

Pages 105–09 *in* BACK FROM THE BRINK: REFINING THE THREATENED SPECIES RECOVERY PROCESS ed by S. Stephens and S. Maxwell. Surrey Beatty & Sons, 1996.

pers. comm.), but certainly its granulate pollen matches that of ancient taxa.

DNA analysis (S. Gilmore and K. Hill, pers. comm.) has supported morphological evidence that the Wollemi Pine is a new genus falling between the previously two known genera *Agathis* and *Araucaria*. Its distinctive morphological features include a spongy, nodular bark with an outer surface of papery scales and a terminal placement of male and female cones on first-order branches (Jones *et al.* 1995).

2. Habitat

The Wollemi Pine is located in two populations in one gorge in the 500 000 ha Wollemi National Park. Sandstone cliffs line this gorge and the pine grows in a narrow band between cliffs and the creek below. Some trees grow on a talus/alluvial slope above the creek in the valley, others grow on ledges on the lower cliffs. The two groves are dominated by the warm temperate rainforest species Coachwood *Ceratopetalum apetalum* and Sassafras *Doryphora sassafras* (suballiance 37 in Floyd 1990), with an understorey dominated by ferns such as *Sticherus flabellatus*.

3. Extant populations

After extensive helicopter and ground searches of surrounding catchments it now seems likely that only two populations of the Wollemi Pine exist, both in the same catchment. One grove contains 23 mature trees (some reaching over 30 m in height) with over 100 seedlings. The second grove contains 17 adults with over 30 seedlings. It is probable that most of the small seedlings will not survive and only a small number of intermediate-sized trees are present at both groves. Individual trees of 1 m diameter may be over 500 years old, based on the number of growth rings counted from cross-sections of smaller stems. Trees also have the capacity to coppice after disturbance such as fire, thus the actual age of an individual may be older than the age of its standing stem.

THREATS

The main threat to the survival of this species is human interference, although access to the groves is not easy. Soon after its discovery fears for the pine's future were aired on the front page of the *Sydney Morning Herald* (Woodford 1995). Trampling of the ground has already possibly led to soil compaction around some of the trees at one of the groves. Seedlings are threatened by trampling because they are difficult to see in the dense fern cover. Trees may be damaged through inappropriate methods of seed collection, or by the collection of living material. The Royal Botanic Gardens (RBG) has used seed traps to catch falling seed, as this method causes minimal disturbance at the groves, but illegal seed collectors could effect the seed resources. Illicit collectors could also damage trees if they climbed them, or lopped branches to obtain the female cones that are held in the uppermost section of the trees. While protocols for disinfecting footwear are in place for researchers, it is possible that pathogens such as *Phytophthora* (and weeds) could be brought into the groves on the footwear, clothes or rucksacks of illegal visitors.

Illegal visits to the groves have been made and are likely to increase as more people learn of the pine's location. The NSW National Parks and Wildlife Service has taken steps to gain the support of the nearest landholders to prevent illegal access to the sites. However, on the 24 April 1996 the *Telegraph-Mirror* newspaper (Benson, S. 1996 — no relation to the author) reported that a member of the New South Wales Parliament released information to his constituents that had previously been published by the anti-wilderness lobby group — the Public Land Users Alliance (1995). This detailed the approximate location of the Wollemi Pine, and therefore has further increased the chances of illegal visitation to the groves that may damage the wild populations.

Wildfire is another threat to the populations and has probably been the main cause of the retraction of the species since the Pleistocene. While there is evidence that the species may coppice after fire, it is likely that intense fires will kill the plants. Potential exists for a single intense wildfire in the future to drive the species to extinction.

MANAGEMENT RESPONSE

1. Memorandum of Understanding: NSW National Parks and Wildlife Service–Royal Botanic Gardens

Topics for research and management requirements were canvassed soon after the discovery of the Wollemi Pine and it became obvious that both the NSW National Parks and Wildlife Service and the Royal Botanic Gardens, Sydney would need to share these responsibilities (Fig. 1). A Memorandum of Understanding (MOU) was entered into by both agencies to formalize responsibilities. Attached to this MOU was a protocol outlining the precautions required before anyone should enter the groves. A joint management committee was established to co-ordinate research and the management of the wild populations. No action can be undertaken without the approval of that committee.

Each participating agency is pursuing sponsorship to support research and management proposals. Surprisingly, as of early 1996 none has been forthcoming, with the exception of grants from the Royal Botanic Gardens and Domain Trust and the Friends of the Royal Botanic Gardens, that have financed the Royal Botanic Garden's horticultural and botanical research programmes on the pine.

2. Research

Research is being conducted by both the NSW National Parks and Wildlife Service and the Royal Botanic Gardens, Sydney and collaborators. In summary, the current research is as follows:

a. in situ *ecological research*

Research at the groves is being undertaken by the National Parks and Wildlife Service and it is investigating: the growth rate of trees and seedlings;

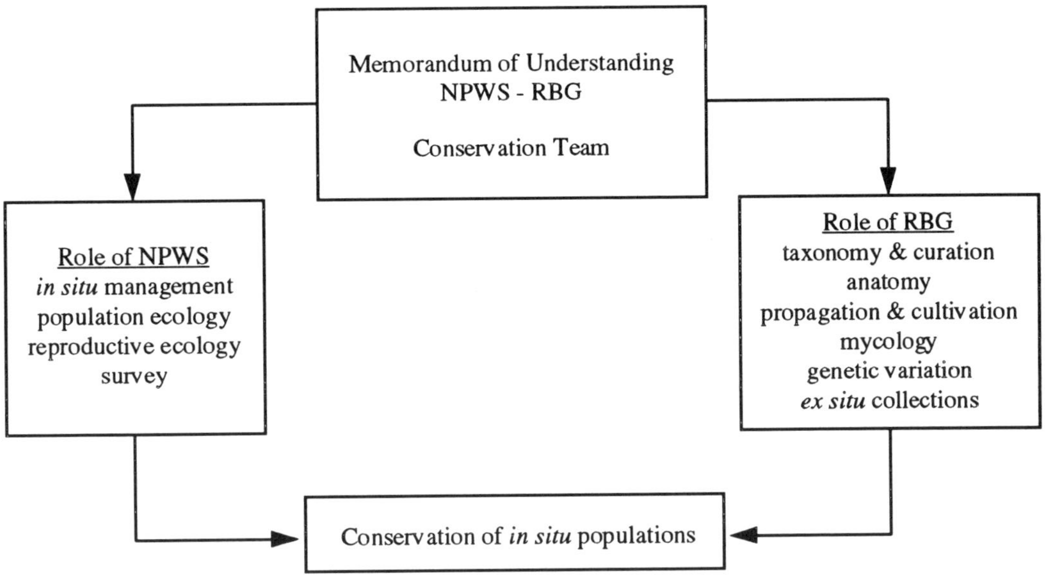

Fig. 1. Respective roles of the NSW National Parks and Wildlife Service (NPWS) and the Royal Botanic Gardens, Sydney (RBG) under a Memorandum of Understanding in studying and conserving the Wollemi Pine.

survival of individuals over time (all stems and most seedlings have been tagged and will be monitored); longevity of seed in the soil; cone production; seed-shed and dispersal mechanisms; and the relationship between the populations and local environmental factors such as soil, climate and associated species.

b. ex situ *research*

Research on the pine at botanic gardens and universities is concentrating on horticulture, morphology, physiology and genetics.

Horticultural research: Researching how to propagate and cultivate the Wollemi Pine is of the greatest urgency because these techniques will lead to a mass production of the plant to meet market demand. Only 5% of the 200 or so seeds produced in each cone are fully developed and viable. In the first year after discovery, only 50 viable seeds were obtained from the wild population. However, subsequently 600 viable seeds were captured in seed traps in 1996. This will allow experimentation to proceed on seed storage viability, and conditions required for seed germination (i.e., light, moisture and temperature). Early findings (Offord *et al.*, in prep.) show that seeds germinate steadily over a period of months indicating there may be a short dormancy period. The nature of this dormancy is subject to current research (C. Offord, pers. comm.).

Research is already well advanced on growing plants from cuttings, and eventually it is envisaged that mass production will use material grown from both seed and cuttings. Cloning of the Wollemi Pine has been achieved using shoot explants placed on a tissue culture nutrient medium, with dormant buds growing out to form new shoots. While rapid tissue culture techniques are being investigated, it may be years before a commercial micropropagation system is established.

Mycological research: Samples of soil, leaf, seed and cone material collected from one of the groves were placed on various growth mediums to investigate the presence of fungi. A total of 52 microfungi and 14 macrofungi (Basidiomycete) taxa have been identified (but not all to species level, B. Summerell, pers. comm.). Most of the microfungi were Deuteromycetes (imperfect) relying only on asexual reproduction. Of the nine *Penicillium* species, five were known and four unknown, but this does not mean the latter are specific to the groves.

The roots of the Wollemi Pine contain a complex nodular structure — more so than other members of the Araucariaceae (B. Summerell, pers. comm.). At least one VA (growing in the roots) mycorrhizal fungus and three ecto-mycorrhizal fungi (growing on the surface of the roots) have been isolated. Future research will attempt to discover whether any of these taxa are specific to the Wollemi Pine and the habitat in which it grows.

A species of *Pestalotiopsis* fungus grows on the leaves of the Wollemi Pine associated with leaf-spots. Professor Gary Strobel, of Montana State University in the USA, has recently isolated the anti-cancer compound, taxol, from this fungus, reported in the *Sydney Morning Herald* by Woodford (1996). This compound may be found in *Pestalotiopsis* on other plant species and it is intended that future research will investigate this.

The root fungus *Phytophthora* has not been discovered at the sites and since this is a potential threat to the species, experiments on the effect of this pathogen on the plant are being planned in cultivation.

Morphological and physiological research: Research has been conducted on the detailed structure of the pine's architecture (Hill, in prep.) and leaf axil anatomy; cuticular structure (Hill and Stanberg, in prep.). Work has also commenced on embryological development, resin and lignin chemistry, and wood and bark anatomy.

Genetic research: A genetic investigation of the Wollemi Pine is being conducted by the Australian National University Division of Botany and Zoology in collaboration with the Royal Botanic Gardens. Threatened species with clonal populations, or those which may be subject to translocation, are suitable for targeting for genetic research (Peakall and Sydes, this volume). The coppicing habit of the Wollemi Pine makes it difficult to distinguish genets from ramets in the two small populations. A genetic study is therefore critical to achieve a cost-effective *ex situ* conservation of the pine. Advanced DNA fingerprinting methods are being applied in order to describe the patterns and levels of genetic diversity within the species, and if suitable genetic markers can be found, this will also enable an assessment of the extent of clonality (R. Peakall, pers. comm.). Preliminary results from one of the populations, indicate below average genetic diversity in the species, however, it is too early to determine whether this reflects extensive clonality or the loss of genetic diversity through inbreeding (R. Peakall, pers. comm.). If it proves to be possible, genetic bar-coding of selected genotypes prior to their commercial release will assist with monitoring trade in the pine.

FUTURE RESEARCH AND MANAGEMENT

A draft species recovery plan has recently been compiled for the Wollemi Pine (NSW National Parks and Wildlife Service 1996). The plan addresses the threats to the wild populations and identifies and costs the most important research topics requiring funding. It would seem that the most important *ex situ* research is to complete the experiments into propagation and cultivation techniques that would facilitate eventual commercial release of the species — studies are also required to establish whether or not a specific mycorrhizal fungus is required for successful growth of the pine; testing for the impact of *Phytophthora* on the species (some other members of the Araucariaceae are susceptible), as if it is found to be susceptible this will have ramifications for the site-access protocol; and investigating the genetic variability in the populations to aid *ex situ* conservation and any long-term requirement for translocation.

Given the degree of soil disturbance in sections of the main grove after only one year of visits by researchers, one could question the cost-benefit of excessive *in situ* research. The most important *in situ* research should aim to discover the environmental determinants required for seed germination and the subsequent growth of seedlings into mature trees. This will involve monitoring individuals through time and measuring microclimatic (particularly light and moisture levels) and edaphic conditions. Laboratory experiments by the Royal Botanic Gardens will test various germination conditions, but it is critical to find out why some seedlings succeed in reaching maturity in the wild and others do not. As mentioned above, seeds germinate fairly quickly with no or only a short dormancy period (C. Offord *et al.,* in prep.). Many seedlings are present at the groves, mainly where more light penetrates the canopy, but most of these will probably die within a short period of time. Few young trees are present and the survival of these is critical to maintaining the population should some mature trees die. As with other rainforest species, gap dynamics may be important in the recruitment of the Wollemi Pine.

The term "recovery" is most applicable to species that require assistance after having declined in recent times due to human interference. This applies to the array of threatened plant and animal species depleted by habitat destruction, competition from exotic or feral species, or unsustainable harvesting (as is the case with some fisheries). With relic species like the Wollemi Pine, the question arises about the suitability of using the term "recovery plan". Recovery from what or to what? Should not the aim be to maintain the existing population and genetic variation? If so, what should be prepared is a plan of management for the species.

The ethical issue of manipulating the future of a species arises when one considers long-term goals. Should we treat a flagship species such as the Wollemi Pine differently from less well known species? Should we allow translocated populations to be established to increase the likelihood of the Wollemi Pine surviving a catastrophe in the wild? At this stage I would say the need to translocate has not been demonstrated but may be appropriate in the future, subject to scientific evidence that there is a need for it, and provided the translocation adhered to the Australian Network for Plant Conservation guidelines (ANPC, in prep.). In the immediate future, groves of the species in botanic gardens will provide the security of an *ex situ* population, should translocation be required in the future.

ACKNOWLEDGEMENTS

I thank David Liddle, Barbara Briggs, Don Blaxell and Cathy Offord for commenting on the text and Brett Summerell for his unpublished mycological information.

REFERENCES

Australian Network for Plant Conservation (in prep.). Guidelines for the translocation of threatened plants in Australia. ANPC: Canberra.

Benson, S., 1996. Jurassic park pine location revealed. *Telegraph-Mirror* 24 April 1996. News Limited: Sydney.

Briggs, J. D. and Leigh, J. H., 1996. Rare or threatened Australian plants. Second edition. CSIRO Publishing: Melbourne.

Floyd, A. G., 1990. Australian rainforests in New South Wales. Surrey Beatty & Sons: Chipping Norton.

Hill, K. D. (in prep.). Architecture of the Wollemi Pine (*Wollemia nobilis*, Araucariaceae), a unique combination of model and reiteration.

Hill, K. D. and Stanberg, L. C. (in prep.). Cuticle micromorphology of *Wollemia nobilis* (Araucariaceae).

Jones, W. G., Hill, K. D. and Allen, J. M., 1995. *Wollemia nobilis*, a new living Australian genus and species in the Araucariaceae. *Telopea* 6(2–3): 173–76.

Macphail, M., Hill, K., Partridge, A., Truswell, E. and Forster, C., 1995. "Wollemi Pine" — old pollen records for a newly discovered genus of gymnosperm. *Geology Today* **11**(2): 48–49.

NSW National Parks and Wildlife Service, 1996. Draft species recovery plan for the Wollemi Pine *(Wollemia nobilis)*. NSW National Parks and Wildlife Service: Hurstville.

Offord, C., Meagher, P. and Porter, C. (in prep.). Factors affecting seed germination and long-term storage viability of *Wollemia nobilis*.

Public Land Users Alliance, 1995. Common ground, winter edition. *Journal of the Public Land Users Alliance:* Glebe.

Tucker, R., 1988. Palms of subtropical Queensland. The Palms and Cycad Society of Australia: Milton, Queensland.

Woodford, J., 1995. Fears for ancient pines after scavengers find secret grove. Page 1 *Sydney Morning Herald* 24/4/1995. John Fairfax and Sons: Sydney.

Woodford, J., 1996. Botanical find of century yields anti-cancer drug. *Sydney Morning Herald* 18/6/1996. John Fairfax and Sons: Sydney.

Population changes from 1990 to 1995 and management of the endangered rainforest palm *Ptychosperma bleeseri* Burret (Arecaceae)

David T. Liddle,[1] Stephen M. Taylor[1] and Darren R. Larcombe[1]

Ptychosperma bleeseri is an endangered endemic palm known from eight rainforests near Darwin in the Northern Territory. Management actions target the threat of habitat disturbance from fire, weeds and introduced animals (buffalo, cattle and pigs). Population data from 1990 and 1995 are presented for adult and juvenile palms, along with more detailed size class data for one population for the period from 1992 to 1995.

There has been a positive response in the survival and growth of juvenile *P. bleeseri* in two populations where introduced animals have been controlled. This response reflects a general increase in the cover of rainforest understorey species through a reduction in trampling and grazing by introduced animals.

A third population has been affected by wildfire twice in the last five years. It has lost more than 75% of adult and more than 50% of juvenile plants. Fire effects are apparent in other components of the rainforest, including formation of canopy gaps and the invasion of weedy grasses. Trends in *P. bleeseri* populations may provide a useful index of the health of the rainforests and illustrate the impact of disturbance by fire and introduced animals. *Ptychosperma bleeseri* provides an example where the management of an endangered species confers a significant conservation benefit for the habitat through addressing commonly occurring disturbance factors which impinge, more generally, on rainforests throughout the Northern Territory.

Key words: Monsoon rainforest, fire, weeds, introduced animals, *Ptychosperma bleeseri*, Arecaceae.

INTRODUCTION

PTYCHOSPERMA BLEESERI is a slender clumping palm endemic to rainforests near Darwin in the Northern Territory. The species is listed as nationally endangered (ANZECC 1993) and a recovery plan was prepared in 1992 (Liddle *et al.* 1992b). Over recent years, management actions have been implemented to protect several of the rainforests from wildfire, weeds and introduced animals, disturbance factors which are widespread in monsoon rainforests in the Northern Territory (Russell-Smith and Bowman 1992). Previous studies of *P. bleeseri* demography and population dynamics have been based on one-off samples or the impact of a single fire event (Duff *et al.* 1992; Barrow *et al.* 1993). In this paper, population data collected over periods from three to five years are used to demonstrate the response of the species to management. Management actions undertaken to protect *P. bleeseri* habitat are described.

Ptychosperma bleeseri habitat

Rainforests in the Northern Territory typically occur as small patches in a mostly eucalypt-dominated savannah. Of the approximately 15 000 rainforest patches (Russell-Smith *et al.* 1992), *P. bleeseri* is known from only eight. All patches containing *P. bleeseri* are wet spring-fed rainforests located on the upland margin of the Adelaide River floodplain or on the Howard River system to the east of Darwin (Barrow *et al.* 1993). The general characteristics of these rainforests are described by Russell-Smith (1991). The extent of the species' range is 30 km long by 20 km wide and the total area of occupancy is 40 ha. In addition to the endangered palm, these eight rainforest patches support two orchids (*Chiloschista phyllorhiza* and *Luisia teretifolia*) and a fern (*Vittaria ensiformis*) that are rare in the Northern Territory (Leach *et al.* 1992).

Threats

The recognized threats to *P. bleeseri* include wildfire, introduced animals (buffalo *Bubalus bubalis*, cattle *Bos indicus*, pigs *Sus scrofa*) and weeds (Liddle *et al.* 1992a; Barrow *et al.* 1993). While these disturbance factors impinge on all *P. bleeseri* rainforests, the pressure from individual factors varies from location to location. Factors likely to be important in the long term, but currently less clearly understood include: changes in the quality and quantity of groundwater through extractive mining; horticulture and rural residential subdivision; and a potential threat through hybridization with ornamental *Ptychosperma* species

[1]Parks and Wildlife Commission of the Northern Territory, P.O. Box 496, Palmerston, Northern Territory, Australia 0831.

Pages 110–13 *in* BACK FROM THE BRINK: REFINING THE THREATENED SPECIES RECOVERY PROCESS ed by S. Stephens and S. Maxwell. Surrey Beatty & Sons, 1996.

introduced to gardens in the local area. Management actions to date, however, have concentrated on the immediate and recognized threats from wildfire, introduced animals and weeds.

MANAGEMENT ACTIONS

Wildfire

Annual fire is a prominent feature of contemporary land use in the tropical savannah adjacent to Darwin. There is a high risk of damage from wildfire to *P. bleeseri* rainforests each dry season unless ameliorated by control burning. Fire management for pastoral purposes in combination with fuel reduction through grazing has provided adequate fire protection at one *P. bleeseri* rainforest (Bankers Jungle) in the last five years. At another (Taminmin), Aboriginal custodians have carried out annual control burning of adjacent vegetation during the early dry season. Of the remaining localities, a program of annual control burning has been progressively implemented at four sites (Black Creek, Crocodile Creek, Whitewood Road and BJ3) to reduce the fuel load on rainforest boundaries. A narrow fire break is established by herbicide application to actively growing grasses in the late wet or early dry season. This sprayed area is burnt and subsequently used as a boundary from which to control burn a wider firebreak.

Weeds

The subdued light conditions created by an intact rainforest canopy inhibit the establishment and growth of recognized environmental weeds of the Northern Territory. Thus the impact of exotic plants on *P. bleeseri* habitat occurs on rainforest margins and in canopy gaps. *Mimosa pigra* is an aggressive weed of the floodplain and riparian habitats, and is well established on floodplains adjacent to several rainforests containing *P. bleeseri*. If left unchecked, *M. pigra* has the potential to develop into dense stands that may seriously impede access to these rainforests, harbour feral animals and invade canopy gaps. Control activities using herbicide and fire have been implemented to minimize the impact of *M. pigra* on the rainforests.

Invasion by exotic grasses such as *Pennisetum polystachion* and *Pe. pedicellatum* pose a serious threat through their ability to modify the fire regime. These species invaded the Whitewood Road rainforest margin and canopy gaps following wildfire. Fuel loads of 27 tonnes of dry matter per hectare have been recorded in a *Pe. polystachion* grassland occupying soils that formerly supported rainforest at Leanyer near Darwin (Panton 1993). In comparison, fuel loads of six to eight tonnes per hectare are present in non-weedy communities which would normally adjoin these rainforests (Bowman and Wilson 1988). The increased fuel loads can produce more intense fires, and if burnt early in the dry season when moisture in the fuel may prevent complete combustion, sufficient material may remain to support a second wildfire within the one season. A cycle may establish whereby increased fire intensity and frequency produces or expands canopy gaps in the rainforest, which allow more fuel to build up. Timely spraying prior to seed set is required to control these grasses. The invasion of these species is reflected in the increased use of herbicides at Whitewood Road where 75 litres of glyphosate-based spray was applied during the 1993–94 wet season, approximately 600 litres in 1994–95 and more than 1 800 litres in 1995–96.

Introduced animals

In the last decade the Brucellosis and Tuberculosis Eradication Campaign that has included the culling of wild stock and feral animals (Lehane 1996), has reduced disturbance to rainforests from grazing and trampling by introduced animals. In addition, with financial support from the Commonwealth Government's Endangered Species and Save the Bush Programs and from the Northern Territory Government, four of the *P. bleeseri* rainforests (Bankers Jungle, Black Creek, Crocodile Creek and Whitewood Road) have been fenced to exclude buffalo, cattle and pigs.

Construction of fences has proved expensive (over $6,000 per kilometre) due to the need to exclude pigs and buffalo, and the irregular shape of the rainforest boundaries requiring frequent corner assemblies. Ongoing maintenance has been substantial due to seasonal inundation of sections of the fences adjoining the floodplains, the necessity to cross watercourses and damage from introduced animals, particularly pigs. While incursions of pigs and buffalo have occurred, a substantial reduction in the number of introduced animals has been achieved within the fenced rainforests.

POPULATION ASSESSMENT

A count of all known populations of *P. bleeseri* was conducted in 1990, with plants recorded in adult and juvenile classes. Where large numbers of juveniles were encountered, an estimate of the minimum population was recorded (Liddle *et al.* 1992a). To monitor the status of *P. bleeseri* and evaluate management practices, a follow up population count was conducted at all known locations in 1995. The 1995 count was highly intensive with lines of observers progressively searching suitable habitat on foot. Each observer counted all *P. bleeseri* within a strip approximately 10 m wide. To minimize the risk of double counting or omission of plants, frequent communication was maintained between adjacent observers and the edge of each pass of a line of observers was defined by a ribbon of tape.

Annual counts were conducted at Bankers Jungle from 1992 to 1995 by walking line transects to assess the whole population within the fenced area. The plants were classified into five height classes (Fig. 1) and all plants were labelled with stainless steel tags for future monitoring.

POPULATION TRENDS

The population counts for 1990 and 1995 are presented in Table 1. While care must be exercised in a direct comparison of total numbers due to the more

Table 1. Ptychosperma bleeseri population counts for 1990 (Liddle *et al.* 1992) and 1995, major management actions and disturbance events.

Population	1990			1995			Management actions and disturbance
	Adult	Juvenile	Total	Adult	Juvenile	Total	
Crocodile Creek	143	1 500+	1 643+	345	4 927	5 272	Fire protection, fenced 1992
Taminmin	10	2 000+	2 010+	25	5 783	5 808	Fire protection, reduced animals
Whitewood Road	137	358	495	33	165	198	Fenced 1993, wildfires 1990 and 1993
Bankers Jungle	24	25	49	13	40	53	Fire protection, fenced 1990 upgrade 1993
Black Creek	2	62	64	2	60	62	Fenced 1993, wildfire 1992
8 km	11	19	30	10	30	40	Reduced animals
7 km	1	0	1	1	2	3	Reduced animals
BJ3	no count	no count	no count	0	4	4	Fire protection, reduced animals
Total	328	3 964+	4 292+	429	11 011	11 440	

intensive sampling in 1995, the population trends reflect patterns of change over the five year period. In 1990, Crocodile Creek and Whitewood Road contained over 85% of the adult species population. By 1995 over 80% of the adult species population was situated at Crocodile Creek alone. A substantial population decline occurred at Whitewood Road with a reduction from 41% of the known adults in 1990 to 7% in 1995.

The stability of the population at Crocodile Creek coincides with a period of little disturbance. On the other hand, the decline at Whitewood Road coincides with the occurrence of two wildfires, the first in September 1990 and the second in August 1993. Barrow *et al.* (1993) detected a decline of at least 22% of adult stems of *P. bleeseri* in the five month period immediately after the September 1990 wildlife. Occurrence of a second fire within a three year period resulted in the loss of over 75% of adult plants and over 50% of juvenile plants. Other impacts of two fires in close succession include the opening of many gaps in the canopy, and the invasion of weeds and pioneer rainforest species such as *Macaranga involucrata, Omalanthus novo-guineenis* and *Trema tomentosa.*

In contrast to the decline at Whitewood Road, there was an increase of adults at Taminmin. On the evidence from over 200 tagged individuals between 1991 and 1994 (Parks and Wildlife Commission of the Northern Territory (PWCNT), unpubl. data), this increase can be attributed to the survival and growth of juvenile plants. There was a substantial loss of adults at Bankers Jungle between 1990 and 1995. Most of this loss occurred between 1990 and 1992. In some plants, death followed damage to the apical shoot (apparently by Little Corellas *Cacatua pastinator*) in combination with a history of disturbance by introduced animals which prevented the survival of basal shoots. Between 1992 and 1995 there was an increase in survivorship and growth of large juvenile plants (0.2–1.5 m) (Fig. 1). This increase in survivorship and growth coincided with the erection of a cattle fence around the rainforest in 1990, which was upgraded to a pig-proof fence in 1993. While fence maintenance has been an ongoing problem with frequent incursions by pigs and buffalo, there has been a dramatic increase in the density of the rainforest understorey and much reduced evidence of damage to trees by trampling and rubbing.

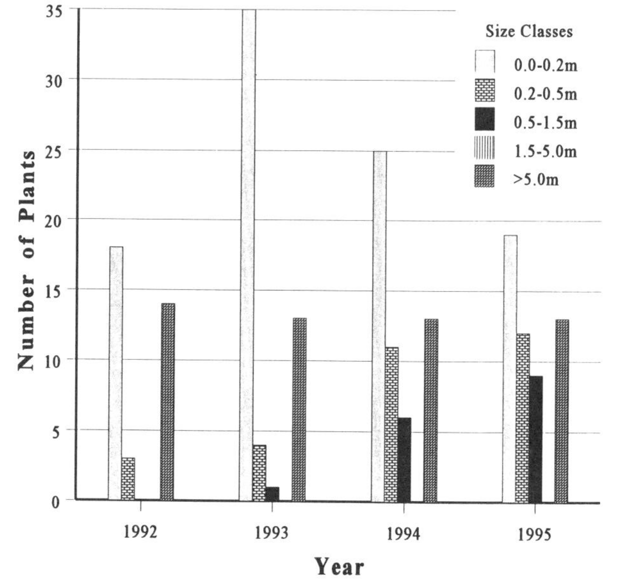

Fig. 1. Size class structure of *Ptychosperma bleeseri* based on total population counts at Bankers Jungle. Plants less than 1.5 m are classified as juveniles and plants equal to or greater than 1.5 m are classified as adults.

CONCLUSIONS

At Bankers Jungle and Taminmin where there has been no wildfire and a reduction in disturbance by introduced animals in recent years, there has been a positive response in the survivorship and growth of intermediate sized plants (large juveniles or small adults). The response by *P. bleeseri* mirrors other components of the rainforest including a marked increase in density of understorey species and a reduction in damage to trees from trampling by introduced animals.

At Whitewood Road, where there has been disturbance from two fires in close succession, the *P. bleeseri* population has undergone a substantial decline. The high level of disturbance is evident in other components of the rainforest, ranging from structural change with an increase in canopy gaps to a shift in species composition with an increase in native pioneer species and exotic weeds. The strong response of pioneer species indicates the rainforest has the capacity to recover but will require an extended fire free period to do so.

At Crocodile Creek there has been little disturbance in recent years and the rainforest retains the largest adult population of *P. bleeseri.*

The trends exhibited by *P. bleeseri* populations reflect the level of disturbance from fire and introduced animals and may provide an indicator of the status or "health" of the rainforest. *Ptychosperma bleeseri* provides an example where the management of an endangered species confers a significant conservation benefit for the habitat through addressing commonly occurring disturbance factors which impinge, more generally, on rainforests throughout the Northern Territory.

ACKNOWLEDGEMENTS

We thank the numerous people who have contributed to the management and monitoring of *Ptychosperma bleeseri.* Collection of population data for Bankers Jungle and the 1995 count was assisted by: M. Anderson, M. Bishop, D. Bullard, L. Bullard, P. Lukitsch, H. Ottley, R. Redfern, M. Smith, A. Stacey, and T. Walmsley of the Northern Territory Palm and Cycad Society; A. Hooper, J. Morgan, D. Murray, and P. Stephenson of the Matilda Quest; A. Bernard, M. Frederiksen, J. Jablonka, and Strider of the Australian Trust for Conservation Volunteers; R. Dempster, R. Liddle, M. Raines, J. Stobo, A. Walter, W. Weatherhead, and B. White of the Top End Native Plant Society; D. Hooper and N. Greenfield of the PWCNT; and conservation volunteers D. Forgie, S. Svensson, J. Walawski and D. Wise. Constructive comments on a draft of this paper were provided by W. J. Freeland, W. J. Panton, S. Stirrat, S. van Cuylenburg and two anonymous referees. The co-operation of local landholders and involvement of staff of PWCNT Parks Division and Wildlife Management Section in management of the rainforests is acknowledged. Funding for the management activities has been provided by the Commonwealth Government through the Endangered Species and Save the Bush Programs, and from the Northern Territory Government.

REFERENCES

ANZECC, 1993. List of Threatened Australian Flora. Endangered Flora Network for the Australian and New Zealand Environment and Conservation Council. Australian Nature Conservation Agency: Canberra.

Barrow, P., Duff, G., Liddle, D. and Russell-Smith, J., 1993. Threats to Monsoon Rainforest Habitat in Northern Australia: the case of *Ptychosperma bleeseri* Burret (Arecaceae). *Aust. J. Ecol.* **18**: 463–71.

Bowman, D. M. J. S. and Wilson, B. A., 1988. Fuel characteristics of coastal monsoon forests, Northern Territory, Australia. *J. Biogeo.* **15**: 807–17.

Duff, G., Wightman, G. and Eamus, D., 1992. Conservation and Management of the Endangered Palm *Ptychosperma bleeseri.* Pp. 69–77 *in* Conservation and Development Issues in North Australia ed by I. Moffatt and A. Webb. North Australia Research Unit, Australian National University: Darwin.

Leach, G. J., Dunlop, C. R., Barritt, M. J., Latz, P. K. and Sammy, N., 1992. Northern Territory Plant Species of Conservation Significance. Northern Territory Botanical Bulletin No. 13, Conservation Commission of the Northern Territory: Darwin.

Lehane, R., 1996. Beating the Odds in a Big Country: The Eradication of Bovine Brucellosis and Tuberculosis in Australia. Commonwealth Scientific and Industrial Research Organization.

Liddle, D. T., Wightman, G. and Taylor, S., 1992a. Conservation Research Statement for the Palm *Ptychosperma bleeseri.* Report to Australian National Parks and Wildlife Service Endangered Species Program No. Project Number 159. Conservation Commission of the Northern Territory: Darwin.

Liddle, D. T., Wightman, G. and Taylor, S., 1992b. Recovery Plan for the Palm *Ptychosperma bleeseri.* Report to Australian National Parks and Wildlife Service Endangered Species Program No. Project Number 159. Conservation Commission of the Northern Territory: Darwin.

Panton, W. J., 1993. Changes in Post World War II Distribution and Status of Monsoon Rainforests in the Darwin Area. *Aust. Geogr.* **24**: 50–59.

Russell-Smith, J., 1991. Classification, species richness, and environmental relations of monsoon rain forest in northern Australia. *J. Veg. Sci.* **2**: 259–78.

Russell-Smith, J. and Bowman, D. M. J. S., 1992. Conservation of monsoon rainforest isolates in the Northern Territory, Australia. *Biol. Conserv.* **59**: 51–63.

Russell-Smith, J., McKenzie, N. L. and Woinarski, J. C. Z., 1992. Conserving Vulnerable Habitat in Northern and North-western Australia: The Rainforest Archipelago. Conservation and Development Issues in Northern Australia. Pp. 63–68 *in* Conservation and Development Issues in North Australia ed by I. Moffatt and A. Webb. North Australia Research Unit, Australian National University: Darwin.

Threatened plant species management in National Parks and Wildlife Act reserves in South Australia

Richard J.-P. Davies[1]

A summary of research and monitoring of threatened plant species in National Parks and Wildlife Act reserves in South Australia is given. Study of a sample of parks indicates that the most frequent threats to populations are weed invasion, goats, bulldozing of firebreaks, rabbits and off road vehicles. Examples are given of populations threatened in parks by the various factors. The causes of inadequate or inappropriate management are 1. paucity of information, 2. inadequate use of existing information, 3. shortages of resources, and 4. the lack of targeted funding. The Department of Environment and Natural Resources (DENR) "Threatened Plant Population Database" has been developed to facilitate information access. The use of volunteers to assist in threatened species management needs to be further encouraged and co-ordinated. The South Australian park system does not adequately reserve all threatened plant species, and a deliberate program of acquisition or negotiation is needed to protect viable populations of each species in parks or Heritage Agreement areas.

INTRODUCTION

BRIGGS and Leigh (1996) list 46 nationally threatened plant species and subspecies as occurring in National Parks and Wildlife Act reserves in South Australia. These species occur in a total of 58 such reserves (Briggs and Leigh 1996; Lang and Kraehenbuehl 1995).

The vulnerability of threatened plant species populations in National Parks and Wildlife Act reserves was indicated in a study of nationally threatened plant species of three regions of the state: "Threatened Plant Species of the Murray Mallee, Mount Lofty Ranges and Kangaroo Island Regions of South Australia" (Davies 1992). Of the species studied, 19 were found to occur in reserves, and 73 populations of these 19 species were located in reserves. Two-thirds of these populations were under apparent threat.

These findings prompted a specific study on threatened species in conservation reserves in South Australia. This study (Davies 1995b) aimed to examine a sample of reserves containing populations of nationally threatened species, determine the factors threatening these populations, formulate management guidelines, and implement these guidelines where practical given the short duration of the project.

The 14 National Parks and Wildlife Act reserves selected were those containing the most significant South Australian populations of several nationally rare or threatened species. Five of the reserves were on Eyre Peninsula, three in the Flinders Ranges, two in the Mount Lofty Ranges, two in the Murray Mallee, and two in the Upper South East.

Populations of 15 nationally threatened species and five nationally rare species were studied, including four threatened species not covered by the previous study. Species studied in detail were *Acacia araneosa, Acacia rhetinocarpa, Caladenia gladiolata, Caladenia xantholeuca, Eremophila barbata, Festuca benthamiana, Pleuropappus phyllocalymmeus, Prasophyllum pallidum, Prasophyllum validum, Prostanthera calycina, Pterostylis arenicola, Pterostylis cucullata, Ptilotus beckerianus, Senecio macrocarpus, Senecio megaglossus, Stipa multispiculis, Swainsona pyrophila, Thelymitra epipactoides, Thelymitra matthewsii* and *Veronica parnkalliana.*

To monitor long-term trends in population size, permanent quadrats and photopoints were set up for 11 of these species, in ten parks. Within each quadrat the numbers of individuals of the target species was counted. In the case of the species being a long-lived perennial, the condition and dimensions of individual plants were also recorded. General vegetation monitoring quadrats were also set up for all these populations to determine whether weed invasion, or changes to the structure or species composition of native vegetation is threatening or potentially threatening populations.

In Gammon Ranges National Park, a set of exclosures were constructed around plants of *Acacia araneosa* to study the long-term impact of goats and rabbits. Similarly, in The Dutchman's Stern Conservation Park, two goat proof exclosures were constructed around plants of *Senecio megaglossus* to study the impact of goats. Quadrats and photopoints were set up inside these exclosures and on comparable

[1]Black Hill Flora Centre, Botanic Gardens of Adelaide, Maryvale Road, Athelstone, South Australia, Australia 5076.
Pages 114–18 *in* BACK FROM THE BRINK: REFINING THE THREATENED SPECIES RECOVERY PROCESS ed by S. Stephens and S. Maxwell. Surrey Beatty & Sons, 1996.

unfenced control areas, to monitor changes to these populations and associated vegetation, attributable to feral animal grazing.

While changes to population size and condition have already become evident for some monitored populations, as described below, some changes are likely only to become evident in the longer term. Davies (1995b) contains baseline monitoring data, to facilitate longer term monitoring.

The goal of this project was primarily to facilitate the preservation and recovery of existing populations *in situ* by reducing threats to their habitats and attending to their specific management needs. This project initiated the minimal disturbance removal of weeds threatening the largest *Pterostylis cucullata* populations in South Australia, in Belair National Park, using volunteer labour. This trial has shown that as few as six persons, each working a total of six hours a year, can significantly push back the weeds and encourage regeneration of natives, as long as the effort is sufficiently targeted.

Where endangered species did not occur in South Australian reserves, attempts were made to reintroduce such species into suitable reserves within their previous range. A trial reintroduction of *Psoralea parva* into Black Hill Conservation Park from a nearby endangered population, appears to be succeeding and continues to be monitored. A trial planting of *Senecio behrianus* into Murray River National Park failed due to the highly altered flooding regime of the river.

THREATS TO POPULATIONS IN NATIONAL PARKS AND WILDLIFE ACT RESERVES

The most frequent threats to threatened species in the 11 National Parks and Wildlife Act reserves studied by Davies (1995b) are presented in Table 1.

Examples of nationally threatened species being destroyed or adversely affected by park management activities abound. In one instance, a small population of *Pterostylis cucullata* was inadvertently bulldozed in Scott Creek Conservation Park during fire track construction, while on another occasion a plant of *Thelymitra epipactoides* was bulldozed in Mount Boothby Conservation Park.

Table 1. Relative frequencies of threats to nationally threatened plant species in 11 National Parks and Wildlife Act reserves in South Australia (after Davies 1995b).

Threats	No. of populations	No. of species	No. of reserves
Weed invasion	20	8	8
Goats	16	2	2
Bulldozing of firebreaks	14	6	7
Rabbits	10	2	2
Off road vehicles	1	1	1
Potential inappropriate fire regime	51	6	8
Recommended changes to hydrology	24	2	1
Potential weed invasion	17	5	5
Potential collectors	7	2	2
Total numbers studied	97	16	11

The siting of a horse trail adjacent to a population of *Pterostylis cucullata* in Belair National Park has increased the chance of new weeds being introduced to the area. The bulldozing of a mound of earth containing weeds into native vegetation, during construction of fire break in Wanilla Conservation Park, threatens the only reserved population of *Ptilotus beckerianus* on Eyre Peninsula.

Certain proposed management activities are likely to adversely affect significant threatened plant populations in parks. The preferred option for the "Upper South East Dryland Salinity and Flood Management Plan" includes the construction of a channel up to eight metres wide across the southeastern section of Messent Conservation Park. The construction of this drain could potentially threaten *Senecio macrocarpus* populations by changing the hydrology of areas containing the species and by encouraging weed invasion. A related scheme called the Wetlands Waterlink could also threaten the species by flooding low lying areas containing the species. This park contains the only known large viable population of this species.

Lack of management activities is also threatening critical reserved populations. Four wheel drive vehicles are threatening *Pleuropappus phyllocalymmeus* in Kellidie Bay Conservation Park. This plant is confined in the park to localized, open, treeless flats, and vehicle access to such areas in the park needs to be prohibited.

In other cases, management actions have not been sufficiently targeted. Despite a program of controlling goats in Gammon Ranges National Park, *Acacia araneosa* continues to decline in the park. Numbers of mature plants in permanent quadrats declined from 33 in 1987 to nine in 1992 due to goats preventing seedlings reaching maturity. Following the construction of a goat and rabbit proof exclosure, mature plants increased dramatically from three in 1992 to 49 in 1995 in the protected area. Where goats (but not rabbits) were excluded a significant increase from one plant in 1992 to 13 in 1995 occurred. This compared with unfenced quadrats where the total number of mature plants remained static, with surviving plants being severely damaged by browsing.

CAUSES OF INADEQUATE OR INAPPROPRIATE MANAGEMENT

Davies (1995b) identified four major factors as contributing to inadequate or inappropriate management of threatened plant species in these reserves. These were:

1. the paucity of information on species' requirements and biology;
2. the inadequate use of existing information largely due to the lack of knowledge of existing computer databases and previous difficulties of access to information;
3. the overall shortages of resources; and
4. the lack of specific fund allocations towards ongoing programs of threatened plant species management in National Parks and Wildlife Act reserves.

The combined effect of these four factors is that long term management goals for threatened plant species do not receive priority.

1. Paucity of information

Forty-six nationally threatened species and subspecies occur in National Parks and Wildlife Act reserves in South Australia. Although studies to date (Davies 1986, 1990, 1992, 1995a, 1995b) have examined the management requirements of 24 of these species, the remaining 22 species require urgent study.

Davies (1995b) examined the management of threatened and rare species in 11 National Parks and Wildlife Act reserves in South Australia. The management of populations of *Prostanthera eurybioides* and *Pterostylis arenicola* in Mount Monster Conservation Park and Coorong National Park, respectively, is being co-ordinated by Black Hill Flora Centre. However, for the majority of the 58 National Parks and Wildlife Act reserves in South Australia containing nationally threatened plant species, no management guidelines have been developed. For many of these reserves, inadequate information exists on the management requirements of the individual populations of these species, to implement informed park management.

Certain threatened species, such as *Thelymitra epipactoides* and *Thelymitra matthewsii*, are known to require specific fire regimes for the perpetuation of populations. Fire management plans written by park managing authorities must take such species into account. The specific fire requirement of other threatened species such as *Senecio macrocarpus* and *Acacia rhetinocarpa* are unclear and the monitoring of such populations is of high priority to determine whether current fire regimes are resulting in long term declines in population size. Some such species, such as *Veronica parnkalliana*, are successional and only survive for a few years following a fire or disturbance. The longevity of these species in the soil seed bank is unknown and requires further research.

2. Inadequate use of existing information

Frequently, inappropriate management decisions could have been averted if existing knowledge had been accessed and taken into account. An opportunity now exists for the more widespread use of such data, with the development during 1994–95, of DENR's "Threatened Plant Population Database" (Davies and Reynolds 1995).

Incorporated into this database was detailed information from five previous studies (Davies 1986, 1990, 1992, 1995a, 1995b) as well as a large amount of information from other professional and amateur botanists, departmental files and other databases.

The database contains two components: a *"Core Information"* component and a *"Visit Information"* component. The *"Core Information"* component includes all information which only needs to be recorded once. Where such information needs to be updated (e.g., land parcels changing due to a subdivision) previous entries are overwritten, corrected or have further information appended.

The *"Visit Information"* component relates to attributes of the population which are likely to change over time. Such data may be recollected during subsequent visits for monitoring purposes. Data from previous visits is retained unaltered to enable trends and changes to be determined (e.g., changes in population size or new threats).

Data stored in the database or available through links to related databases are as follows:

1. A unique reference number for each population of rare or threatened plant species stored in the database.
2. Scientific and common names and threat codes for each species, and indices of reliability of identification.
3. Location description for each population including:
   a. distance and direction from nearest map locations;
   b. AMG co-ordinates, latitude and longitude and map sheet;
   c. regions (as per Briggs and Leigh (1995) and Adelaide Herbarium);
   d. detailed location descriptions;
   e. indices of reliability;
   f. details of any quadrats in the DENR "SURVEY SYSTEM";
   g. reserve information where applicable;
   h. road numbers and nearest road markers, where in roadside vegetation.
4. Cadastral, land tenure and management authority information for land containing each population, and dates when such information was last updated.
5. Information on habitat including associated vegetation, soil type, rock cover, and landform with indices indicating the reliability of such information.
6. When and by whom populations were surveyed.
7. Monitoring information *for a particular visit* including:
   a. a unique visit number;
   b. the numbers of plants in the population;
   c. the age and size structure of the population;
   d. statistics on reproductive stages;
   e. dimensions and size of area covered by population;
   f. condition of and threats to population;
   g. area and condition of associated native vegetation;
   h. fire history and population response;
   i. quadrats read and photopoints photographed;
   j. herbarium specimens collected.
8. Details of management practices undertaken, results of past management, and suggestions for future management.

9. Descriptions of permanent photo points and quadrats set up to monitor population, and storage locations of photo point photos and quadrat data.
10. References for information recorded in database.

The database presently contains information on over 980 South Australian populations of 79 nationally threatened plant species. This database is maintained by the Biological Survey and Research Section (Resource Management Branch), DENR.

The challenge still remains to facilitate access to this data for on-the-ground park managers. Should this data be too widely available, the risk exists that certain populations will be destroyed by collectors. If access is too restricted, populations will continue to be inadvertently damaged by ill-informed management decisions.

While use of this database can help prevent known populations of threatened species from being damaged, it should not be assumed that it contains details of every threatened species population in every park. Sites for proposed visitor facilities, roads, and tracks in reserves should be surveyed by experienced field botanists before work commences to ensure that unknown threatened species populations will not be affected.

3. Shortages of resources

The relatively small human population of South Australia in comparison to the relatively large area of parks is often cited as the reason for the sparsity of government resources to manage threatened flora in parks. While there is some truth in this argument, the historic emphasis on funding tracks and amenities in parks at the expense of vegetation management to maintain biodiversity, is also a major factor. A disproportionate amount of staff time is also spent managing threatened fauna at the expense of threatened flora, due to fauna's greater emotional appeal.

Weed invasion is one of the greatest threats to flora in the temperate regions in South Australia. Weed control by park staff has traditionally targeted areas of heavy infestation, high visual impact, or high fire risk, rather than areas of high botanical value. Bulldozers and broad acre spaying have frequently been used rather than minimal disturbance methods such as cut and swab, and handpulling.

To date, key volunteer groups have initiated and undertaken much of the work associated with the management of threatened plant species in parks in South Australia. This work has been of a high standard and it is desirable that they continue to be encouraged. The Threatened Plant Action Group has an ongoing program of removing weeds at the sites of *Pterostylis cucullata* populations in Belair National Park. Likewise the "Friends of Scott Creek Conservation Park" have an ongoing program removing weeds around populations of *Caladenia gladiolata* and *Caladenia rigida* in Scott Creek Conservation Park. *Prostanthera eurybioides* is now regenerating in Mount Monster Conservation Park in response to extensive rabbit control works by the "Friends of Upper South East Parks".

The Threatened Plant Action Group, the Nature Conservation Society of South Australia, and Friends of Parks groups have assisted in the monitoring of *Acacia rhetinocarpa, Acacia araneosa* and *Senecio megaglossus* in Ferries McDonald Conservation Park, Gammon Ranges National Park and The Dutchmans Stern Conservation Park respectively.

However, Taylor (1995) cites instances of bush regenerators on other projects removing *Dianella revoluta, Bursaria spinosa*, and *Senecio odoratus* from reserves, thinking that they were controlling Watsonia, African Boxthorn and South African Daisy respectively. Taylor argues that this indicates a need for expert guidance in such bushland management. The need for volunteers and staff to be trained, interested and painstakingly careful is heightened when working with threatened species.

It is therefore recommended that skilled field botanists experienced in minimal impact bushland weeding and monitoring, be employed by park managing authorities to co-ordinate weeding in parks, by volunteer groups and staff. The drawing up of weed management plans for heavily invaded reserves and areas of high biological significance should be one of the tasks undertaken by these botanical advisers. Technical assistants experienced in minimal impact bushland weeding should also be employed by these authorities to undertake the weeding of sites of particularly high biological significance in parks, where suitable volunteers do not exist.

The employment of such staff is necessary to:

1. ensure continuity of effort and implementation according to a planned schedule;
2. provide training of volunteers and technical staff to a high standard and match tasks to proficiency level;
3. maintain enthusiasm among volunteers;
4. ensure monitoring evaluation and reassessment of program.

INADEQUACY OF RESERVE SYSTEM

Twenty-one of South Australia's nationally threatened plant species are not conserved in any National Parks and Wildlife Act reserves or Heritage Agreement areas. A deliberate program of acquisition or negotiation is required to protect viable populations of each species.

One such species is *Agrostis limitanea*. Previously listed as extinct, this species is only known from one population, on a narrow council reserve between a road and a pipeline. The Threatened Plant Action Group is controlling weeds at the site, and has facilitated the fencing of the population. However, until the site is legally protected, the survival of *Agrostis limitanea* in the wild is uncertain.

Other examples exist where critical populations are only partially protected in conservation reserves, with the majority of plants occurring on unprotected adjacent land. For example, large populations of *Senecio megaglossus* and *Ptilotus beckerianus* occur on private grazing land adjacent to The Dutchmans Stern and Wanilla Conservation Parks respectively. The protection of such populations is important since they greatly enhance the viability of reserved populations. Their inclusion in the adjacent reserves or protection under a Heritage Agreement is of high priority.

CONCLUSION

Much research into the management requirements of reserved populations of South Australia's threatened plant species is still required, along with greater resources to enable the implementation of the findings of such research. However, this research must be targeted and the findings readily communicated to park managers. More use must be made of existing data on such populations, in particular that in the Department of Environment and Natural Resources "Threatened Plant Population Database".

Active threatened species management in parks requires staff and volunteers to be trained, interested and painstakingly careful. To this end, further botanist positions need to be created specifically to facilitate such activities.

The South Australian reserve system does not contain viable populations of many threatened plant species, and a deliberate program of acquisition or negotiation is needed to protect viable populations of each species in reserves or Heritage Agreement areas.

REFERENCES

Briggs, J. D. and Leigh, J. H., 1996. Rare or threatened Australian plants. Second edition. CSIRO Publishing: Canberra.

Davies, R. J.-P., 1986. Threatened plant species of the Mount Lofty Ranges and Kangaroo Island regions of South Australia. Conservation Council of South Australia Inc: Adelaide.

Davies, R. J.-P., 1990. Gammon Ranges National Park — vegetation monitoring and rare and threatened plant species survey, 1987. Nature Conservation Society of South Australia Inc: Adelaide.

Davies, R. J.-P., 1992. Threatened plant species of the Murray Mallee, Mount Lofty Ranges and Kangaroo Island regions of South Australia. Conservation Council of South Australia Inc: Adelaide.

Davies, R. J.-P., 1995a. Threatened plant species management in the Arid Pastoral Zone of South Australia. Pastoral Management Branch, DENR: Adelaide.

Davies, R. J.-P., 1995b. Threatened plant species management in *National Parks and Wildlife Act* reserves in South Australia. Black Hill Flora Centre: Adelaide.

Davies, R. J.-P. and Reynolds, T., 1995. Threatened plant population database for South Australia, Resource Management Branch, DENR and IDAB, DHUD — Draft Users Manual.

Lang, P. J. and Kraehenbuehl, D. N., 1995. Plants of particular conservation significance in South Australia's agricultural regions. July 1995 update of unpublished database. Resource Management Branch, Department of Environment and Natural Resources.

Taylor, R., 1995. Involving Friends Groups in weed management programs. Pp. 48–49 *in* Weeds of conservation, seminar and workshop papers 5th–6th April, 1995 ed by D. Cooke and J. Choate. Department of Environment and Natural Resources, Animal and Plant Control Commission, South Australia: Adelaide.

Defining priorities for achieving practical outcomes from the genetic studies of rare plants

Rod Peakall[1] and Marita A. Sydes[1]

With the aid of case studies, priorities for achieving practical conservation management outcomes from the genetic studies of rare plants are defined. In *Acronychia littoralis,* genetic analysis did not support the hypothesis that the species was a hybrid. Rather *A. littoralis* was found to be genetically very similar to another species, highlighting the need for further taxonomic research. It is now apparent from a genetic study of *Haloragodendron lucasii,* that clonality is extensive with fewer than 10 genetic individuals represented among thousands of stems. Thus the location of new populations is now the highest management priority and is currently being attempted by an integrated conservation initiative. *Zieria prostrata* has been the subject of an *ex situ* and translocation program. Genetic study has revealed that the initial *ex situ* collection may fail to adequately conserve genetic diversity and has raised the possibility of mislabelled plants. Consequently, a postponement of the translocation program has been recommended. With respect to achieving practical conservation outcomes in future studies, we rank the genetic study of clonal plants, plants targeted for translocation, and plants of uncertain taxonomic status a high priority, particularly when conducted within a hypothesis testing framework. In contrast, we rank other areas, including many of those proposed for genetic study in existing recovery plans, a lower priority. Clearly a more critical evaluation and justification of the need for conservation genetic research on a case by case basis is essential for more effective recovery plans.

Key words: Plant conservation, genetics, genetic markers, reserve design, clonality.

INTRODUCTION

THE conservation of rare and threatened plants has become a national imperative. Within Australia, 74 plant species are now presumed extinct (ANCA 1994), largely through habitat loss, grazing and other agricultural practices. The remaining native plant species face increasing levels of endangerment by continuing habitat reduction and destruction (Leigh *et al.* 1988; ESAC 1992; Table 1). Currently, of the 20 000 or so Australian vascular plant species, 226 are listed as endangered and 660 as vulnerable (ANCA 1994). Those species now listed as endangered face the risk of extinction, while vulnerable species are likely to become endangered, if no action is taken to reverse the species decline.

Given the large and increasing number of threatened species, and the often costly actions required to maximize the chance of a species' survival, conservation managers must continually reassess priorities. Ideally, among other considerations, conservation status should form the basis for defining which species to target for a recovery plan. Currently, some progress is being made towards an objective evaluation of priority plant taxa (Jensz, this volume). However, defining priorities does not end here. Within the recovery plan itself it is further necessary to define priorities for management and research. Furthermore, as action is implemented or research conducted it may be necessary to reconsider priorities. A cost effective process of species recovery will thus be an iterative one, in which priorities are adjusted with changing information and circumstances.

Genetic research is often proposed as a research action in recovery plans for plants. In New South Wales alone, 57% (17/30) of all recovery plans published to the end of 1994 suggested that genetic studies should be undertaken as part of the species recovery program. However, as for other proposed research actions, the practical importance of the intended genetic research is often not well defined or justified. In this chapter we evaluate the relevance of genetics to the conservation of plants, and critically discuss and define priorities for achieving practical outcomes from the genetic studies of rare plants.

[1]Division of Botany and Zoology, Australian National University, Canberra, Australian Capital Territory, Australia 0200.

Pages 119–29 *in* BACK FROM THE BRINK: REFINING THE THREATENED SPECIES RECOVERY PROCESS ed by S. Stephens and S. Maxwell. Surrey Beatty & Sons, 1996.

Table 1. A summary of the presumed causes of plant extinction in Australia, and past and present threats for Australian endangered plants. Numbers of species based on Leigh *et al.* 1988.

Threats	Extinct Presumed	Endangered Past	Present
Agriculture/Grazing	97	228	191
Forestry/Mining	2	19	31
Roadworks	1	8	54
Urbanization	7	22	23
Fire	4	5	29
Other	3	30	97
Unknown	15	9	—

Priority conservation actions

Clearly, habitat loss and habitat alteration have been and continue to be the major causes of plant species decline in Australia (Table 1). The dedication of money to the recovery of populations or individuals of a species for which the land tenure and management are not well established may thus be futile. Land acquisition and land protection are therefore the first critical steps towards conservation. Appropriate land management will also be essential and particularly important if the habitat is affected by disturbances such as grazing by introduced herbivores, excessive vehicular or pedestrian traffic, or where fire control and weed management are necessary.

For many rare plant species the successful implementation of land protection and land management will create little need for further active conservation management. However, it seems recovery plans are more often focused on additional active management options such as *ex situ* conservation and translocation rather than land protection and management. Research tasks that are largely only relevant when implementing these additional active management options are also frequently emphasized. At times these research and management options may be implemented prematurely, in some cases even before adequate surveys have been completed.

Among the research tasks suggested in recovery plans, genetic study is frequently proposed, but often there is little justification for this proposal. Table 2 shows the breakdown of funds requested in the recovery plans of 12 threatened plants of New South Wales which proposed genetic research as one of the research objectives. Note, that up to $15,000 per species was requested for genetic research. Overall, if these recovery plans were funded as requested, the funds for genetic research would be equivalent to 7% of the total budget. As a percentage of the research budget, genetics as a single tool represents 15% of the requested funds. Although these figures do not indicate the actual funding received, they do highlight the importance placed on genetic studies in the species recovery process.

Why is genetics relevant to conservation?

Given the perceived importance of the genetic study of rare species it is essential to critically evaluate how and why genetics is relevant to conservation. The answer in part can be found in Australia's National Strategy for Conservation which consists of both short and long-term goals (ESAC 1992). In the short term, the goal is to prevent the extinction of species and to ensure that species survive and flourish in their natural habitats. As noted previously, for many plants this may often be achieved simply by land protection and land management. In the long term, the goal is larger and aims to preserve genetic diversity so as to ensure the species retains its potential for evolutionary change. Genetic considerations are clearly relevant to this goal, and this is a major reason for the vast conservation genetics literature, including monographs on plant conservation genetics (e.g., Falk and Holsinger 1991). Although it is rarely stated, this also appears to be the basis for the emphasis given to genetics studies in many recovery plans where it is perceived that in order to maximize the preservation of diversity it is necessary to understand the patterns and extent of genetic diversity.

What is meant by "genetic diversity"? Genetic diversity is the sum of genetic variation in a given unit (e.g., individual, population or species). Virtually all species exhibit visible morphological variation among individuals. Some of this variation is caused by the environment, the remainder represents the underlying genetic variation. Morphological characters such as flower colour can exhibit a clear genetic basis, but

Table 2. A summary of the funding requested within recovery plans for genetic research compared with the total research and management funding requests for 12 plant species. References for species recovery plans in Table 5.

Species	Genetics	Research	Management	Total
Allocasuarina defungens	12 000	43 000	26 000	69 000
Kunzea rupestris	3 000	40 000	23 000	63 000
Haloragodendron lucasii	15 000	60 300	43 150	103 450
Zieria prostrata	4 000	40 000	41 300	81 300
Acronychia littoralis	3 000	63 578	153 304	216 882
Elaeocarpus williamsianus	5 000	56 814	21 973	78 787
Fontainea oraria	5 000	63 768	42 457	106 225
Diploglottis campbellii	5 000	54 546	73 094	127 640
Allocasuarina portuensis	15 000	32 100	23 660	55 760
Cynanchun elegans	5 650	75 458	124 990	200 448
Pimelea spicata	7 000	60 300	55 260	115 560
Pterostylis gibbosa	13 000	27 600	55 390	82 990
Total	92 650	617 464	683 578	1 301 042
Average	7 721	51 455	56 965	108 420

more often environmental effects will mask the underlying genetic variation. In order to measure genetic diversity it is therefore important to look beyond the obvious morphological differences found among the units of interest.

Genetic markers are the means by which the assessment of genetic diversity is made. They are simply inherited characters that can be measured and quantified. To be useful for measuring genetic diversity, genetic markers need to be variable in the unit of interest, be it at the level of the individual, population or species. Floral colour, for example, can be used to assess genetic diversity if it shows these properties. Populations of a plant whose flowers are all of the same colour are clearly less diverse for the genes contributing to flower colour than populations where there are several flower colours represented. By combining results across a number of genetic markers it is possible to estimate and describe the patterns of genetic diversity within and among units, although a very large number of genetic markers are required for accurate estimates of total genetic diversity. Unfortunately, there are very few morphological characters like flower colour that can be used as genetic markers, hence laboratory based markers must be used.

Allozymes have been widely used in genetic studies of rare plants because of their ease of use, their low cost and rapid assay. These genetic markers provide an indirect measure of genetic diversity by visualizing variation in enzymes — the products of genes (Morell *et al.* 1995). However, because the number of enzyme systems that can be assayed is limited (usually less than 20 in plants) allozymes may underestimate total genetic diversity (Schaal *et al.* 1991), and in some rare plants may detect little or no genetic variation (e.g., Waller *et al.* 1987; Lesica *et al.* 1988; Soltis *et al.* 1992; Crawford *et al.* 1994; Kress *et al.* 1994; Peakall 1994a). This may be a characteristic of rare plants existing in small populations and probably indicates a low level of genetic diversity within the species.

For many studies of rare plants, the resolution achieved with allozymes will be adequate to answer the question(s) of interest. However, in the cases where little allozyme variation is found, more detailed genetic study cannot proceed unless more informative markers are used. Recently, new molecular procedures have been developed that allow direct quantification of genetic variation at the DNA level (see Morell *et al.* 1995 for review). These genetic markers promise to be important and powerful tools in conservation, not only because they enable more accurate estimates of genetic diversity, but because their high resolving power allows them to be used to address specific, often previously intractable hypotheses. These hypotheses need not be concerned *per se* with patterns of genetic diversity, but rather may address taxonomic, ecological or other non-genetic questions of relevance to conservation managers. To date, the practical application of genetic markers in this context has been largely overlooked, but may greatly extend the utility of genetic studies to conservation in the future.

Outcomes from genetic studies

Irrespective of whether one is concerned about conserving maximum genetic diversity or using genetics as a tool to address specific questions, it will not be possible or necessary to conduct genetic research on all rare plant taxa. In order to achieve an effective and cost efficient recovery of rare species, the challenge is to identify those cases where genetic studies are likely to produce practical outcomes for conservation managers. This may be best achieved by assessing the outcomes of genetic studies already conducted. Six possible outcomes of genetic research are suggested in Table 3. Of these only two are directly relevant to management (5 and 6) while a third (3) may ultimately have management implications. Below we describe three studies we have undertaken in collaboration with the NSW National Parks and Wildlife Service (NSW NPWS). In all cases, the genetic study was initiated because the recovery plans recommended genetic studies in their research phase. For each case we critically evaluate the outcome against the list in Table 3.

A rare plant with uncertain taxonomic status (Peakall 1994a) — *Acronychia littoralis*

The endangered tree *Acronychia littoralis* is known from 18 locations in northern New South Wales, most consisting of only one or two individuals. *A. littoralis* displays a number of characteristics intermediate between two other coexisting species, *A. imperforata* and *A. wilcoxiana,* and clarification of a possible hybrid status was a key concern of the NSW NPWS (Hunter *et al.* 1991a). Using both allozyme and DNA based genetic markers the hypothesis that *A. littoralis* was a hybrid between *A. imperforata* and *A. wilcoxiana* was explored.

The allozyme study was limited by a lack of detectable genetic differences between the taxa. However, for the three variable genetic markers that provided some information it was apparent that the data did not support the hybrid hypothesis. However, more resolution was necessary to establish whether or not *A. littoralis* is a distinct taxon. DNA markers (RAPDs) were therefore used to further clarify the level of genetic similarity between individuals of each of the three taxa. In addition, single samples of *A. baeuerlenii* and *A. pauciflora*, two morphologically distinct taxa, were included as outgroups. The DNA markers provided a vast improvement in resolution compared to the allozymes. Again no support for the hybrid hypothesis was found, instead the consistent pattern to emerge from the study was that *A. littoralis* and *A. wilcoxiana* were more closely related to each other than either was to *A. imperforata.*

A closer inspection of the data suggested that *A. littoralis* and *A. wilcoxiana* may not be distinct species. A genetic similarity of 70% distinguished the two outgroup taxa from each other. Similarly, genetic similarity values of 65–76% were observed between *A. imperforata* when compared with both *A. littoralis* and *A. wilcoxiana*. In contrast, the genetic similarity between *A. wilcoxiana* and *A. littoralis* (84 to 86%) was almost as great as that within *A. littoralis* (88 to 94%).

Table 3. Six possible outcomes from the genetic study of rare plants.

Possible outcomes of genetic studies
No new insights gained
1. The results obtained reinforce current knowledge, but do not provide justification for further research of a change in management.
2. The results obtained reinforce current knowledge and provide additional justification for further research that may have management implications.
3. The results obtained reinforce current knowledge and provide additional justification for a change in management.
New insights gained
4. The results obtained provide new insights that are not directly relevant to management.
5. The results obtained provide new insights with potential management implications but further research required before any change in management.
6. The results obtained provide new insights that lead directly to a change in management.

Based on these findings it was recommended that as a priority the status of *A. littoralis* be re-investigated by a thorough taxonomic revision of the species complex. Should these findings support the genetic evidence it may be appropriate to review the conservation priority of this taxon. In the long term, the outcome of continuing research may lead to a change in the management plan, particularly if the conservation priority of *A. littoralis* is altered on taxonomic grounds. This case may thus equate to outcome number 5 (Table 3). However, in the description of the species, Hartley and Williams (1983) suggested that *A. littoralis* was very closely related to *A. wilcoxiana*. They also noted that the geographic distribution and morphology suggested that *A. littoralis* may be a hybrid, although they subsequently discounted this hypothesis on the grounds that the pollen appeared to be fertile.

Nevertheless, the recovery plan noted that the inability to germinate the seed and other factors still raised doubt about the taxonomic status of *A. littoralis* (Hunter *et al.* 1991a). A recommendation for further taxonomic study could thus have been made on the basis of existing knowledge (or doubts in knowledge). In which case, the genetic study merely reinforced current knowledge and provided additional justification for further research (outcome 2). However, it is worth noting that in this case genetics had the potential to resolve the taxonomic status of *A. littoralis* if it was in fact a hybrid.

A rare plant limited by sexual reproduction (Sydes, unpubl.) — *Haloragodendron lucasii*

In 1994 the endangered shrub *Haloragodendron lucasii* was known from only two populations in suburban Sydney (Nash and Matthes 1994). One population was small (<40 individuals) while the other was extensive. At the larger population *H. lucasii* formed dense stands, presumably by vegetative spread of plants via underground stems. This meant that estimating the actual number of genetic individuals was impossible. The NSW NPWS had also noted that despite producing many flowers each season mature fruit was not produced by *H. lucasii.* These observations lead to the hypothesis that fruit set was limited by extensive clone size, self incompatibility and restricted pollen dispersal.

In the subsequent year, when this study was initiated, a third population was found and observations in the field and laboratory showed that the flowers of *H. lucasii* at all known sites were male sterile. Genetic markers were used to determine the extent of clonality in this species and obtain an estimate of the actual number of genetic individuals at all the known sites. Despite screening more than 50 representative samples with variable allozyme and DNA markers, only five genotypes were found at the three sites. Given the ease at which genetic individuals were identified in this species it is likely that each of these five genotypes represent single clones. The most extensive putative clone covers an area of 180 m × 20 m and consists of thousands of ramets. Based on growth rates observed in cultivation, this clone is likely to exceed 450 years in age.

Thus at present, the census size of the species grossly overestimates the true number of genetic individuals and none of the known individuals appear capable of sexual reproduction. Collectively, the findings indicate that the species urgently requires active management despite the apparently large number of plants in the field. First and foremost, other populations must be located in the hope of finding more secure populations and possibly male fertile clones. The imperative of finding new populations had not been previously recognized. To achieve this, in 1995 an integrated conservation initiative was established involving co-operation between the Australian National University (ANU), NSW NPWS, local council and volunteer groups to systematically search for new populations (see Sydes *et al.,* this volume). One new population, located well within a national park was found as a result. Ongoing genetic analysis of this site indicates it consists of two clones, but both also appear to be male sterile. Should male fertile clones be found in the future, it may be possible to assess the value of artificially introducing sexual reproduction into these male sterile populations. On the other hand, if no male fertile individuals are located, all that can be done for the conservation of the species is to preserve the few remaining genetic individuals in their habitat. This is a case where genetic study has revealed the critical state of the species and the urgent need for action (outcome 6).

A rare plant requiring translocation (Peakall 1994b) — *Zieria prostrata*

Zieria prostrata is a prostrate endangered shrub with a highly restricted distribution. Only about 500 individuals are known from three headlands spanning a mere three kilometre stretch of coastline near Coffs Harbour on the north coast of New South Wales. Formerly, it also occurred at a disjunct headland 24 km south of its extant distribution. The species is easily cultivated by cuttings and consequently a population enhancement and re-introduction program was proposed in the recovery plan (Griffith 1992a). By 1994, an *ex situ* collection was established and population enhancement at two sites was undertaken prior to any genetic study. Subsequently, in late 1994 a genetic survey was requested to evaluate the patterns of genetic diversity within and among the natural populations. The sole remaining individual collected from the now extinct site was also included in the analysis.

Although DNA based markers (RAPDs) were used in this study, little genetic variation was detected within the species. Of the total genetic diversity detected, 34% represented divergence among the populations, with the remaining 66% being within populations. Thus the results for *Z. prostrata* tended towards those of a selfer with lower levels of variation being detected within rather than between populations (cf. Hamrick *et al.* 1991). The results indicate extensive genetic differentiation among the three *Zieria* populations, although they are geographically very close together. Gene flow between the populations would thus appear to be restricted. The divergence between populations may be further accentuated by local adaptation or genetic drift. These findings emphasize the importance of all populations in the conservation of the genetic diversity of the species.

With respect to the *ex situ* program for *Z. prostrata*, it appears that the original collection was based on just a few individuals from two populations (Clemesha, pers. comm. 1996) and is unlikely to have represented the genetic diversity of the species adequately. However, because permanent tagging of the sampled and translocated plants was not performed, this will never be known for certain. Future translocation attempts must ensure that multiple individuals from all sites are used. The genetic analysis has also suggested the possibility of mislabelled samples in the *ex situ* collection; the plant believed to be the only individual rescued from the now extinct site was genetically identical to several plants from the largest population. Given significant differentiation between the proximate extant sites, we might expect individuals from a disjunct site to have a high probability of being genetically distinct with the genetic markers used. The lack of information on the *ex situ* collection and concerns about labelling have forced the population re-introduction program to be postponed pending further study.

Until we better understand the possible consequences of mixing genetically distinct populations, it has also been recommended that mixing of the populations be avoided when enhancing populations. On the other hand, a planned reintroduction program that evaluates the consequences of mixing or not mixing populations, is likely to yield the most useful long-term outcome for this species. In 1996, a three year study commenced at the Australian National University that will evaluate and implement a population enhancement and re-introduction program for this species. However, unlike the previous translocation program, population enhancement may only be implemented at a new fourth site within the species' range where less than 10 individuals are found. Thorough surveys indicate that the remaining three populations are larger than originally thought and enhancement appears unnecessary, especially since the land protection has been achieved and excessive disturbance has been minimized. Population reintroduction, however, remains an important consideration.

The recommendation to preserve and adequately represent all populations in the *ex situ* collection of *Zieria prostrata* may at first seem to be an important practical outcome of the genetic study. However, given that only four populations are known, all populations require preservation irrespective of the genetic considerations. Indeed, even with all four populations adequately protected and managed, the species would retain its endangered status under the risk assessment scheme currently being applied to plants in New South Wales (Chalson and Keith 1995). Thus in this context, the genetic study merely provided additional justification for preserving all populations (outcome 3). On the other hand, the suspension of the *ex situ* collection and translocation program, pending further study, lead directly to a change in management (outcome 6).

LESSONS FROM THE GENETIC STUDIES

Conservation of genetic diversity can frequently be achieved without formal genetic study

The case of *Zieria prostrata* highlights a critical point that is frequently overlooked. The best way to conserve genetic diversity is to preserve all populations; in which case we do not need formal genetic studies. Indeed for many endangered plants, the number of populations is already so low that we cannot afford to lose a single population. It follows that concerns about conserving genetic diversity only become relevant when it is not possible to conserve all extant populations adequately or when other management actions, such as *ex situ* conservation and translocation programs, may inadvertently lead to losses of genetic diversity.

For species represented by many populations, genetic surveys can provide one criterion for consideration in the design of a representative reserve system. Several genetic studies of Australian plants discuss the possible use of genetic data in reserve design and selection. The majority of these studies have been on rare members of the genus Eucalyptus (Moran and Hopper 1983; Sampson *et al.* 1988, 1989; Peters *et al.* 1990; Prober *et al.* 1990), but studies have also been completed on *Banksia* (Coates and Sokolowski 1992;

Sampson *et al.* 1994), *Acacia* (Coates 1988), *Atherosperma* (Shapcott 1994) and some herbaceous species (Leeton and Fripp 1991; Coates 1992). Some general patterns are evident from these studies. First, when significant genetic differentiation was found among regions, it is emphasized that in the very least the largest and most genetically diverse populations in each region should be preserved (Coates 1988; Peters *et al.* 1990; Prober *et al.* 1990; Leetin and Fripp 1991; Coates and Sokolowski 1992). Second, populations containing rare alleles were also considered a conservation priority, leading to the recommendations that reserves, even if small in size, should be selected to incorporate this component of genetic diversity (Sampson *et al.* 1988; Peters *et al.* 1990). In all cases, the authors emphasized the importance of preserving as many populations as possible, especially when the populations are typically small, since the genetic diversity in small populations was rarely representative of that in the species as a whole (Moran and Hopper 1983; Coates 1988; Sampson *et al.* 1988).

While genetic surveys may be a desirable basis for making informed decisions about reserve options for a species, it may be possible to maximize the preservation of genetic diversity by simply incorporating well established genetic principles into reserve design. One consequence of the many published descriptions of genetic diversity within plants is the finding that the extent and distribution of genetic diversity conforms more or less to one of two common patterns. In predominant outcrossers, on average 80% of the species genetic diversity will be represented within populations, whereas in predominant selfers, only 50% of the species diversity will be found within populations, the remainder existing as genetic differences among populations (Hamrick *et al.* 1991). To maximize the conservation of genetic diversity, more populations of selfers would therefore need to be conserved than for outcrossers. Without a formal genetic study, one could therefore err on the side of caution by assuming the patterns of genetic diversity within a rare species equates to that of a selfer, thus suggesting that as many as possible populations should be preserved.

Genetic considerations are clearly relevant to reserve selection. However, formal genetic studies will only have practical outcomes if reserve design is an option, that is, if there are sites to choose amongst based on genetic and other criteria. If other factors, such as the feasibility of managing the land override genetic considerations for reserve selection, then genetic surveys and even the application of genetic principles to reserve design are merely academic, albeit interesting exercises.

Potentially clonal endangered plants are priority targets for genetic research

Many plants have the potential to reproduce asexually. This is most often achieved as vegetative reproduction by stolons, rhizomes or tubers but also by agamospermy; the asexual production of seeds. Regardless of its basis, asexually reproducing species exhibit two levels of organization: the genetic individual the "genet" or "clone" and the asexual module the "ramet". Because connections between ramets are easily lost (or absent in the case of agamosperms), and genets and ramets may be intermixed, a census based on counts of recognizable individuals will invariably over-estimate the number of genetically distinct individuals. The case of *Haloragodendron lucasii* exemplifies this point clearly: a census size of more than 1 000 presumed individuals equated to less than 10 genetic individuals. Given the complex within population organization of clonal species and the usual absence of phenotypic markers, a genetic study, ideally using the most sensitive genetic markers available, is essential to elucidate the extent of clonality within plants.

There are at least three important implications of extensive clonality in rare plants: (1) the true status of the plant may go undetected, (2) in self-incompatible species seed set may fail when clone size exceeds the average pollen dispersal distance while in self-compatible species increased levels of selfing may occur, and (3) the extent and patterns of genetic diversity can neither be predicted or measured until clonal structure is known: applying general principles may not suffice to adequately preserve the genetic diversity of the species.

One consequence of the latter point is that a very large number of populations may need to be preserved to adequately represent the genetic diversity of a clonal species. This point was highlighted in the genetic study of the clonal populations of *A. anomala* (Coates 1988). *A. anomala* is a rare Western Australian wattle that occurs in two main regions, each consisting of several populations. One region of the species is predominantly outcrossing, while in the other region reproduction is predominantly by vegetative growth. A genetic study was conducted and the results were then used to devise an optimal reserve design to maximize the genetic diversity conserved in this species. In the case of the outcrossing populations, genetic diversity could be adequately protected in two or three large reserves. On the other hand, because all the clonal populations were fixed for different genotypes it was recommended that all populations should be protected. Clearly, alternative reserve designs that did not consider the underlying genetic patterns would have very likely failed to maximize the preservation of genetic diversity in this species.

Similarly, the *ex situ* collections for clonal species may need to be much more extensive in order to represent the genetic diversity of the species adequately. The case of *H. lucasii* illustrates the unintentional inadequacy of an *ex situ* collection of a species in which clonality was unrecognized. Although an attempt to sample broadly within the largest known population was made, it is likely that only a single (male sterile) genetic individual is represented in the Australian National Botanic Gardens collection (Sydes, pers. obs.).

In other cases, recognition of the extent of clonality may actually simplify active management. For example, in 1995 road works in New South Wales threatened a population of the endangered

Acronychia littoralis. Translocation of the plants at risk was supported by funds from the Roads and Transport Authority. At the same time, a genetic survey of the population was undertaken (Peakall 1995). No genetic variation was detected within the population, although it differed genetically from nearby populations, suggesting it may be a single clone. This hypothesis was further supported by a lack of seed set within the population over a two year period, despite numerous flowers. Based on this information, recreating a new artificial population genetically representative of the old was straightforward. On the other hand, the potential for extensive clonality within the species and the possibility that this may limit seed set warrants further research that is likely to have important conservation management implications (Peakall 1995). Clarifying the taxonomic status of *A. littoralis* remains the highest research priority for this taxon (see above).

It is apparent that the genetic study of endangered clonal plants rates amongst the highest research priorities. The challenge is to identify those endangered species that have the potential to be clonal. In some cases the potential for vegetative reproduction is obvious given the habit of the plant, but in others small scale excavations may be necessary. A lack of seed set in what appears to be an otherwise healthy population may provide a clue about possible clonality, although this could equally indicate a lack of pollinators. The potential for asexual reproduction may be more difficult to detect in agamospermous species where bagging and emasculation of flowers is necessary to confirm agamospermy. High levels of seed set in the absence of pollinators may however provide a clue to agamospermy. Information about the extent of asexual reproduction in the genus or family may also be a useful indication of the possible mechanisms and likelihood of clonality in the species of interest.

Endangered plants targeted for translocation are priority targets for genetic research

One goal of translocation, be it population enhancement or reintroduction is to ensure that new or modified populations exhibit similar levels of genetic diversity to the original population(s). For this to occur it is first necessary to adequately represent genetic diversity in any *ex situ* collection which will form the source of plants for subsequent translocation. Guidelines for conservation collections of endangered plants have been established by the Center for Plant Conservation at the Missouri Botanic Gardens (Falk and Holsinger 1991). These guidelines attempt to maximize the sampling of genetic diversity for typical species, without the requirement for formal genetic surveys. Brown and Briggs (1991) also provide advice on sampling strategies and currently the Australian Network for Plant Conservation is preparing national guidelines for the translocation of threatened plants in Australia.

For the reasons outlined in preceding sections, general guidelines for translocation may fall short of their genetic goals for clonal species and consequently genetic study remains essential in such cases. Genetic studies may also be crucial when re-establishing populations from *ex situ* collections that are based on just a few individuals or for which the history of the collection is poorly known. In these cases genetic diversity may be lost inadvertently, while in other cases translocation may fail simply because insufficient genetic diversity exists. For example, when re-establishing a population of self-incompatible plants that unknowingly originated from a single plant or clone.

The example of *Zieria prostrata* clearly illustrates the problems associated with a poorly planned translocation program. In such cases, a genetic study may enable the salvaging of at least some of the program, although as a consequence of the retrospective genetic study of *Z. prostrata* a completely new start was recommended. In other cases, this may not be feasible, and genetic study may be useful in defining the best possible strategies with the existing collection. The use of genetics in the planning phase of reintroduction projects is highlighted in Rieseberg *et al.* (1989). In this study genetics was used to identify hybrids between the endangered *Cerocarpus traskiaea* and a common congener. Identified hybrids were subsequently removed from the reintroduction program. Another example is provided by Rossetto *et al.* (1995) who incorporated genetic data in their strategy for conserving the highly endangered *Grevillea scapigera* which is dependent on advanced propagation and translocation techniques.

Genetic considerations are vital when designing and implementing a translocation program, although by following appropriate guidelines formal genetic analysis may not be necessary for typical species. Clonal plants, however, remain a high priority for genetic study so as to avoid undesirable and unknown duplication of clones in the *ex situ* collection and translocation program. Additionally, as in the case of reserve selection, genetic studies may aid the decision making process about which populations to represent within an *ex situ* collection when it is not possible to sample all populations. For endangered species with few populations, it remains advisable to sample from every population.

Endangered plants of uncertain taxonomic status are priority targets for genetic research

Although we emphasized earlier that the priority conservation strategies of land protection and management should precede active management and its associated research, there is one exception: defining the taxonomic status of rare taxa. This is critical because hybrids and subspecies may not warrant the same level of protection as distinct species, or at the very least they may be ranked a different priority.

Taxonomic uncertainty can take several forms ranging from a lack of information which has prevented formal taxonomic delineation, to disagreement amongst taxonomists about taxonomic rank or position. Regardless, genetic markers provide an additional set of characters that may aid in the resolution of the taxonomy of the taxa in question. However, as in the case of *Acronychia littoralis,* the outcome of genetic study may merely reinforce the known existing understanding of the taxonomic relations.

While genetic studies of endangered taxa of uncertain taxonomic status cannot replace the formal taxonomic process, when conducted collaboratively or in consultation with the relevant taxonomists, they may aid in the resolution of taxonomic uncertainty. However, it is essential to justify the case for a genetic analysis within the context of a broader taxonomic study or revision.

Practical outcomes from genetic studies are maximized by a hypothesis testing framework

The cases of *Acronychia littoralis* and *Haloragodendron lucasii* illustrate genetic studies that were conducted within a hypothesis testing framework. In neither case was there direct concern about the patterns of genetic diversity *per se,* but rather genetic markers were used as tools to evaluate hypotheses on taxonomic status and clonality respectively.

Genetic studies conducted within a hypothesis testing framework are more likely to have practical outcomes than descriptive studies of the patterns of genetic diversity. As such, in defining priorities for research, directed genetic studies of rare plants should be ranked higher than descriptive genetic studies.

OTHER PRACTICAL APPLICATIONS OF GENETIC STUDIES OF RARE PLANTS

Identifying factors limiting reproductive success

Of the myriad factors limiting the continuing survival of endangered plants, factors limiting reproductive success are likely to be important. By combining reproductive biology and genetic studies within a hypothesis testing framework, it may be possible to identify problems limiting reproductive success. Nevertheless, genetic studies require justification within the context of the need to explore the basis of suspected limitations to reproduction: genetic study cannot be justified just because it may identify limiting factors.

Genetic barcoding to discourage illegal trade

The recent discovery of the Wollemi Pine *Wollemia nobilis* (Jones *et al.* 1995; Hill 1995) made international headlines. The antiquity of this species, and likelihood that it can be grown in cultivation may make this species a target for illegal collection and trade. If unchecked this could seriously threaten its prospects of survival in the wild. We are currently developing high resolution genetic markers to "fingerprint" every genetic individual. Based on this knowledge, a cost effective *ex situ* collection could be maintained that would minimize unnecessary duplication of clones.

New genetic technology also offers the exciting possibility of "barcoding" selected genotypes prior to their commercial release on national and international markets. Thus any illegal trade in the species could be confirmed by DNA fingerprint analysis. Accompanied by an extensive international publicity program, this may provide a powerful way to discourage illegal collection and trade. Genetic studies of this type are both costly and time consuming, but are clearly justified by the almost certain practical outcomes that will enhance the conservation of this fascinating species.

DO PROPOSED GENETIC STUDIES MATCH PRIORITY AREAS?

In the preceding discussion, the use of genetics as a research priority was evaluated with the aid of several case studies. It is clear that a distinction can be made between studies that have immediate implications for the conservation management of a species and those that simply contribute to a growing and important body of knowledge. Due to the lack of resources to fund all requested genetic studies, it is essential that those studies that have direct application to conservation management are ranked the highest priority by conservation agencies. Table 4 outlines our view of the priority areas for achieving practical outcomes from the genetic study of rare plants. It must be stated that ranking some areas a low priority does not mean that no interesting information would be gained from a genetic study. Nor are we discouraging the use of genetics in the process of endangered species recovery. Rather, we wish to encourage critical evaluation and convincing justification of the need for genetic research on a case by case basis.

Table 4. A summary of the high and low priority areas for achieving practical outcomes from the genetic study of rare plants. Note that the numbering is not intended to represent ranks within the high and low priority areas.

High priority genetic research

1. TRANSLOCATION
   - 1.1 *Ex situ* collection/history poorly known
   - 1.2 Impossible to adequately sample all populations
   - 1.2 Major genetic differences among populations suspected
2. RESERVE DESIGN
   - 2.1 Choice of populations for reservation possible
3. ASEXUAL REPRODUCTION
   - 3.1 Extensive clonality suspected
   - 3.2 Sexual reproduction suspected to be limited by clonality
4. SPECIES OF UNCERTAIN TAXONOMIC STATUS
   - 4.1 Morphological data insufficient for taxonomic clarification
5. OTHER
   - 5.1 Limits to reproductive success suspected
   - 5.2 Target for illegal trade

Low priority genetic research

6. TRANSLOCATION
   - 6.1 Species only known from a single population
   - 6.2 All populations can be adequately sampled
   - 6.3 Minimal genetic differences among populations suspected
7. RESERVE DESIGN
   - 7.1 All populations need to be preserved
   - 7.2 No choice of populations for reservation possible
8. SPECIES OF UNCERTAIN TAXONOMIC STATUS
   - 8.1 Available data sufficient for taxonomic clarification
9. OTHER
   - 9.1 Genetic diversity assessment without a hypothesis testing framework

Table 5. A summary of the number of known populations and the basis for proposing genetic research in the recovery plans of 18 endangered plants of New South Wales.

Species	No. pops	Objective	Reference
Allocasuarina defungens	8	Asexual reproduction	Benwell 1993
Kunzea rupestris	9	Asexual reproduction	Cohn 1993
Haloragodendron lucasii	2	Asexual reproduction	Nash and Matthes 1994
Grevillea caleyi	19	*Ex situ*/translocation	Scott *et al.* 1995
Allocasuarina portuensis	1	*Ex situ*/translocation	Matthes and Nash 1994
Microstrobos fitzgeraldii	7	*Ex situ*/translocation	Jones 1995
Acronychia littoralis	17	Uncertain taxonomy	Hunter *et al.* 1991a
Phaius australis and *P. tancarvilliae*	18	Uncertain taxonomy	Benwell 1994
Cynanchum elegans	3	Descriptive	Matthes and Nash 1993a
Diploglottis campbellii	14	Descriptive	Hunter *et al.* 1991b
Elaeocarpus williamsianus	2	Descriptive	Hunter *et al.* 1991c
Fontainea oraria	1	Descriptive	Hunter *et al.* 1991d
Grevillea beadleana	2	Descriptive	Benson 1991
Pimelea spicata	18	Descriptive	Matthes and Nash 1993b
Pterostylis gibbosa	3	Descriptive	Anon. 1994
Thesium australe	12	Descriptive	Griffith 1992b
Zieria prostrata	3	Descriptive	Griffith 1992a

To further illustrate the need for more critical evaluation and justification, we examine the reasons given in recovery plans for the use of genetics. Table 5 lists 18 endangered plants of New South Wales for which genetic studies were proposed as one of the research actions of the recovery plan. Consistent with our high priority, genetic study was proposed for three species based on the likelihood of asexual reproduction. In three species, planned translocation was the basis for proposing genetic studies, but as for *Zieria prostrata* it appears that genetic studies were planned after the *ex situ* collection had commenced in *Microstrobos fitzgeraldii*. In two other plants, genetic study was suggested as an aid to resolving uncertain taxonomic status. For the remaining nine species, descriptive genetic studies were proposed but in all cases there was little or no justification. We suspect that genetic studies were proposed in the belief that they would aid in maximizing the protection of genetic diversity. If so, it is of interest to note that in most cases very few populations of the species are known (Table 5). Therefore, protecting the few remaining populations has a very high priority irrespective of genetic considerations — in which case genetic diversity is conserved by default. Based on the available information, we consider these cases to have a low priority for genetic research.

It is important to note that although we rate some of the above studies as a low priority based on the justification provided, in some cases there may well be good reason to undertake a genetic study — the reasons have simply not been recognized. For example, both *Elaeocarpus williamsianus* and *Fontainea oraria* appear to have the potential for extensive clonality which may lead to low levels of seed set in the few remaining populations. A genetic investigation thus has a higher priority than apparent from the recovery plan. The contrasting case of *Thesium australe* is also worth noting. The recovery plan suggested a survey of genetic variation across the wide ecological range of the species (Griffith 1992b). Yet, in redefining priorities for the research and management of this species the recovery team in New South Wales has recently postponed a formal genetic study (Andrew Steed, pers. comm. 1996). We concur with the recovery team action, while such a study is of academic interest, the species does not fit any of the high priority areas proposed in Table 4.

CONCLUSIONS

In our exploration of the practical outcomes of the genetic study of rare plants, we have identified both high and low priorities for genetic research (Table 4). We are in no way decrying the academic merit of any genetic study, irrespective of its practical outcome. Neither are we calling for fewer genetic studies of rare plants. Rather we are calling for more genetic studies, but within a hypothesis testing framework on species that fit the priority areas. Such studies will most certainly contribute in a tangible way to the cost effective achievement of both the short and long-term goals of conservation. The task of defining priorities for all research and management actions remains a continuing challenge for conservation biologists and managers. Defining priorities for genetic research is but one component of this critical need.

ACKNOWLEDGEMENTS

We wish to thank Siegy Krauss, Peter O'Hanlon and Tricia Hogbin for valuable comments on the manuscript, and the NSW National Parks and Wildlife Service for providing financial support to undertake the genetic studies described here.

REFERENCES

ANCA, 1994. *Endangered Species Protection Act* 1992. An Outline. Australian Nature Conservation Agency: Canberra.

Anon., 1994. *Pterostylis gibbosa* Conservation Research Statement and Species Recovery Plan. Prepared by Quality Environmental Management Pty Limited. Australian Nature Conservation Agency Endangered Species Program No. 250.

Benson, J., 1991. *Grevillea beadleana* Species Recovery Plan. Australian National Parks and Wildlife Service Endangered Species Program Project No. 155.

Benwell, A., 1993. *Allocasuarina defungens* Species Recovery Plan. Australian National Parks and Wildlife Service Endangered Species Program.

Benwell, A., 1994. Swamp orchids *Phaius australis, Phaius tancarvilliae* Recovery Plan. Australian Nature Conservation Agency Endangered Species Program No. 371.

Brown, A. H. D. and Briggs, J. D., 1991. Sampling strategies for genetic variation in *ex situ* collections of endangered plant species. Pp. 99–122 *in* Genetics and Conservation of Rare Plants ed by D. A. Falk, K. E. Holsinger. Oxford University Press: New York.

Chalson, J. M. and Keith, D. A., 1995. A risk assessment scheme for vascular plants: Pilot application to the flora of New South Wales. NSW National Parks and Wildlife Service: Sydney, New South Wales.

Coates, D. J., 1988. Genetic diversity and population genetic structure in the rare Chittering grass wattle, *Acacia anomala* court. *Aust. J. Bot.* **36**: 22–27.

Coates, D. J., 1992. Genetic consequences of a bottleneck and spatial genetic structure in the triggerplant *Stylidium coroniforme* (Stylidiaceae). *Hered.* **69**: 512–20.

Coates, D. J. and Sokolowski, R. E. S., 1992. The mating system and patterns of genetic variation in *Banksia cuneata* A.S. George (Proteaceae). *Hered.* **69**: 11–20.

Cohn, J., 1993. Conservation Research Statement and Recovery Plans (Research and Management) for *Kunzea rupestris.* Australian National Parks and Wildlife Service Endangered Species Program Project No. 251.

Crawford, D. J., Stuessy, T. F., Cosner, M. B., Haines, D. W., Wiens, D. and Penaillo, P., 1994. *Lactoris fernandeziana* (Lactoridaceae) on the Juan Fernandez Islands: allozyme uniformity and field observations. *Conserv. Biol.* **8**: 277–80.

ESAC, 1992. An Australian National Strategy for the Conservation of Australian Species and Communities Threatened with Extinction. Endangered Species Advisory Committee, Australian National Parks and Wildlife Service: Canberra.

Falk, D. A. and Holsinger, K. E., 1991. Genetics and Conservation of Rare Plants. Pp. 1–283. Oxford University Press: New York.

Griffith, S. J., 1992a. *Zieria prostrata* Species Recovery Plan. Australian National Parks and Wildlife Service Endangered Species Program Project No. 195. NSW National Parks and Wildlife Service: Sydney.

Griffith, S. J., 1992b. *Thesium australe* Species Recovery Plan. Australian National Parks and Wildlife Service Endangered Species Program Project No. 196. NSW National Parks and Wildlife Service: Sydney.

Hamrick, J. L., Godt, M. J. W., Murawski, D. A. and Loveless, M. D., 1991. Correlations between species traits and allozyme diversity: Implications for conservation biology. Pp. 75–86 *in* Genetics and Conservation of Rare Plants ed by D. A. Falk, K. E. Holsinger. Oxford University Press: New York.

Hartley, T. G. and Williams, J. B., 1983. A new species of *Acronychia* (Rutaceae) from Australia. *Brunonia.* **6**: 251–55.

Hill, K., 1995. The Wollemi Pine. Watch out, look around you. *Friends of the Royal Botanic Gardens Sydney Newsletter.* **November–January 1995–96**: 8–9.

Hunter, J., Jay, A., Nicholson, N., Nicholson, H. and Horton, S., 1991a. *Acronychia littoralis* Species Recovery Plan. Australian National Parks and Wildlife Service Endangered Species Program Project No. 155.

Hunter, J., Jay, A., Nicholson, N., Nicholson, H. and Horton, S., 1991b. *Diploglottis campbellii* Species Recovery Plan. Australian National Parks and Wildlife Service Endangered Species Program Project No. 155. NSW National Parks and Wildlife Service: Sydney.

Hunter, J., Jay, A., Nicholson, N., Nicholson, H. and Horton, S., 1991c. *Elaeocarpus williamsianus* Species Recovery Plan. Australian National Parks and Wildlife Service Endangered Species Program Project No. 155. NSW National Parks and Wildlife Service: Sydney.

Hunter, J., Jay, A., Nicholson, N., Nicholson, H. and Horton, S., 1991d. *Fontainea oraria* Species Recovery Plan. Australian National Parks and Wildlife Service Endangered Species Program Project No. 155. NSW National Parks and Wildlife Service: Sydney.

Jones, W., 1995. The biology and management of the Dwarf Mountain Pine *(Microstrobis fitzgeraldii)* in New South Wales. Species Management Report No. 13. NSW National Parks and Wildlife Service.

Jones, W. G., Hill, K. D. and Allen, J. M., 1995. *Wollemia nobilis*, a new living Australian genus and species in the Araucariaceae. *Telopea* **6**: 173–76.

Kress, W. J., Moddax, G. D. and Roesel, C. S., 1994. Genetic variation and protection priorities in *Ptilimnium nodosum* (Apiaceae), an endangered plant of Eastern United States. *Conserv. Biol.* **8**: 271–76.

Leeton, P. and Fripp, Y. J., 1991. Breeding system, karyotype and variation within and between populations of *Rutidosis leptorrhynchoides* F. Muell. (Asteraceae: Inuleae). *Aust. J. Bot.* **39**: 85–96.

Leigh, J., Boden, R. and Briggs, J., 1988. Extinct and Endangered Plants of Australia. World Wildlife Fund: Australia.

Lesica, P., Leary, R. F., Allendorf, F. W. and Bilderback, D. E., 1988. Lack of genetic diversity within and among populations of an endangered plant, *Howellia aquatilis. Conserv. Biol.* **2**: 275–82.

Matthes, M. and Nash, S., 1993a. *Cynanchum elegans* Conservation Research Statement and Species Recovery Plan. Australian Nature Conservation Agency Endangered Species Program No. 311. NSW National Parks and Wildlife Service: Sydney.

Matthes, M. and Nash, S., 1993b. *Pimelea spicata* Conservation Research Statement and Species Recovery Plan. Australian Nature Conservation Agency Endangered Species Program No. 311. NSW National Parks and Wildlife Service: Sydney.

Matthes, M. and Nash, S., 1994. Conservation Research Statement and Recovery Plan for *Allocasuarina portuensis.* Australian Nature Conservation Agency Endangered Species Program. NSW National Parks and Wildlife Service: Sydney.

Moran, G. F. and Hopper, S. D., 1983. Genetic diversity and the insular population structure of the rare granite rock species, *Eucalyptus caesia* Benth. *Aust. J. Bot.* **31**: 161–72.

Morell, M. K., Peakall, R., Appels, R., Preston, L. R. and Lloyd, H. L., 1995. DNA profiling techniques for plant variety identification. *Aust. J. Exp. Agric.* **35**: 807–19.

Nash, S. and Matthes, M., 1994. Conservation Research Statement and Species Recovery Plan: *Haloragodendron lucasii.* Australian National Conservation Agency Endangered Species Program No. 250. NSW National Parks and Wildlife Service: Sydney.

Peakall, R., 1994a. Genetic analysis of four endangered rainforest plants. Report to the NSW National Parks and Wildlife Service.

Peakall, R., 1994b. Genetic analysis of the endangered *Zieria prostrata:* Implications for its conservation. Report to the NSW National Parks and Wildlife Service.

Peakall, R., 1995. The extent of clonality in a roadside population of the rare and endangered plant *Acronychia littoralis.* Report to the Roads and Transport Authority of New South Wales.

Peters, G. B., Lonie, J. S. and Moran, G. F., 1990. The breeding system, genetic diversity and pollen sterility in *Eucalyptus pulverulenta*, a rare species with small disjunct populations. *Aust. J. Bot.* **38**: 559–70.

Prober, S. M., Tompkins, C., Moran, G. F. and Bell, J. C., 1990. The conservation genetics of *Eucalyptus paliformis* L. Johnson et Blaxwell and *E. Parvifolia* Cambage, two rare species from southeastern Australia. *Aust. J. Bot.* **38**: 79–95.

Rieseberg, L. H., Zona, S., Aberbom, L. and Martin, T. D., 1989. Hybridization in the island endemic, Catalina Mahogany. *Conserv. Biol.* **3**: 52–58.

Rossetto, M., Weaver, P. K. and Dixon, K. W., 1995. Use of RAPD analysis in devising conservation strategies for the rare and endangered *Grevillea scapigera* (Proteaceae). *Mol. Ecol.* **4**: 321–29.

Sampson, J. F., Collins, B. G. and Coates, D. J., 1994. Mixed mating in *Banksia brownii* Baxter ex R.Br. (Proteaceae). *Aust. J. Bot.* **42**: 103–11.

Sampson, J. F., Hopper, S. D. and James, S. H., 1988. Genetic diversity and the conservation of *Eucalyptus crucis* maiden. *Aust. J. Bot.* **36**: 447–60.

Sampson, J. F., Hopper, S. D. and James, S. H., 1989. The mating system and population genetic structure in a bird-pollinated mallee, *Eucalyptus rhodantha*. *Heredity* **65**: 189–99.

Schaal, B. A., Leverich, W. J. and Rogstad, S. H., 1991. Comparison of methods for assessing genetic variation in plant conservation biology. Pp. 123–34 in Genetics and Conservation of Rare Plants ed by D. A. Falk, K. E. Holsinger. Oxford University Press: New York.

Scott, J., Marshall, A., and Auld, T. D., 1995. *Grevillea caleyi* Conservation Research Statement and Species Recovery Plan. Australian Nature Conservation Agency Endangered Species Program No. 456. NSW National Parks and Wildlife Service: Sydney.

Shapcott, A., 1994. Genetic and ecological variation in *Atherosperma moschatum* and the implications for conservation of its biodiversity. *Aust. J. Bot.* **42**: 663–86.

Soltis, P. S., Soltis, D. E., Tucker, T. L. and Lang, F. A., 1992. Allozyme variability is absent in the narrow endemic *Bensoniella oregona* (Saxifragaceae) *Conserv. Biol.* **6**: 131–34.

Waller, D. M., O'Malley, D. M. and Gawler, S. C., 1987. Genetic variation in the extreme endemic *Pedicularis furbishiae* (Scrophulariaceae) *Conserv. Biol.* **1**: 335–40.

An area-based multiple species approach to threatened flora conservation and management in the Merredin area of Western Australia

Andrew Brown[1], David Coates[2], Mike Fitzgerald[3] and Claire Welbon[3]

The vascular flora of Western Australia is extremely rich, particularly in the largely cleared and highly populated areas of the South-west land division. Here, about 8 000 of the 11 000 named species are found, including most of the State's 282 declared threatened flora and 1 700 poorly known priority 1–4 flora. Recovery plans and interim recovery plans are being written and implemented for 41 taxa of declared threatened flora, of which 38 are ranked as critically endangered. However, because of limited resources, individual recovery plans are unlikely to be developed for most flora in the lower endangered and vulnerable categories, and it is unlikely that surveys will be carried out for individual taxa of poorly known priority flora in anything but an opportunistic *ad hoc* manner. Alternative strategies are therefore required. An approach taken in Western Australia is to address the conservation and management of declared threatened and poorly known flora on a geographical area basis. This area-based approach has advantages over single taxon approaches as it has the potential to be more efficient in its use of resources, is more cost-effective and many taxa are covered by a single recovery team. It also provides broader options for management in relation to threatening processes. In this paper we outline the implementation and operation of the first area-based threatened flora recovery program in Western Australia. This case study covers the problems encountered, lessons learned and solutions developed in the first 20 months of managing many threatened flora over a large, extensively cleared area of the South-west agricultural region.

INTRODUCTION

THE vascular flora of Western Australia is extremely rich, particularly in the largely cleared and highly populated areas of the South-west land division. Here, about 8 000 of the 11 000 named Western Australian species are found, including most of the State's 282 declared threatened flora (threatened flora are broken down into the categories "Critically Endangered", "Endangered" and "Vulnerable" depending on perceived degree of threat) and 1 700 poorly known priority 1–4 flora (an explanation of these codes is outlined under the heading "CALM'S Priority Flora List", see p.136). Causes of threatened status are many and varied but are largely due to massive past land clearance for agriculture and housing development. In many areas much of the remaining natural vegetation is fragmented and is represented in small reserves and narrow road corridors. As many poorly known priority flora occur in these areas, have a low number of extant individuals and appear to have a restricted occurrence or susceptibility to various threats, they may qualify for declared threatened status. These taxa do, however, require urgent survey in order to determine their conservation status (Hopper *et al.* 1991). Once this has been done the true figure for threatened taxa is likely to rise to more than 500.

At the time of writing, recovery plans (RPs) and interim recovery plans (IRPs) are being written and implemented for 41 taxa of declared threatened flora, of which 38 are listed as critically endangered, (for an explanation of what an IRP is see the last paragraph under the heading "Ranking", p.135). However, because of limited resources, individual recovery plans are unlikely to be developed for most endangered and vulnerable flora, and it is unlikely that surveys will be carried out for individual taxa of poorly known priority flora in anything but an opportunistic *ad hoc* manner. Alternative strategies are therefore required if all of Western Australia's declared threatened and poorly known priority flora are to be considered in the recovery process.

[1]Threatened Species and Communities Unit, Department of Conservation and Land Management, P.O. Box 51, Wanneroo, Western Australia, Australia 6065.
[2]Science and Information Division, Department of Conservation and Land Management, Western Australian Herbarium, P.O. Box 104, Como, Western Australia, Australia 6152.
[3]Department of Conservation and Land Management, Wheatbelt Region, P.O. Box 332, Merredin, Western Australia, Australia 6415.

Pages 130–37 *in* BACK FROM THE BRINK: REFINING THE THREATENED SPECIES RECOVERY PROCESS ed by S. Stephens and S. Maxwell. Surrey Beatty & Sons, 1996.

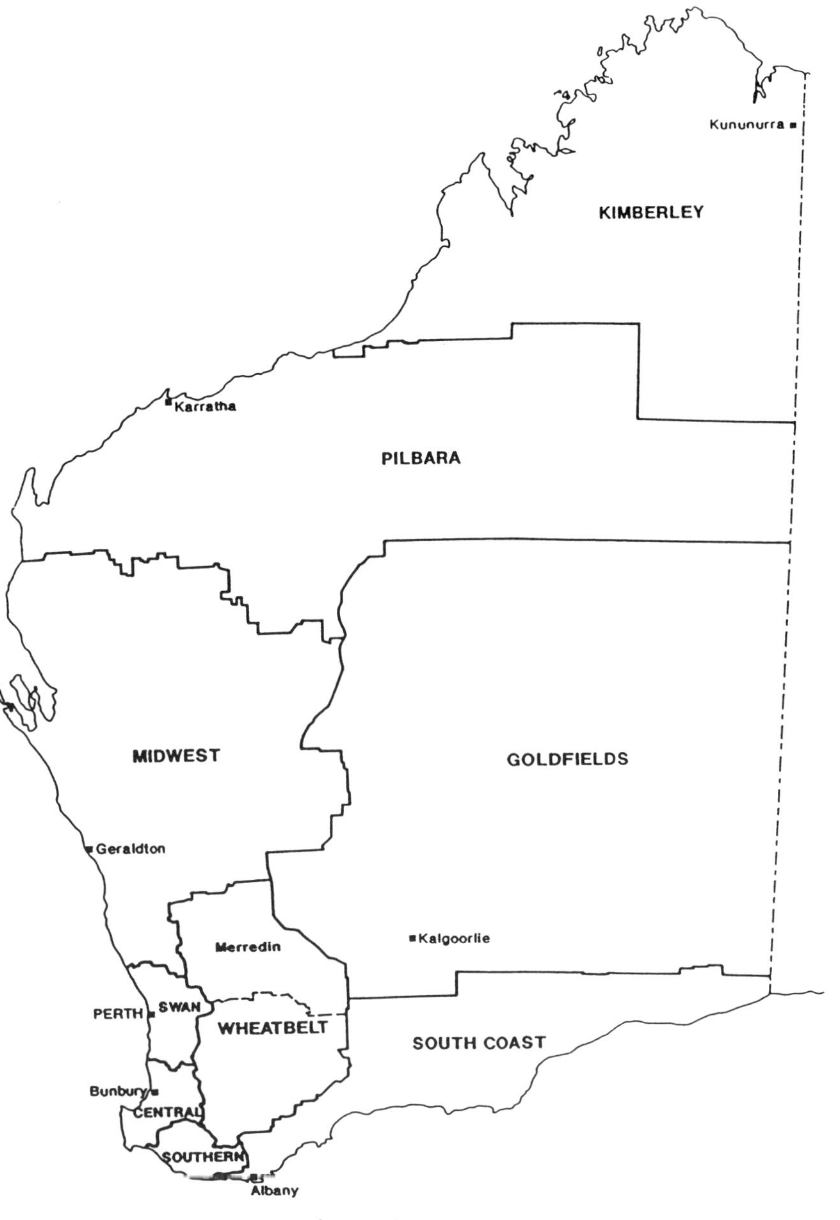

Fig. 1. CALM regions with the Merredin District.

An approach taken in Western Australia is to address the conservation and management of declared threatened and poorly known flora on a geographical area basis. In this State these areas correspond to the Department of Conservation and Land Management's (CALM) administrative regions or districts. There are nine CALM regions and 22 districts (Fig. 1).

The area-based approach to the conservation and management of declared threatened and poorly known priority flora has successfully developed over the last five years and now forms an integral part of flora conservation in Western Australia (see CALM Policy Statement No. 44 in references). To date, it has focused on the South-west Botanical Province, due to its floristic richness, high level of endemism and the widespread land clearing that has occurred in that area. However, it is envisaged that the whole of the State will be covered within the next 5 to 10 years.

Declared threatened flora are concentrated in the South-west Botanical Province in the northern and southern sand health plant communtities, and in the shrublands and heathlands of more inland areas (Wheatbelt). In contrast the higher rainfall forest areas of the extreme South-west have far fewer threatened taxa. Preparation of area-based threatened flora recovery programs has therefore initially focused on areas outside the forests. These recovery programs consider all declared threatened and poorly known priority taxa within a specified geographical area which, in this case, is based on CALM administrative districts or regions. Each program prioritizes individual declared threatened taxa and threatening processes for conservation action. To date, four such recovery programs have been completed covering parts of the northern sand heaths and forests (Kelly *et al*. 1990, 1993), the Merredin District in the Wheatbelt (Mollemans *et al*. 1993) and part of the southern sand heaths (Robinson and Coates 1995). Seven other area-based recovery programs are in preparation and are expected to be completed by the end of 1997.

This area-based approach has many advantages over single taxon approaches as it has the potential to be more efficient in its use of resources, is more cost-effective and many taxa are covered by a single recovery team. It also provides broader options for management in relation to threatening processes, although less detailed attention can be given to any single taxon in the area the program covers.

In this paper we outline the implementation and operation of the first of these area based threatened flora recovery programs, which was developed for CALM's Merredin District. This case study will cover the problems encountered, lessons learned and solutions developed in the first 20 months of managing many threatened flora over a large (77 000 km^2), extensively cleared area of the South-west agricultural region.

THE MERREDIN DISTRICT

Situated within the central Wheatbelt of Western Australia some 100 kilometres east of Perth (Fig. 1), the Merredin District covers an area of approximately 63 000 km^2 within the northern half of CALM's Wheatbelt Region. It is a heavily cleared part of Western Australia which encompasses its main inland grain growing areas. Within its boundaries lie 16 local government authorities (LGAs). It covers four botanical districts within two natural phytogeographic regions and a major interzone. Broadly, vegetation patterns in the district are comprised of *Acacia* shrublands in the north, *Eucalyptus* shrublands (mallee) in the south and east and *Eucalyptus* forests and woodlands in the south and west.

Settlement of the District commenced in the 1860s with limited pastoralism and sandalwood harvesting. This was followed in the 1880s by gold mining in its eastern part and in the 1890s by agricultural expansion. Agricultural land use now occupies approximately 80% of the District's area. In the remainder, the main land uses include pastoralism, mining and nature conservation. Currently, the human population is around 20 000 with the majority dependant on agriculture for their livelihood.

The main operational focus of the three CALM Merredin District staff (a District Manager, a Reserves Officer and a Wildlife Officer) is in the nature conservation area. This devotes a substantial portion of the District's resources to the management of the conservation estate (national parks, conservation reserves and timber reserves). Other areas covered include fire management, wildflower industry regulation, threatened species management and land acquisitions for the purposes of conservation. Tourism and recreation activities complete the major components of the District's operational portfolio.

As at May 1996, the District contains 190 nature reserves managed by CALM, with a total area of 662 532.9 ha. This equates to about 10.5% of the total area of the district. However, the reserves vary greatly in size and are unevenly distributed. The two largest reserves in fact make up 78% of the District's conservation estate.

Much of the other natural vegetation in the agricultural zone is also highly fragmented. It usually occurs along narrow transport corridors or in isolated small parcels on private and public land and is biased towards non-arable landscape elements such as the margins of salt lakes, granite outcrops and lateritic breakaways. In some LGA's, where up to 97% of the original vegetation has been removed, the conservation estate occupies a significant proportion of the remnant vegetation.

However, in the non-agricultural northeastern and eastern fringes of the District, native vegetation is almost continuous, disrupted only by the occasional settlement and mine site. The largest nature reserves in the District are located in this area (Karroun Hill = 309 678 ha, Jilbadgi = 208 866 ha).

THE MERREDIN DISTRICT THREATENED FLORA RECOVERY PLAN

In 1991, with the aid of funds from the Australian National Parks and Wildlife Service (ANPWS), now the Australian Nature Conservation Agency (ANCA), surveys of threatened and poorly known plants in CALM's Merredin District commenced. This resulted in a detailed threatened flora recovery program for the District (Mollemans *et al.* 1993).

This area-based recovery program collated the available biological and management information on all declared threatened, and poorly known priority one, two and three taxa in CALM's Merredin District, excluding the Wongan-Ballidu shire which will be the subject of a separate threatened flora recovery program.

At the time of publication, some 33 declared threatened flora were known from the district, 10 were presumed extinct and a further 40 were listed as priority one, 36 as priority two and 23 as priority three. Some 355 populations of declared threatened and poorly known flora were spread over a variety of land tenures (see Table 1).

The objective of the area-based recovery program was to ensure and enhance, by appropriate management, the continued survival in the wild of populations

Table 1. Merredin District declared threatened flora population and land tenure information (Dec. 1995).

Tenure	No. of pops	% of total pops
Conservation reserve	63	17.6
Private property	88	24.8
Railways	14	3.9
Main Roads Reserve	9	2.5
SHIRE road verge	95	26.7
Non-vested Crown land	18	5.1
Shire other land	10	2.9
Other reserves	8	2.3
Water reserve	30	8.5
Not known	20	5.7
Total	355	100

populations of declared threatened flora and other plants in need of special protection, and aimed to achieve this by:

- providing a useful reference for CALM staff and other land managers for the day to day management and protection of threatened flora and populations of other taxa which are poorly known and potentially at risk;
- directing departmental resources within the District to those taxa most urgently in need of attention;
- assisting in the identification of threatened flora and other species potentially at risk, and their likely habitats;
- fostering an appreciation and increased awareness of the importance of protecting and conserving threatened flora and other taxa potentially at risk or in need of special protection; and
- listing a number of strategies for management and research including: survey, liaison with landowners, fire management, monitoring, land acquisition, fencing, weed control, propagation and areas of research required.

IMPLEMENTATION OF THE MERREDIN DISTRICT THREATENED FLORA RECOVERY PLAN

At the time of the Merredin District Threatened Flora Recovery Plan's publication, CALM's Merredin District had a staff of just three — a District Manager, a Reserves Officer and a Wildlife Officer, all of whom were fully committed to established works programs. District staff were able to address some of the population monitoring work and urgent threatened flora management issues, but the large size of the district (345 km × 240 km), the large number of taxa (33 threatened and 99 poorly known) and the many threats to their conservation status, including weeds, severely degraded habitat, salinity, grazing and accidental destruction, combined with the limited resources available in the District, meant that they could not develop the broad strategies and recovery actions recommended in the program.

In 1993 CALM sought funding from the Australian Nature Conservation Agency (ANCA) to implement the programme, and in January 1994 this funding led to the appointment of an Assistant Conservation Officer (threatened flora co-ordinator) to the District. This person's primary role was to implement the recommendations made in the program, under the guidance of the Merredin District Threatened Flora Recovery Team.

The recovery team

The recovery team held its first meeting in January 1994. It is relatively small, comprising two CALM staff, an externally funded Assistant Conservation Officer, an ANCA representative from Canberra who usually does not attend meetings due to distance, two private land owners, one of whom is also a representative of the Wild-flower Society and a Shire representative. It is chaired by CALM's Merredin District Manager, Mike Fitzgerald.

The problem with such a small group was one of ownership (many species spread over a very large area) and an inability to adequately spread the substantial workload or get the message across to the community at large. Local government and community groups were each represented by a single member on the team. While both provided invaluable service, the ability of the team to communicate with the groups they represent was limited. There are 16 LGAs in the district, four wildflower and wildlife groups, more than 20 Land Conservation District Committees (LCDs) and several independent but highly competent individuals that needed to be included in the team's information dissemination.

The team is currently being expanded to include a wider representation of relevant non-government groups such as naturalist clubs, wildflower groups, Land Conservation District Committees and local communities. To help the work of the community representatives, and to facilitate better contact and co-ordination with these important groups, the team is considering setting up subsidiary advisory groups in outlying areas (i.e., Wongan Hills) to help spread its message and workload. Advisory groups will regularly report back to the main team and, from time to time, members of these groups will attend recovery team meetings as observers or to discuss particular issues.

It is important that members of recovery teams and subsidiary advisory groups be chosen carefully as experience with several other recovery teams has shown that poorly selected representatives, who are not directly involved in recovery actions or management of land on which the threatened species occurs, or who are hostile to the conservation ethos of the group due to competing interests, provide little active and useful input.

A further trap is one where a few members of a team do most of the talking and work. Recovery team chairs should also be mindful of involving members representing other groups in planning, discussion and active management. Experience in several early recovery teams elsewhere has shown that there could be a tendency for decisions to be made by government representatives prior to meetings who then simply use the meeting as a forum to inform the non-government members of those decisions, without first having involved them in discussions.

Development of a recovery schedule

At its first meeting, the recovery team reviewed progress made since publication of the Merredin District Threatened Flora Recovery Plan and adopted a schedule of works. The schedule was based on six strategies that have been central to the success of the recovery program to date. They were:

1. Prioritization of recovery actions for threatened flora.
2. Good information management.
3. Regular liaison with other key agencies.
4. Regular community liaison.
5. Regular monitoring of populations.
6. Additional survey.

Table 2. Additions and deletions (between 1992 and 1995) to the list of declared threatened flora in the Merredin District. Status in 1995 shown. CE = critically endangered; E = endangered; V = vulnerable; PE = presumed extinct; P1 = priority 1; P2 = priority 2; P3 = priority 3; P4 = priority 4.

Additions	Deletions
Rediscovered	**Status downgraded**
Hemiandra rutilans (rediscovered but now PE again)	*Acacia merrickiae* (P4)
Stylidium merrallii (V)	*Caladenia cristata* (P4)
Recently discovered	*Daviesia oxylobium* (P4)
Grevillea pythara (CE)	*Gastrolobium callistachys* (P4)
Recently found in the District	*Grevillea prostrata* (P4)
Gastrolobium graniticum (V)	*Prostrantha magnifica* (P4)
Upgraded	*Sowerbaea multicaulis* (P4)
Acacia sciophanes (E)	*Stylidium neglectum* (P3)
Grevillea christineae (E)	**District boundary change**
Myriophyllum lapidicola (CE)	*Tetratheca aphylla* (V)
Tetratheca deltoidea (CE)	*Tetratheca harperi* (V)
	Tetratheca paynterae (E)

The work schedule developed by the team listed, in chronological order, the actions that were to be carried out and the first task the Assistant Conservation Officer had was to update the 12 month old Merredin District Threatened Flora Recovery Plan.

The recovery plan detailed the management and research requirements for all 33 threatened and 99 poorly known priority 1–4 flora found in the Merredin District. However, in the several years since the writing of the plan, there had been a number of changes in the status of those taxa as shown in Table 2.

The total number of declared threatened flora had not changed substantially from the 33 in 1992 to the 31 in 1995 (Table 3) but, during that period, 11 taxa had been de-listed and eight new taxa added, including three critically endangered and two endangered (Table 2). The main cause of these changes was due to the increased survey and management attention that threatened and poorly known flora were receiving in the District. Surveys of poorly known taxa resulted in several being added to the list once their threatened status was confirmed, while several that were listed as threatened were found to be well represented on conservation reserves or were more common than previously thought. These were downlisted to P4. District boundary changes also resulted in three taxa that were in the Merredin District being placed in the Goldfields Region.

Due to the substantial changes in the composition of the species requiring recovery actions, the existing ranking of actions for declared threatened taxa contained in the program required review.

Table 3. Number of declared threatened flora in the Merredin district (1992 and 1995).

	In 1992	In 1995
Critically endangered	4	7
Endangered	4	5
Vulnerable	25	18
Total	33	31

Ranking

In the published Merredin District Threatened Flora Recovery Plan the conservation status of each threatened taxon was assessed and recommendations for protection, research and management made (Mollemans *et al.* 1993). On the basis of these recommendations, each taxon was scored on a scale of 1 to 3 under 19 categories recognized as potential threats or management requirements. The scores were summed for each of the 33 threatened taxa and for each threat/requirement category.

It was intended that recovery work be conducted, programmed or deferred according to priority, available funds and existing resources and workloads. This would then enable resources and staff within the Merredin District to be allocated where they were most urgently required.

As some time had elapsed since the plan was written in 1992 and a number of changes had occurred for both the species listed and the threatening processes effecting them, it was thought appropriate that the taxa be re-ranked. A new more accurate ranking method had been developed (CALM Policy Statement No. 50) and it was decided to use it, rather than the older method outlined above.

All declared threatened flora that were known to occur in the Merredin District were ranked into one of four categories by scoring each against 12 criteria. The management categories, their associated score ranges and the number of taxa in each category are shown in Table 4.

Several important lessons were learnt during ranking which highlighted the desirability of using this process when implementing area-based recovery programs.

Table 4. The number of critically endangered, endangered, vulnerable and conservation dependent taxa in the Merredin District.

Category	Score	No. of taxa
Critically endangered	≥60	9
Endangered	45–59	10
Vulnerable	21–44	17
Conservation dependent	≤20	0

Firstly, if taxa are not ranked there is a strong human tendency to deal with easily rectified management problems rather than those that are essential but where the probability of success is less certain and more difficult to accomplish. There is also a temptation (often due to pressure from outside groups) to deal with popular, well known or attractive taxa rather than those with fewer patrons. Having an objective approach to priority setting by ranking has helped the Merredin District Threatened Flora Recovery Team hold its focus.

Secondly, the variations in the raw rankings each quarter can be used as a rough performance (recovery) indicator. By the completion of the recovery team's first year, the average raw ranking score had fallen by 20%. This essentially means that the overall status of the declared threatened flora in the Merredin District had improved. It is expected that by the completion of the second year the average score will have fallen by about a further 8%.

The main cause of the drop in ranking is due to active management and the increased scrutiny the declared threatened flora in the Merredin District were receiving. In the team's first year, many recovery actions were implemented (for examples of these see "Recovery actions" below) and following intensive surveys, some taxa were found to be more common than previously thought and new populations were found in conservation reserves.

At the same time this ranking was being done by the recovery team in March 1995, the first comprehensive ranking of all Western Australia's declared threatened flora had been completed by a panel of expert scientists. Thirty eight taxa were ranked critically endangered, seven of which were known to occur in the Merredin District. As a matter of urgency, Interim Recovery Plans (IRPs) have been prepared for them. The implementation of the recovery recommendations contained in the IRPs are being overseen by the Merredin District Threatened Flora Recovery Team. IRPs are designed to address the most urgent threatening processes effecting populations of threatened flora and run for three years only. For the long-term management of threatened species. IRPs will be developed into full recovery plans following that three year period.

Recovery actions

Since implementation of the Merredin District Threatened Flora Recovery Plan commenced, a number of recovery actions are in progress or have been completed for threatened species. These include:

Wongan Cactus *Daviesia euphorbioides:* Two populations of *D. euphorbioides* are being treated for weed control in a trial currently conducted by CALM's Science and Information Division. The objective of the trial is to determine the effect of weeds on population dynamics of the species and the best methods of control.

Pythara Grevillea *Grevillea pythara:* This species is also part of the CALM Science and Information Division trial on weed control. Spot treatment of weeds was undertaken in late August 1996, and its effect will be assessed over time.

Colourful Snakebush *Hemiandra rutilans:* Although the one extant plant of this species died shortly after being found, cutting material was taken and has been grafted on to Westringia rootstock. Several plants are now held by Kings Park and Botanic Garden. The area of bushland in which the species occurred has been protected from grazing by a fence. The soil has been smoke treated to encourage germination of soil-stored seed. This area also contains populations of declared threatened species *Stylidium scabridium* and *Verreauxia verreauxii.*

Chiddarcooping Myriophyllum *Myriophyllum lapidicola:* Due to the dry winters that have resulted in little water runoff into the rock pool habitat of this species, it had not been relocated at the single known site in recent years. However, following a recent (August 1996) field trip, plants were again located in a single rock pool. The area is a water reserve and the appropriate authorities have been notified of the species' presence. The area will be fenced to protect it from rabbits and kangaroo grazing and trampling.

Wyalcatchem Foxglove *Pityrodia scabra*: Plants propagated from seed, cuttings and tissue culture have been translocated back into the area of the known population on private land as well as a nearby reserve. This was done by CALM and Kings Park and Botanic Garden staff and volunteers. The trial has proved most successful with some 60% of plants surviving the first summer and many flowering in the first year. Fencing, rabbit and weed control were critical to the success of the trial. The sites are being regularly monitored by CALM and volunteers.

Feral animal control has been implemented in a number of areas containing populations of threatened plant taxa, including both reserve areas and private land.

Poorly known priority 1–3 taxa

While the main focus of the Merredin District Threatened Flora Recovery Team has been on declared threatened flora it is also extremely important that more information be obtained for poorly known priority 1–3 taxa that are known from few populations, all of which appear to be under threat. An explanation of these codes is outlined below. For these taxa the main operational focus was in gaining additional population information, mainly through surveys. Priorities for selecting species for survey have been based largely on the number of known populations, the time since they were last surveyed, the status and size of land that they occur on, the health of associated vegetation and the quality and reliability of the information produced at the time of past surveys.

To date, surveys have been undertaken, and the conservation status reviewed, for 45 poorly known priority taxa. From the information gained during this study, one taxon has now been upgraded to declared threatened status and it is likely that another five will be upgraded following further survey work. Recommendations will also be made to de-list 24 taxa, as they are now known to be secure in the wild.

CALM's Priority Flora List

CALM maintains a priority flora list to determine priorities for survey of plants of uncertain conservation status. The list comprised 1 408 taxa (at October 1992) that are poorly known and in need of high priority survey or are adequately surveyed but in need of monitoring. The poorly known taxa are possibly at risk but do not meet the survey requirements for gazettal as Declared Rare Flora (DRF), as outlined in Policy Statement No. 9 (see references). Only those plants considered to be threatened on the basis of thorough survey or presumed extinct can be included on the DRF schedule.

The priority flora list is divided into the following categories according to the degree of perceived threat (for more information see Hopper *et al.* 1990).

Priority one — poorly known taxa

Taxa which are known from one or a few (generally <5) populations which are under threat, either due to small population size, or being on lands under immediate threat, road verges, urban areas, farmland, active mineral leases, etc., or the plants are under threat, for example, from disease, grazing by feral animals, etc. May include taxa with threatened populations on protected lands. Such taxa are under consideration for declaration as "rare flora", but are in urgent need of further survey.

Priority two — poorly known taxa

Taxa which are known from one or a few (generally <5) populations, at least some of which are not believed to be under immediate threat, that is, not currently endangered. Such taxa are under consideration for declaration as "rare flora", but are in urgent need of further survey.

Priority three — poorly known taxa

Taxa which are known from several populations, at least some of which are not believed to be under immediate threat, that is, not currently endangered. Such taxa are under consideration for declaration as "rare flora" but are in need of further survey.

Priority four — rare taxa

Taxa which are considered to have been adequately surveyed and which, while being rare (in Australia), are not currently threatened by any identifiable factors. These taxa require monitoring every 5–10 years.

Information management

Information management is one of the most demanding work areas for those working to recover declared threatened taxa found in the Merredin District. A number of different challenges arose during the first 20 months.

Initially, the Threatened Flora Recovery Plan was the main source of information for each taxon the recovery team had to manage. However, with the dramatic changes in status of a number of taxa, and the development of new policies and methods, the plan, as originally published, is well out of date. To offset this, the District is attempting to maintain an up to date database on all taxa, with the necessary population, threat and management information.

Liaison with local government authorities

In the Merredin district 30% of the declared threatened flora populations occur on narrow degraded road verges managed by local government authorities. It was considered by the team that there was an enormous potential for accidental damage to those populations. An extensive and effective liaison and education program was developed to reduce the chances of this occurring. This involved developing LGA information kits for each of the 16 shires in the district. The information kits provided:

- 1:100 000 scale wall maps illustrating all populations in that LGA.
- Detailed site maps for each roadside population.
- Training sessions on rare flora issues.

In addition, a threatened flora road marking program was undertaken in each LGA to make it easy to identify areas with declared threatened plants and prevent accidental damage during operations such as road maintenance.

This has proved highly successful and the kits provide a good example of the success of area-based recovery programs. Rather than providing management for a single species at a time, this involved managing a group of populations of many species in a specific area. A number of points have been learned from this exercise which will be used in future management. They include:

- The development of standardized threatened flora information kits within CALM.
- Standard training in threatened flora issues for all LGA employees.
- The recommendation that threatened flora information be included in LGA staff induction packages.

Given the amount of time spent liaising with and educating LGAs, it has become clear there is a need for one or two LGA representatives on the recovery team to assist in this process. This is planned for 1996.

Resources, community education and participation

One of the major problems encountered in the Merredin District was managing a large number of plant taxa and many populations over a wide area with so few resources. In many cases funding was not the major problem, rather, there was a lack of sufficient field staff to either survey and monitor populations or initiate recovery actions. To overcome this problem it was clear volunteers would be essential for the success of this recovery program.

Potential groups such as wildflower societies were targeted. The Assistant Conservation Officer gave information and motivation talks and threatened flora volunteers were recruited. Kits providing information on threatened flora and poorly known priority taxa that were thought to occur in the voluntary groups' areas were handed out. These included species descriptions and photographs. Very active volunteers were given two or three species of high priority to monitor and survey and have also been involved in assisting in various recovery actions (see "Recovery actions", p. 135).

The success of the volunteer program cannot be understated. Volunteers have been responsible for the discovery of populations of a number of declared threatened taxa (e.g., *Hemiandra rutilans* and *Stylidium merrallii*). In addition, they were also responsible for rediscovering a presumed extinct taxon (*Leucopogon marginatus*) and are involved in the monitoring of many declared threatened flora populations. Volunteers also act as ambassadors, educating the public about the importance of threatened flora and general conservation issues.

All volunteer programs are time consuming to manage, as they require regular contact, advice, encouragement and most of all appreciation. The volunteer group in the Merredin district is now largely self-supporting with people in different areas being responsible for those species that are found in their area.

The recovery team intends to increase its volunteer pool by targeting local Land Care District Committees (LCDCs) which will give the team greater access to private property remnant vegetation. With approximately 20 LCDCs in the Merredin district it is intended to add an LCDC representative to the recovery team. It is hoped volunteers will become more and more involved in recovery work.

To be effective, a volunteer program of this size, with the potential to greatly increase, requires co-ordination. The recovery team is therefore considering the appointment of a part-time volunteer co-ordinator.

Taxonomy

A problem encountered by the recovery team has been the uncertain taxonomic status of a number of the species listed in the Merredin District Threatened Flora Recovery Plan.

For example, two species in the recovery plan were listed as presumed extinct (*Leucopogon marginatus* and *Phlegmatospermum drummondii*) but had been included under other names at the Western Australian Herbarium. A great deal of time and effort was spent looking for these species when in fact they had already been found. In addition, *Hemiandra rutilans*, which was initially regarded by the recovery team as critically endangered is currently under taxonomic review. It is suspected that the species description may have been misinterpreted and the taxon may be simply a form of a more common widespread species.

Clearly, it is difficult to prioritize management actions in these circumstances and much time can be spent on the recovery or survey of some taxa only to find out that they are synonyms or forms of more widespread and common species. Resources may be channelled to some species to the detriment of others. If the taxonomy is not sorted out then the consequences for declared threatened taxa may be severe.

The Western Australian Herbarium has now provided staff to sort out the taxonomy of many difficult taxa, and the recovery team will be lobbying for the development and maintenance of a policy to give threatened flora taxonomy a much higher priority.

CONCLUSIONS

The development of area-based recovery plans is a viable alternative to individual species recovery plans. It allows a more efficient use of resources, is more cost-effective and many taxa are covered by a single recovery team. In the case of the Merredin District, 32 individual recovery teams would be required on an individual species basis, an alternative that is quite impractical given the limited resources and personnel. It also provides broader options for management in relation to threatening processes, that is, where single threatening processes are effecting both many populations and many different taxa within a geographic area.

There were, and still are some teething problems in one recovery team managing and implementing recovery actions for a large number of taxa and many populations over a very large area with so few resources. However, these are being overcome through the bringing together of land owners, land managers, shires and other government departments in a unified approach to recovery, the involvement of enthusiastic and knowledgeable volunteers and the development of new strategies.

More thorough surveying and the implementation of successful recovery actions during the first 20 months of implementation has resulted in the down-listing of 11 taxa previously considered to be threatened and, with better information, the addition (to threatened status) of nine poorly known taxa which may otherwise have become extinct.

REFERENCES

Hopper, S. D., van Leeuwin, S., Brown, A. P. and Patrick, S. J., 1990. Western Australia's Endangered Flora. Department of Conservation and Land Management. Como, WA.

Kelly, A. E., Coates, D. J., Herford, I., Hopper, S. D., O'Donaghue, M. and Robson, L., 1990. Declared rare flora and other plants in need of special protection in the Northern Forest Region. Department of Conservation and Land Management. Western Australia Wildlife Management Program No. 5.

Mollemans, F. H., Brown, P. H. and Coates, D. L., 1993. Declared rare flora and other plants in need of special protection in the Merredin District. Department of Conservation and Land Management. Western Australia Wildlife Management Program No. 9.

Policy Statement No. 9 Conservation of Threatened Flora in the Wild, May 1991. Department of Conservation and Land Management.

Policy Statement No. 44, 1992. Wildlife Management Programs Department of Conservation and Land Management Western Australia.

Policy Statement No. 50, 1994. Setting Priorities for the Conservation of Western Australia's Threatened Flora and Fauna Department of Conservation and Land Management Western Australia.

Robinson, J. and Coates, D. J., 1995. Declared rare and poorly known flora in the Albany District. Department of Conservation and Land Management. Western Australia Wildlife Management Program No. 20.

Native grasslands and threatened species in the Australian Capital Territory: steps towards protection and recovery

Sarah B. Sharp[1] and David J. Shorthouse[1]

Only about 1 000 hectares (5%) of the original natural temperate grasslands in the Australian Capital Territory remain in more or less natural condition. A *Recovery Plan for the Australian Capital Territory Lowland Native Grasslands* being implemented by the ACT Parks and Conservation Service guides a systematic approach to determine the distribution and ecological requirements of grassland communities and selected threatened species, and to develop monitoring and management strategies for remnant grassland sites.

Significant areas of native grassland and habitat for the threatened Striped Legless Lizard *Delma impar* will be protected as part of revised plans for the new Gungahlin Town Centre. The location and site plans for new offices are being modified to retain important grassland habitats, and management of peri-urban and rural grasslands is taking greater account of the need to protect conservation values.

The focus for grassland recovery in the Australian Capital Territory is shifting towards research into the long-term management requirements and rehabilitation of degraded sites, including weed control, recovery of forbs, conservation of threatened fauna and the role of grazing stock. Active community interest in native grasslands is encouraged and community involvement in conservation activities is anticipated.

Key words: Natural temperate grasslands, recovery, community management, endangered communities.

INTRODUCTION

THE original native grassland communities of south-eastern Australia have almost completely disappeared following sustained use for agricultural purposes and urban development. Commonwealth, State, Territory and local governments are now co-operating to ensure conservation of remaining native grasslands and their component species, many of which are also threatened with extinction. One element of this task is a research and recovery program undertaken by the ACT Parks and Conservation Service with funds from the Commonwealth's Endangered Species Program. The recovery program is complex, spanning the Australian Capital Territory and adjacent parts of the Southern Tablelands region of New South Wales including both the native grassland ecological community and its threatened flora and fauna. This paper aims to record how the research and recovery program has developed to meet the challenge, and to describe how the information generated through research is beginning to influence land use planning and management practices.

Prior to European settlement, natural temperate (lowland native) grasslands occurred in three major discrete areas in the Australian Capital Territory, with a total area of approximately 20 000 ha. The remaining area is now highly fragmented, with the most recent surveys indicating that only approximately 1 000 ha of naturally treeless grasslands still remain in more or less natural condition. This represents about 5% of the pre-European extent of this community. In addition, approximately 500 ha of very open grassy woodland and lowland secondary grassland (grassy woodland from which trees have been cleared) with a similar groundstorey composition also exists. Remaining natural temperate grasslands are also found throughout southern New South Wales, Victoria, Tasmania and eastern South Australia occurring as fragmented communities of similar size, but the estimated extent of remaining sites is 0.5% of the original 2 million hectares, and this is poorly represented in reserves (McDougall and Kirkpatrick 1994).

In addition to the significant reduction and fragmentation of grasslands in the Australian Capital Territory, there has been marked alteration of their composition and structure due mainly to management practices and invasion by weeds. Threatening processes include continuing urban expansion, weed invasion, clearing for agricultural purposes and inappropriate management regimes.

[1]Wildlife Research Unit, ACT Parks and Conservation Service, P.O. Box 1119, Tuggeranong, Australian Capital Territory, Australia 2901.

Pages 138–49 *in* BACK FROM THE BRINK: REFINING THE THREATENED SPECIES RECOVERY PROCESS ed by S. Stephens and S. Maxwell. Surrey Beatty & Sons, 1996.

The dominant plant species are perennial native tussock grasses, with few or no shrubs, and trees either absent or in very low numbers (Moore 1964). The dominant grasses are *Themeda triandra, Danthonia* spp., *Stipa* spp., *Bothriochloa macra* and *Poa* spp. Forbs, particularly Asteraceae, comprise about 70% of species present. The grasslands virtually always contain exotic grasses and forbs (Groves and Williams 1981).

The Australian Capital Territory natural temperate grasslands contain a number of significant species. Three plant species (Sharp 1994) and three vertebrate species (Osborne *et al.* 1995) are listed as nationally threatened and three invertebrate species are considered to be threatened (Driscoll 1994). A number of other plant species are regarded as regionally uncommon (Benson 1994; Sharp 1994), including the Purple Peas *Swainsona sericea, S. monticola*, Yam Daisy *Microseris lanceolata* and Emu Foot *Psoralea tenax*.

The Interim Biogeographic Regionalization for Australia (Australian Nature Conservation Agency (ANCA) 1995) identifies two bioregions that cover the Australian Capital Territory: the Australian Alps, covering the southwestern third of the Australian Capital Territory, and the Southeastern Highlands. Although significant parts of the Australian Capital Territory are already included in conservation reserves, natural temperate grasslands remain as the conspicuous exception, reflecting an equally poor situation within the whole Southeastern Highlands bioregion.

The urgent need to survey and document the remaining grasslands in the Australian Capital Territory and to address their conservation requirements was identified in a proposal for a *Recovery Plan for Lowland Native Grasslands in the Australian Capital Territory* prepared by the Wildlife Research Unit (WRU) of the ACT Parks and Conservation Service in 1991. In 1992 the Plan was approved for funding for five years with a budget of $450,000 under the Commonwealth's Endangered Species Program administered by the ANCA, with additional resources provided by the ACT Parks and Conservation Service.

Distribution and characteristics of natural temperate grasslands in the Australian Capital Territory

Natural temperate grasslands are found in northern Australian Capital Territory, generally below 600 m altitude in valleys or lower slopes with less than 600 mm rainfall. It is thought that these areas were maintained as open non-wooded communities through a combined effect of cold air drainage, cracking clay soils, low nutrient content of the soils and regular burning by Aborigines (Chan 1980; Groves and Lodder 1991). Naturally treeless grasslands merge with grassy woodlands on slopes, and secondary grasslands occur where the trees have been cleared. Grasslands also incorporate some significant wetland areas, such as the Ginninderra and Horse Park wetlands (National Capital Development Commission 1988).

The Australian Capital Territory grasslands contain five distinct vegetation associations which occur as a result both of intrinsic site factors and land use since European settlement (Benson 1994; Sharp a, in prep.). In particular, drainage patterns related to slope and landform, soil characteristics and intensity of past and current agricultural practices appear to influence which associations occur at a site. These factors influence the species present, their life and growth forms and the amount of invasion by exotic plants, and indicate probable responses to management practices (Sharp a, in prep.). The associations may occur in isolation in smaller sites or may form a continuum in larger sites. Native species diversity varies considerably within the associations, as a result of varying intensities of management.

A comparison of the extent of grasslands prior to European settlement (Pryor 1938) with the present distribution (Fig. 1), indicates a severe decline in the distribution and fragmentation of grasslands in the Australian Capital Territory. There are also other areas of rough native pasture, containing some native grasses with few native forbs, and generally dominated by exotic species.

In the 1970s, 40 remnants were mapped by Chan (1980) (although it is known that this was an incomplete record of the total natural temperate grassland distribution). Over the past 20 years, 14 of these sites have been lost and a further 14 sites have been partially destroyed. These sites have been either resumed for urban development, been replaced by exotic pasture or become severely degraded through invasion by exotic species (WRU, unpubl. data). If this documented change is indicative of the loss of grasslands over the last two decades, 70% of the Australian Capital Territory's natural temperate grassland sites have been either totally or partially lost. Since 1991, thorough surveying has revealed a total of 50 sites remaining, ranging in size from less than 1 ha to several sites greater than 100 ha. Current land uses include open space, special purpose sites (e.g., defence and telecommunication facilities that require extensive buffer zones), rural lease and roadsides (Sharp 1995).

There has also been a marked decline in the vertebrate fauna associated with grasslands (Osborne *et al.* 1995). This decline is related to the land use since European settlement, and includes both loss of species (such as the Australian Bustard *Ardeotus australis* and the Plains Wanderer *Pedionomus torquatus*) and decline in abundance of others (including the Striped Legless Lizard *Delma impar* and the Eastern Lined Earless Dragon *Tympanocryptis lineata pinguicolla*) (Osborne *et al.* 1995).

OUTLINE OF THE RECOVERY PLAN

The *Recovery Plan for Lowland Native Grasslands in the Australian Capital Territory* (WRU 1992), funded by ANCA in 1992 for the period to 1997, identified the actions required to reduce the threat to natural temperate grassland communities and component species in the Australian Capital Territory, in the context of national conservation priorities for natural temperate grasslands. The recovery plan is being implemented by the Wildlife Research Unit of the ACT Parks and Conservation Service assisted by a recovery

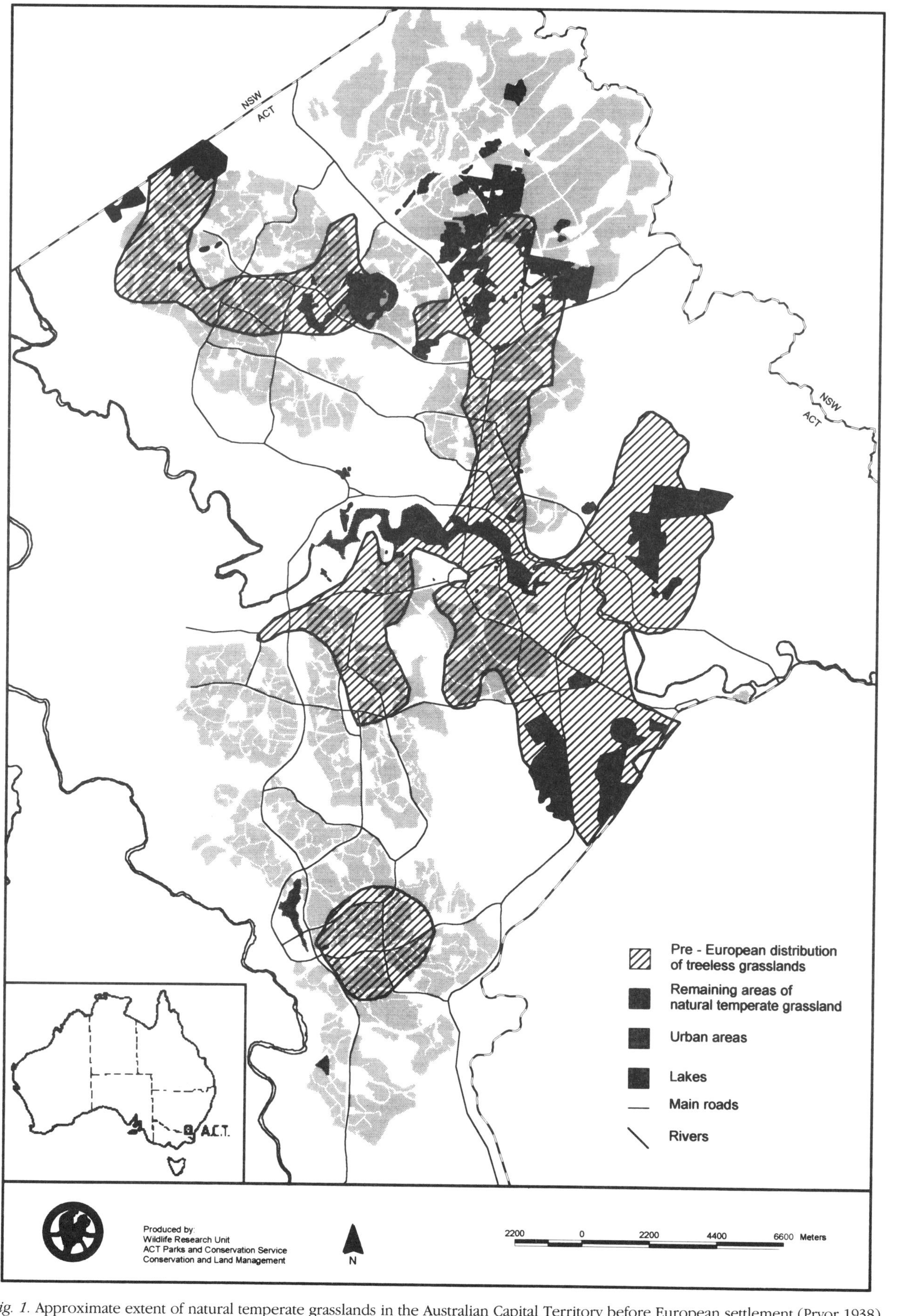

Fig. 1. Approximate extent of natural temperate grasslands in the Australian Capital Territory before European settlement (Pryor 1938) and current distribution.

team, with members drawn from ACT Parks and Conservation Service, ANCA, NSW National Parks and Wildlife Service, University of Canberra, CSIRO, Australian Capital Territory Planning Authority and the National Capital Authority. A Grasslands Project Officer (funded by both ANCA and ACT Parks and Conservation Service) is employed to carry out actions listed in the recovery plan and to administer and assist with other associated projects being implemented through research grants to other organizations and individual research scientists.

The actions were necessarily broad generalizations for studies on species and the community, rather than specific actions for conservation and management. This was because of the minimal level of knowledge in 1992 about the Australian Capital Territory grassland community and its component species, and the extent of studies that would be required to cover all aspects of recovery of the community. Three categories of action were identified: those actions to be carried out by the Grasslands Project Officer (50% of budget); those actions related to threatened species survey and ecological studies (40% of budget); and some lower priority actions which would be carried out subject to receipt of project proposals (10%). A full list of projects funded under the second and third categories and the experimental research carried out by the Grasslands Project Officer is in Appendix 1.

A broad spectrum of researchers met in 1994 to review the outcomes of research and actions at that time, and to recommend priorities for research and survey during the final two years of funding. The 1994 review also provided a renewed focus on any gaps or deficiencies in the actions outlined in the recovery plan. All priority actions set out in the recovery plan have been addressed to the extent that the projects offered and the limited funds available have allowed.

Summary of actions of the recovery plan research and implementation phase

The key actions of the recovery plan that have been carried out are as follows:

1. Quantitative surveys and mapping of remnant grassland sites in the Australian Capital Territory (1991 to present)

The distribution and boundaries of remaining areas of grasslands in the Australian Capital Territory have been documented and quantitatively surveyed using all available information, including past flora and fauna surveys (e.g., Chan 1980; Benson 1994), air photo interpretation and extensive communications with scientists and field naturalists. Additionally, at sites with suitable habitat, surveys of the significant invertebrates and vertebrates and uncommon and threatened plant species have been, or are being, undertaken. Survey information is being analysed and the data has been entered into a geographic information system (GIS) for use by the Australian Capital Territory Government and others.

2. Research and management of Australian Capital Territory grasslands

Patterns and dynamics in natural temperate grasslands

The Grasslands Project Officer has undertaken an ecological study of the grassland community using pattern analysis to identify floristic associations and assess which attributes, including soil and site variables, invertebrate abundance and diversity, are related to these associations (Sharp a,b, in prep.). Hypotheses have been developed about the dynamics of the floristic associations and how they respond to gap replacement. Results indicate a response by both native and exotic species to gap creation, in particular by the forbs (Sharp c, in prep.). These studies, together with those of others which address floristic associations, in particular Benson (1994), provide a basis for developing management strategies.

The impacts of burning on plants in Australian Capital Territory grasslands

A study is under way in mown grasslands which aims to determine how grasslands respond to seasonal (spring and autumn) burning (again on the basis of the floristic associations). Changes in floristics after seasonal burning are also being assessed in secondary grassland and grassy woodland sites along a railway.

Seed viability of plants *in situ*

Eight native and two exotic grassland plants were used to study seed viability over a two year period in soil under field conditions. Results indicate that there is a widely varying response, with some species (for instance the Yam Daisy *Microseris lanceolata* and the exotic grass *Holcus lanatus*) germinating within 12 months, with few or no viable seeds surviving after two years. Other species retained high viability after two years, including the exotic grass *Briza maxima* (Lunt 1996).

Control of exotic weeds using herbicides

Studies of the tolerance of native grasses to herbicides and methods of direct application of selective and non-selective herbicides have been undertaken (Nazer and Carder 1996). This study suggests some herbicides which may have lower impact on native grasses. The effects of these herbicides on native forbs are not known.

National workshop and public seminar, 1993

A national workshop and public seminar on management of relict lowland grasslands was jointly funded by the recovery plan, ACT Parks and Conservation Service and University of Canberra, and the proceedings have been published (Sharp and Rehwinkel 1995).

A management plan for temperate grasslands

A management plan was prepared for ANCA (WRU 1994). It incorporates requirements for both protection and management and provides recommendations for each known grassland site in the Australian Capital Territory. The management plan will provide the basis for the development of an action plan for natural temperate grassland conservation (see p. 144). The action plan will address long-term and strategic management, planning and protection of the native grasslands in the Australian Capital Territory.

3. Long-term monitoring in selected grassland sites

A study has been designed to measure seasonal and long-term dynamic changes in grasslands over a 20-year period and complements other management studies. A range of sites, including sites which have been recently burnt, have been selected to reflect different land uses, management practices and floristics. Changes in floristic cover and frequency, and structure in the grasslands are being monitored in permanent quadrats.

4. Database compilation and maintenance

All site and plant data are being entered on to a computer database to facilitate easy access to and analysis of data. A database has been developed for the Southern Tablelands region using interrelated spreadsheets covering literature references, studies, sites, species and management regimes (Wellington and Larwill 1996). This is being used to identify the major gaps in information and to assist in the development of a strategic approach to further work on grassland distribution and species diversity in the region.

5. Liaison with other organizations and individuals

Considerable effort and time is taken to liaise with planners, land holders, the general public and community groups to increase awareness of the conservation values of grasslands and to provide advice on the protection and management of these sites. Liaison with other grassland researchers is also maintained, through attendance at workshops, conferences, and interstate representation of the Australian Capital Territory on grassland conservation matters. These links have been particularly important in forming an appropriate regional perspective for actions taken in the Australian Capital Territory under the recovery plan and in linking with other activities. The emergence of a local Friends of Grasslands community group has been a particular focus for liaison and mutual assistance.

Other studies relevant to the recovery of grasslands and component species in the Australian Capital Territory have also been carried out at universities in Canberra, Hobart and Melbourne and at CSIRO, in a separate process to the recovery plan, through funding provided by several organizations. Close links are retained with those organizations to ensure a co-ordinated approach.

6. Threatened species surveys and ecological research

For the purposes of the recovery plan, the recovery team decided that grassland species considered by Sharp (1994) and Driscoll (1994) as being nationally threatened, whether officially listed or not, would be considered for inclusion for priority research.

APRASIA PARAPULCHELLA PINK-TAILED LEGLESS LIZARD

A recovery plan prepared by Osborne and Jones (1995) sets out research requirements and actions required for this species which is listed as nationally endangered. Research on the ecology of the species is being undertaken at the University of Canberra.

DELMA IMPAR STRIPED LEGLESS LIZARD

Extensive surveys in over 150 locations in the Australian Capital Territory (Gungahlin, Yarramundi, Majura Valley and Jerrabomberra Valley) have been carried out over a five year period by the WRU (Rauhala *et al.* 1995). This was partially funded in 1993 through the implementation of the recovery plan.

The recovery plan has also funded studies on the ecology, genetics and physiology of *D. impar* (Osmond 1994), its diet (Nunan 1995) and the role of habitat and historical processes in the distribution of the species (Dorrough 1995).

TYMPANOCRYPTIS LINEATA PINGUICOLLA EASTERN LINED EARLESS DRAGON

Surveys for this species, listed as nationally vulnerable, have been funded through the recovery plan (Smith 1994; Langston 1996), by the WRU and through other studies in the region (Mills 1994) and Osborne in Rowell (in prep.). Additionally, there have been genetic studies on the sub-species (Smith 1994) which have provided the basis for a new description of the taxon and a proposal for its elevation to species status (Smith and Osborne in prep.). Extensive examination of trapping methods has also been carried out (Smith 1994; Langston 1996).

KEYACRIS SCURRA A MORABINE GRASSHOPPER

A survey for the species has been conducted in the Australian Capital Territory (Rowell and Crawford 1995). A study of population density in relation to vegetation composition and past management is currently being undertaken.

SYNEMON PLANA GOLDEN SUN MOTH

A survey of potential habitat in the Australian Capital Territory carried out in 1993/94 (Edwards 1994) added to the known distribution of the Golden Sun Moth, bringing the total number of known sites in and near the Australian Capital Territory to 16. However, eight of these sites are considered to be of low conservation value for the species (Edwards, pers. comm.).

SWAINSONA RECTA SMALL PURPLE PEA

A study of the impacts of seasonal burning on the germination, establishment and seeding of *S. recta* plants is being undertaken. This was partially funded through the implementation of the recovery plan.

RUTIDOSIS LEPTORRHYNCHOIDES BUTTON WRINKLEWORT

In the survey for uncommon and declining plants in the grasslands (Crawford and Rowell 1996) a new extensive population of *R. leptorrhynchoides* has been discovered in a small area (less than 1 ha) in the

Majura Field Firing Range. An additional population of about 30 plants has been found at the Campbell Park Offices, also in the Majura Valley (Sharp, unpubl. data).

7. *Other research funded through the recovery plan*

Studies on the potential use of selected invertebrate taxa (using collembola, ants and beetles) as indicators for other invertebrate groups have been undertaken (Melbourne 1993; Greenslade 1994).

Funds have also been allocated to the preparation of a poster on grassland invertebrates, assisting the establishment of the Friends of Grasslands community group, and towards preparation of a field guide to temperate grasslands of the Southern Tablelands.

COMPLEMENTARY ACTIONS BY AUSTRALIAN CAPITAL TERRITORY GOVERNMENT AND OTHERS

The Grasslands Project Officer and other ACT Parks and Conservation Service staff have increasingly been required to engage in a wide range of activities not originally conceived as part of the recovery program, as survey, research and management work progressed and a range of Australian Capital Territory agencies and community groups became more aware of the developing information base and availability of advice. This has been an important, though time consuming, development, greatly assisting in the overall effort directed at the conservation of grasslands. However, it places great strains on the resources available and can result in unsatisfied expectations and, potentially, can divert resources from the intended recovery activities.

The major additional activities are:

1. Involvement of the community in grassland conservation

Grasslands Management Group

This group contains representatives from both community and government organizations that are practically involved with native grassland management in the Australian Capital Territory. The Management Group includes on-the-ground managers, from Australian Capital Territory and New South Wales Government agencies, the rural community and community groups. It acts as a network, exchanging information and linking community interest and concerns directly into the recovery process.

Friends of Grasslands and other community groups

A number of community organizations are developing an interest in the care and protection of natural temperate grasslands. The newest group, "Friends of Grasslands" has joined the longer established Parkcare and Landcare groups, the Conservation Council of the South-East Region and Canberra and the National Parks Association in providing a focus for community interest and advocacy for conservation. Community groups are able to focus on the Australian Capital Territory and surrounding New South Wales region, to play a political role in the community not appropriate to a government agency, to provide information and education to a wider group of people, and to generate a greater sense of local ownership of the region's biodiversity.

The ACT Parks and Conservation Service and the Grasslands Project Officer assist groups where possible, by providing advice and, at times, material support to on the ground land management activities such as weed control, trail construction and vegetation rehabilitation. The ACT Parks and Conservation Service is also responding to increasing numbers of general inquiries on grasslands from members of the public.

2. Provision of advice to government and its agencies

Surveys and ecological studies of reptiles of grasslands

Three threatened species of lizard, *Delma impar* Striped Legless Lizard, *Tympanocryptis lineata pinguicolla* Eastern Lined Earless Dragon and *Aprasia parapulchella* Pink-tailed Legless Lizard have been the focus of a sustained survey effort by the ACT Parks and Conservation Service. These species and others, including some invertebrates, have at times been regarded as "flagships" for the conservation of grasslands, while information on grassland communities and their conservation requirements is being gathered. While protection measures may be focused on particular threatened species, it is essential that an ecological community and habitat approach underpins conservation management.

Assessment of planning and development proposals for sites with potential grassland conservation interest

Information on the distribution and conservation values of native grasslands and their component species is provided regularly to other government agencies for use in planning, development and management of land in the Australian Capital Territory.

State of the environment reporting

Information on temperate grasslands has been made available to the Australian Capital Territory Commissioner for the Environment for use in preparing the first two State of the Environment Reports in 1994 and 1995. The 1995 Report included a special section on grasslands and the Commissioner has recommended that legislation requiring written approval for the clearance of or damage to grasslands on any land assessed on the basis of scientific principles be enacted to prevent the loss of further sites, and that planning laws be amended to specifically prohibit development on grasslands (ACT Commissioner for the Environment).

Production of GIS maps of grasslands and species compatible with other GIS information

All available data on the distribution of grasslands, including associations, dominant species and conservation ratings and the distribution of species have been entered into geographic information systems for access by relevant government organizations. This will be utilized in management planning and consideration of land use issues.

3. Flora and Fauna Committee and legislation for threatened species and communities

Coincidentally with the recovery plan work, the Australian Capital Territory Government moved to establish a legislative framework for identifying and protecting species and ecological communities threatened with extinction. This initiative has provided a timely mechanism through which grasslands and their component species may receive added recognition followed by statutory protection and conservation management.

The primary legislation in the Australian Capital Territory for protection of the natural environment, the *Nature Conservation Act 1980,* was amended in 1994 to establish the Australian Capital Territory Flora and Fauna Committee. The Committee is an independent body comprising seven experts in biodiversity and/or ecology and is required to identify native species and ecological communities that are threatened with extinction, and processes that threaten the survival of native species and communities in the Australian Capital Territory region.

The process by which items are declared is one of nomination, assessment by the Committee, and formal recognition by the Minister. Declaration initiates a management response by the Conservator of Flora and Fauna, who is required to prepare an action plan for each item. The action plan will include consideration of conservation issues and proposed measures for the conservation of the declared species or community, or management of the threatening process.

Natural temperate grassland has been recently listed as an endangered ecological community and *Tympanocryptis lineata pinguicolla, Synemon plana, Swainsona recta* and *Rutidosis leptorrhynchoides* have been listed as endangered species and *Delma impar* has been listed as a vulnerable species. Consequently, action plans are required to be prepared. These plans will draw heavily on the results of research and survey programs initiated under the recovery plan. The ACT Parks and Conservation Service will co-ordinate implementation of such plans including, where required, off-reserve conservation measures.

AUSTRALIAN CAPITAL TERRITORY GRASSLANDS IN A BIOREGIONAL PLANNING CONTEXT

The Australian Capital Territory lies within the Southern Tablelands region of New South Wales, bordering the Australian Alps region and encompassing part of the Murrumbidgee River. The Australian Capital Territory and sub-region is the largest, and until recently, the fastest growing inland region in Australia in terms of population. The division of responsibility between State, Territory and national governments has not, in the past, assisted in the development of a co-ordinated approach to natural resource planning and management. More recently, however, greater consultation and co-ordinated activities have set agencies on a new path which should lead to improved knowledge of the region and planning for conservation.

Interest in a regional approach to planning for the sub-region and identifying the significant natural resources and conservation needs of the area is being shown by local government councils, private development interests and governments. The four spheres of government (Federal, State, Territory and local) have come together to establish a strategy for managing future regional development (Anon. 1995). Private enterprise organizations have also expressed an interest in this approach because of uncertainty surrounding development opportunities, which is generated by a lack of information about the regional or national significance of the region's ecological resources.

A focus of concern is the remaining areas of native vegetation, particularly natural temperate grasslands and woodlands, and plants and animals that are now considered to be threatened with extinction. The lack of knowledge of the distribution and relative conservation status of these species and ecological communities can be a significant impediment to the planning of urban and other development in Canberra and Queanbeyan. This situation has the potential to arise in other parts of the region as local councils deal with development proposals.

CASE STUDIES

The following case studies are representative of the ways in which information from the research and survey phase of the recovery plan is being used to protect and bring remnant natural temperate grasslands and their component species under conservation management.

1. Gungahlin Town Centre

The Striped Legless Lizard *Delma impar* is recognized under Commonwealth, New South Wales, Victorian and Australian Capital Territory legislation as being vulnerable to extinction. The species is largely restricted to areas of natural temperate grasslands, but in the Australian Capital Territory it has also been located in lowland grasslands with a high content of exotic species.

One of the more significant populations of the species is found in the Gungahlin area of the Australian Capital Territory (Australian Capital Territory Government 1994), where a new town is being developed to accommodate some 100 000 residents. Consideration by the Australian Capital Territory Legislative Assembly of a draft plan for the Gungahlin Town Centre ceased in December 1994, pending the results of additional surveys for the lizard.

In 1995 the Australian Capital Territory Planning Authority co-ordinated a comprehensive review of planning policies for the Town Centre and some adjacent areas of Gungahlin. An independent study also developed principles to guide the development of a conservation strategy for natural temperate grassland and component species in Gungahlin, and to advise on the conservation significance of the grasslands in Gungahlin (Williams *et al.* 1995).

This study concluded that areas of native grassland in Gungahlin, and other parts of the Australian Capital Territory, together with extensive populations of *Delma impar* and threatened invertebrates are of major regional significance, representing in some cases, communities unique to the Canberra-Queanbeyan region, and all of which are poorly protected.

The independent report recommended protection for three areas for the conservation of both native grasslands and *Delma impar*. These areas comprise about 170 ha of natural temperate grassland and about 330 ha of associated adjacent areas of lesser grassland conservation value, but which encompass the main areas where *Delma impar* is found in the Australian Capital Territory (Williams *et al.* 1995).

From the town planning perspective these recommendations were to have a major effect on the location and form of the Gungahlin Town Centre, residential development and provision of the infrastructure required to service Gungahlin. The consultants undertaking the review adopted and held to the view that *"rather than regarding the nominated conservation areas as problematic for the urban plan... the presence of the endangered species... and their consequent conservation reserves (are) an important and positive element of the recognition of the unique character of the Gungahlin area from which the entire urban plan can take its form"*.

The preliminary assessment of the new planning proposals, required under the *Land (Planning and Environment) Act 1991* found that while the establishment of three conservation areas for native grasslands and *Delma impar* results in substantial loss of residential and Town Centre land, the revenue foregone as a result is substantially offset by savings in reduced provision of infrastructure (Australian Capital Territory Planning Authority 1995). The revised plan provides for an integrated system of urban recreational spaces, cycle paths and landscaped waterways which will provide a range of social and environmental benefits.

In December 1995 the Australian Capital Territory Legislative Assembly accepted the new variation to the Territory Plan. Community reaction to the revised plan is positive, and the Australian Capital Territory Government announced, as part of its 1995–96 budget, the establishment of reserves to protect natural temperate grasslands and the Striped Legless Lizard in Gungahlin, and support for research into the animal's biology and habitat needs.

2. Office development site

The site for a proposed office development adjacent to possibly one of the most culturally significant avenues in Canberra's central national area, Anzac Parade, includes a remnant native grassland which is assessed as having high conservation value. The small, slightly elevated area is dominated by *Danthonia* spp., and *Stipa* spp., and contains only a small number of exotic species. Several locally uncommon plants have been found in this site. Previous surveys have recorded the endangered Golden Sun Moth *Synemon plana* as present on the site.

Early negotiations with Australian Capital Territory Government agencies and the consulting civil and structural engineers have identified part of the site which can be left undeveloped while not diminishing the overall development potential of the site. Assisted by the topography of the site, the remnant grassland can readily be protected from construction activities and site run-off. It will be necessary to ensure that future office workers are made aware of the significance of the site and the need to avoid uncontrolled or inappropriate use and that management of the area reflects the conservation values of the site. Long-term monitoring on this site will assist in determining the impacts on the site.

3. Landscape enhancement through tree planting

Forward tree planting programs associated with new residential developments have, historically in Canberra, been largely driven by the need to revegetate previously cleared agricultural land and to provide a treed landscape. On the western fringe of Belconnen, the suburb of Dunlop has fragmented a native *Themeda triandra* grassland which had not been subjected to significant pasture improvement or land clearing. Early proposals for tree-planting, however, discounted the conservation value of the remaining native cover and proposed extensive areas of tree plantings for landscaping and to act as windbreaks.

A survey of the area has now identified the location and extent of the more significant remnant native grassland patches in the area and the degree to which other parts are degraded and require rehabilitation through tree planting. Collaboration in the field between conservation staff and the landscape designers responsible for forward tree planting programs has resulted in a planting proposal acceptable to all parties, and one expected to create a visually appealing peri-urban landscape. Small groupings of trees (up to 20 per group) will be planted across the site, but only in those areas identified in the field survey as degraded or along the fire trail to be constructed on the urban edge. Existing trees and significant areas of the *Themeda* grassland will be retained in the proposed landscape setting which represents a positive step forward in the management of Canberra's urban edge.

4. Conservation grazing

A strategy being developed by the ACT Parks and Conservation Service will see limited, controlled grazing of some grassland conservation sites through short-term agreements with farmers wishing to participate. It is generally accepted that in order to retain diversity in native grasslands, there needs to be control of biomass, produced predominantly by the native grasses and weeds, so that herbaceous species can germinate and reproduce. Where grasslands have been, or currently are being, grazed it may be appropriate to continue a grazing regime, although for conservation objectives rather than for rural production purposes. Some species and species associations are more common, or only found, in areas which are grazed, and may require continued grazing if they are to be maintained *in situ*. In other cases, grazing may be the best way known to control weeds, or to retain a level of biomass control where alternatives such as burning or mowing may not be appropriate. However, to graze areas which have not been grazed for a long period of time is not considered an option.

Currently it is not known if it is preferable to maintain a continuous, light grazing regime or to graze intermittently, either lightly or heavily (as in cell or pulse grazing), avoiding peak flowering and seeding times for native plants or avoiding breeding seasons for particular animal species. It may be that particular sites require different regimes, on the basis of the different species mix present, type of exotic infestation, drainage or other factors. The conservation grazing program will be monitored to determine if the conservation objectives for particular areas are being met.

CONCLUSIONS: RECOVERY OF GRASSLANDS IN THE AUSTRALIAN CAPITAL TERRITORY

The foundation for long-term recovery of natural temperate grasslands in the Australian Capital Territory is being laid through a process which includes integrating research results, developing and implementing a protection strategy and new legislation, and the involvement and co-operation of both the government and community. Implementation of the actions outlined in the recovery plan has been largely responsible for a very significant increase in understanding by the general public, government agencies and politicians of their responsibility to protect the remaining grasslands and component species. There has also been increased understanding of the ecological requirements of the community and its component species. The funding provided by ANCA has provided essential seed money which has been supplemented by funds from other government and non-government organizations. Not surprisingly the resulting total effort is much more than the sum of its parts, with corresponding results and synergies.

Information from the Grasslands Recovery Plan has played a pivotal role in the outcome of the case studies. The Australian Capital Territory-wide grassland surveys have identified all grassland sites and assessed those which have significant conservation value and warrant protection. This information became available at a critical time in the development of plans for the new Gungahlin Town Centre. Research results are being applied immediately to the new areas where control of weeds and management of grazing are two issues urgently requiring a sound basis for ongoing management. New research can be undertaken in these protected environments where long-term monitoring programs are established.

Quantitative ecological surveys of grasslands have increased our knowledge of the distribution of uncommon plant species and the ecology of several threatened plants and animals. The improved understanding of the ecological dynamics of the community is now reflected in recommendations for management of specific grassland sites.

A number of advantages arise from the integrated approach taken by the recovery team in managing the recovery plan. The full-time Grasslands Project Officer provides a focal point for Australian Capital Territory government agencies and the general community on a wide range of grassland issues; and because this position is partly funded by a Commonwealth agency, co-ordination between Commonwealth and Australian Capital Territory Governments is facilitated, particularly in respect to Commonwealth owned or controlled sites that have significant grassland values. The Grasslands Project Officer position has been a vital means of dealing with the complex issues and range of players that are associated with conservation of an endangered ecological community and associated threatened species within an urban and peri-urban context.

A number of government agencies and community groups have been involved in the strategic approach, through the recovery plan implementation, to the recovery of an ecological community and a range of species. This has enabled parties to develop proactive approaches to potentially destructive processes, such as urban development, weed invasion, site disturbance and to provide timely advice and guidance on how to mitigate these impacts and to manage sites for conservation. Increased understanding by the general public and agencies of the need to integrate protection of individual threatened species with that of their grassland habitat assists this approach.

The recovery plan provides a strategic framework for research and implementation of research results. The knowledge gained can also be extrapolated to other areas within the region and nationally. Future research requirements are being identified and these will focus primarily on the need to determine the dynamic responses of species and the community to management practices.

Lessons learned by the recovery team over the past few years indicate that a more integrated regional approach, taken from the beginning of the project, could have promoted a more consistent focus on the ecological boundaries of the community and species, rather than allowing political boundaries to have undue influence. This could have assisted with decision making by participating governmental agencies.

The original recovery plan also did not sufficiently integrate the research phase with the urgent need to implement protection and management measures. Management advice had to be given before research results became available. However, there is little that can be done about this, other than acknowledge that it is inevitable, and develop a strategy for dealing with it.

While the recovery program has enabled much relevant and highly important research to be undertaken, insufficient attention was paid to the importance of ensuring that reports were peer reviewed and published, thus making them fully available and accepted scientific literature. This has been left up to the individual researchers to complete, and in many cases, this has not been done. Consultancies for research should include a requirement for reports to be of a standard that can, where relevant, be published immediately, and the recovery team should assist with ensuring that this is done.

In reviewing the final stage of the research phase of the recovery plan, the recovery team has identified as its top priority the completion of the following specific studies:

- continued negotiations towards protection of sites and implementation of management for conservation;
- continued surveys in the region for grassland communities;
- completion of surveys in both the Australian Capital Territory and the Southern Tablelands region for uncommon and threatened species, including invertebrates, and assessment of their status;
- continued research to assist management of communities/associations and species; and
- encouragement of studies relating to the enhancement of habitat, including rehabilitation and weed control.

Application of the results from research projects sponsored by the recovery plan and long-term monitoring requires a commitment by the Australian Capital Territory and Commonwealth Government agencies and lessees responsible for the grassland sites and threatened species found in them. Additionally, all areas requiring special protection need to be identified and an attempt made to implement this protection while resources are available under the recovery plan.

Finally, it is important to ensure that on-going programs and processes are in place that are not reliant solely on ANCA funding. This will be assisted by developing a long-term conservation strategy, which includes the identification of Commonwealth and Territory agency responsibilities. Liaison within and between lessees, the community and government agencies is vital for recovery. The strategy will include education programs, and involvement of both government agencies and community groups in research and implementation of conservation management and protection.

Management of both the native grasslands and their component threatened species in the Australian Capital Territory is setting a new challenge for Australian Capital Territory and Commonwealth Government agencies as development of the urban area progresses and land stocks are reduced. Research initiated by the Grasslands Recovery Plan will be continued in the grassland areas and the results applied to their management.

ACKNOWLEDGEMENTS

The authors would like to thank the Australian Nature Conservation Agency (Endangered Species Program) and the ACT Parks and Conservation Service for their funding and support for the native grassland recovery program. The recovery team has provided advice, support and encouragement throughout the implementation of the recovery plan.

Mark Dunford, Wildlife Research Unit, ACT Parks and Conservation Service, prepared Figure 1 and assisted with field data collection. Thanks are also due to other colleagues at the Wildlife Research Unit who have undertaken site surveys, mapped sites, carried out floral and faunal surveys and research relevant to the recovery plan. Recommendations for protection and management of sites is a result of combined input from colleagues at the Unit and researchers funded under the recovery plan. Thanks are due also to the referees for comments on the draft manuscript.

REFERENCES

ACT Commissioner for the Environment, 1995. Australian Capital Territory State of the Environment Report, 1995. Office of the Commissioner for the Environment: Canberra.

Anon., 1995. Australian Capital Territory and Sub-region Planning Strategy, draft for community comment.

Australian Capital Territory Government, 1994. Draft conservation strategy for the Striped Legless Lizard *Delma impar* in the Australian Capital Territory. Prepared by a working party convened by the Minister for the Environment, Land and Planning October 1994.

Australian Capital Territory Planning Authority, 1995. Preliminary assessment Gungahlin Town Centre and Central Area. Preliminary assessment for public notification.

Australian Nature Conservation Agency, 1995. An interim biogeographic regionalization for Australia: a framework for setting priorities in the national reserves system co-operative programme ed by R. Thackway and I. D. Creswell, version 4.0.

Benson, J. S., 1994. The native grasslands of the Monaro region: Southern Tablelands of New South Wales. *Cunninghamia* 3(3): 609–50.

Chan, C., 1980. Natural grasslands in Canberra: their distribution, phenology and effects of mowing. MSc Thesis, Australian National University.

Crawford, I. and Rowell, A. M., 1996. Uncommon or declining vascular plants of lowland native grasslands of the Australian Capital Territory. Unpublished report to the ACT Parks and Conservation Service, June 1996.

Dorrough, J., 1995. The distribution of *Delma impar* in lowland grasslands of the Australan Capital Territory: the role of habitat and historical processes. Unpublished report to the ACT Parks and Conservation Service, August 1995.

Driscoll, D., 1994. Invertebrates of lowland native grasslands in the Australian Capital Territory: conservation and research strategies for a Recovery Plan. Technical Report 9, ACT Parks and Conservation Service.

Edwards, T., 1994. Survey of lowland grassland sites in Australian Capital Territory for the Golden Sun Moth, *Synemon plana*. Unpublished report to the ACT Parks and Conservation Service, October 1994.

Greenslade, P., 1994. A study of the use of indicator groups in assessing the conservation value of native grassland sites in the Australian Capital Territory and New South Wales. Unpublished report to the ACT Parks and Conservation Service, May 1994.

Groves, R. H. and Lodder, M., 1991. Flora of the Australian Capital Territory's natural grasslands. *In* The Australian Capital Territory's Native Grasslands ed by R. Falconer. Proceedings of a workshop held by the Conservation Council of the South-East Region and Canberra.

Groves, R. H. and Williams, O. B., 1981. Natural Grasslands. Pp. 293–316 *in* Australian Vegetation ed by R. H. Groves. Cambridge University Press: Cambridge.

Langston, A. C., 1996. The ecology and distribution of *Tympanocryptis lineata pinguicolla* (Southern Lined Earless Dragon) in the Australian Capital Territory and adjacent sub-region. Unpublished report to the ACT Parks and Conservation Service.

Lunt, I. D., 1996. Germination responses and longevity in the field of ten grassland herbs. Unpublished report to the ACT Parks and Conservation Service.

McDougall, K. and Kirkpatrick, J. B., 1994. Synthesis and recommendations. Pp. 158–61 *in* Conservation of lowland native grasslands in southeastern Australia ed by K. McDougall and J. B. Kirkpatrick. World Wide Fund for Nature: Australia.

Melbourne, B., 1993. The assessment of insect biodiversity. A case study: ants and carabid beetles in grasslands of the Australian Capital Territory. Unpublished report to the ACT Parks and Conservation Service, December 1993.

Mills, K. and Associates, 1994. Fauna survey and assessment "The Poplars, Queanbeyan, New South Wales". Unpublished report to Queanbeyan City Council.

Moore, C. W. E., 1964. Distribution of grasslands. *In* Grasses and Grasslands ed by C. Barnard. Macmillan: London.

National Capital Development Commission, 1988. Sites of significance in the Australian Capital Territory. Technical Paper 56.

Nazer, C. and Carder, J., 1996. Directed application of a range of selective and non-selective herbicides. Unpublished report to the ACT Parks and Conservation Service, June 1996.

Nunan, D., 1995. *Delma impar:* general insectivore or dietary specialist? Unpublished report to the ACT Parks and Conservation Service, July 1995.

Osborne, W. and Jones, S., 1995. Recovery Plan for the Pink-tailed Worm Lizard *(Aprasia parapulchella).* Technical Report 10, ACT Parks and Conservation Service.

Osborne, W., in prep. Assessment of reptile habitat on the Monaro grasslands. Appendix. *In* A study of the native grasslands of the Monaro district. A. Rowell. Report to the NSW National Parks and Wildlife Service and New South Wales Department of Planning (Heritage Assessment Program).

Osborne, W., Kukolic, K. and Jones S., 1995. Management of threatened vertebrates in native grasslands: a case of grasping at straws? Pp. 89–96 *in* Management of relict lowland grasslands. Proceedings of a workshop and public seminar, September 24 and 25, 1993. Conservation Series No. 8 ed by S. Sharp and R. Rehwinkel. ACT Parks and Conservation Service: Canberra.

Osmond, H., 1994. *Delma impar*, the Striped Legless Lizard, genetics, physiological limits and activity patterns. Unpublished report to the ACT Parks and Conservation Service, December 1994.

Pryor, L., 1938 Vegetation map of the Australian Capital Territory. Dept of the Interior: Canberra.

Rauhala, M. A., Shorthouse, D. J. and Ingwersen, F., 1995. The Striped Legless Lizard *Delma impar* in the Gungahlin, Majura and Jerrabomberra Valleys. Internal Report 95/2 ACT Parks and Conservation Service.

Rowell, A. and Crawford, I., 1995. Survey of the Morabine Grasshopper *Keyacris scurra* (REHN) in the Australian Capital Territory. Unpublished report to ACT Parks and Conservation Service, June 1995.

Sharp, S. and Rehwinkel, R. (eds), 1995. Management of relict lowland grasslands. Proceedings of a workshop and public seminar, September 24 and 25, 1993. Conservation Series No. 8, ACT Parks and Conservation Service.

Sharp, S. B., 1994. Lowland native grasslands in the Australian Capital Territory and surrounding region: a review and research strategy for a Recovery Plan. Technical Report 8, ACT Parks and Conservation Service: Canberra.

Sharp, S. B., 1995. Distribution of grasslands and floral diversity in natural temperate grasslands in the Australian Capital Territory. Pp. 80–88 *in* Management of relict lowland grasslands. Proceedings of a workshop and public seminar, September 24 and 25, 1993. Conservation Series No. 8, ed by S. Sharp S and R. Rehwinkel. ACT Parks and Conservation Service: Canberra.

Sharp, S. B., in prep a. Vegetation diversity and patterns in natural temperate grasslands in the Australian Capital Territory and sub-region.

Sharp, S. B., in prep. b. Invertebrate biodiversity and its relationship with floristic and other site attributes in natural temperate grasslands in the Australian Capital Territory sub-region.

Sharp, S. B., in prep. c. Dynamic response of natural temperate grassland in the Australian Capital Territory sub-region to gap replacement, litter removal and soil disturbance in two grassland communities.

Smith, W., 1994. The ecology and taxonomy of the Southern Lined Earless Dragon *Tympanocryptis lineata pinguicolla.* Unpublished report to the ACT Parks and Conservation Service, December 1994.

Smith, W. and Osborne, W., in prep. Taxonomic re-description of *Tympanocryptis lineata pinguicolla.*

Wellington, B. and Larwill, S., 1996. A survey of available information concerning the distribution and conservation values of native lowland grasslands in the Southern Tablelands of New South Wales and the Australian Capital Territory. Unpublished report to the ACT Parks and Conservation Service, January 1996.

Wildlife Research Unit, 1991. Recovery Plan for Lowland Native Grasslands in the Australian Capital Territory. ACT Parks and Conservation Service.

Wildlife Research Unit, 1992. Lowland native grassland ecosystems in the Australian Capital Territory. Progress Report to the Australian National Parks and Wildlife Service, Endangered Species Unit, November 1992.

Wildlife Research Unit, 1994. Management plan for Australian Capital Territory lowland native grasslands. Report to the Endangered Species Unit, Australian Nature Conservation Agency, March 1994.

Williams, D. G., Osborne, W., Georges, A. and Holloway, S., 1995. Principles and strategic options for the conservation of native grasslands and their threatened fauna in Gungahlin, Australian Capital Territory Applied Ecology Research Group, University of Canberra, Unpublished report to Australian Capital Territory Planning Authority.

FURTHER READING

National Capital Development Commission, 1984. The Ecological Resources of the Australian Capital Territory. Technical Paper 42.

APPENDIX 1

NATURAL TEMPERATE GRASSLANDS IN THE AUSTRALIAN CAPITAL TERRITORY IMPLEMENTATION OF THE RECOVERY PLAN

Reports to the ACT Parks and Conservation Service, Wildlife Research Unit.

Carder, J., Nazer, C. and Arney, M., 1996. Evaluation of the Weed-bug for its ability to selectively control tall growing weeds in native grasslands. Unpublished report by Parks and Public Places Asset Management, City Services to the ACT Parks and Conservation Service, January 1996.

Crawford, I. and Rowell, A. M., 1996. Uncommon or declining vascular plants of lowland native grasslands of the Australian Capital Territory. Unpublished report to the ACT Parks and Conservation Service, June 1996.

Dorrough J., 1995. The distribution of *Delma impar* in lowland grasslands of the Australian Capital Territory: the role of habitat and historical processes. Unpublished report to the ACT Parks and Conservation Service, August 1995.

Driscoll D., 1994. Invertebrates of lowland native grasslands in the Australian Capital Territory: Conservation and research strategies for a Recovery Plan. Technical Report 9, ACT Parks and Conservation Service.

Edwards T., 1994. Survey of lowland grassland sites in Australian Capital Territory for the Golden Sun Moth, *Synemon plana*. Unpublished report to the ACT Parks and Conservation Service, October 1994.

Greenslade P., 1994. A study of the use of indicator groups in assessing the conservation value of native grassland sites in the Australian Capital Territory and New South Wales. Unpublished report to the ACT Parks and Conservation Service, May 1994.

Langston, A. C., 1996. The ecology and distribution of *Tympanocryptis lineata pinguicolla* (Southern Lined Earless Dragon) in the Australian Capital Territory and adjacent sub-region. Unpublished report to the ACT Parks and Conservation Service, August 1996.

Lunt, I. D., 1996. Germination responses and longevity in the field of ten grassland herbs. Unpublished report to the ACT Parks and Conservation Service.

Melbourne, B., 1993. The assessment of insect biodiversity. A case study: ants and carabid beetles in grasslands of the Australian Capital Territory. Unpublished report to the ACT Parks and Conservation Service, December 1993.

Nazer, C. and Carder, J., 1996. Directed application of a range of selective and non-selective herbicides. Unpublished report to the ACT Parks and Conservation Service, June 1996.

Nunan D., 1995. *Delma impar*: general insectivore or dietary specialist? Unpublished report to the ACT Parks and Conservation Service, July 1995.

Osborne, W. and Jones, S., 1995. Recovery Plan for the Pink-tailed Worm Lizard *(Aprasia parapulchella)*. Technical Report 10, ACT Parks and Conservation Service.

Osmond, H., 1994. *Delma impar,* the Striped Legless Lizard, genetics, physiological limits and activity patterns. Unpublished report to the ACT Parks and Conservation Service, December 1994.

Rauhala, M. A., Shorthouse, D. J. and Ingwersen, F., 1995. The Striped Legless Lizard *Delma impar* in the Gungahlin, Majura and Jerrabomberra Valleys. Internal Report 95/2 ACT Parks and Conservation Service.

Rowell, A. and Crawford, I., 1995. Survey of the Morabine Grasshopper *Keyacris scurra* (REHN) in the Australian Capital Territory. Unpublished report to ACT Parks and Conservation Service, June 1995.

Sharp, S. B., 1994. Lowland native grasslands in the Australian Capital Territory and surrounding region: literature review and research strategy for a Recovery Plan. Technical Report 8, ACT Parks and Conservation Service.

Sharp, S. and Rehwinkel, R. (eds), 1995. Management of relict lowland grasslands. Proceedings of a workshop and public seminar, September 24 and 25, 1993. Conservation Series No. 8, ACT Parks and Conservation Service.

Smith, W., 1994. The ecology and taxonomy of the Southern Lined Earless Dragon *Tympanocryptis lineata pinguicolla*. Unpublished report to the ACT Parks and Conservation Service, December 1994.

Wellington, B. and Larwill, S., 1996. A survey of available information concerning the distribution and conservation values of native lowland grasslands in the Southern Tablelands of New South Wales and the Australian Capital Territory. Unpublished report to the ACT Parks and Conservation Service, January 1996.

Wildlife Research Unit, 1994. Management Plan for Australian Capital Territory lowland native grasslands. Unpublished report to the Endangered Species Unit, Australian Nature Conservation Agency, March 1994.

Projects in Progress

Effects of fire on the understorey composition of a temperate grassy woodland, Stage 1 and 2. John Briggs, NSW National Parks and Wildlife Service.

Effects of burning on the endangered plant, *Swainsona recta*, John Briggs, NSW National Parks and Wildlife Service.

Taxonomic re-description of *Tympanocryptis lineata pinguicolla*, Warwick Smith and Will Osborne, University of Canberra.

Alternative trapping methods for the striped legless lizard, *Delma impar*, Tony Corrigan, ACT Parks and Conservation Service.

Keyacris scurra in lowland native grasslands in the Australian Capital Territory: population density in relation to vegetation composition and past management, Alison Rowell and Isobel Crawford.

Vegetation diversity and patterns in natural temperate grasslands in the Australian Capital Territory and sub-region, Sarah Sharp.

Invertebrate biodiversity and its relationship with floristic and other site attributes in natural temperate grasslands in the Australian Capital Territory sub-region, Sarah Sharp.

Dynamic response of natural temperate grassland in the Australian Capital Territory sub-region to gap replacement, litter removal and soil disturbance in two grassland communities, Sarah Sharp.

A comparison of burning and mowing in two natural temperate grassland sub-communities in the Australian Capital Territory, Sarah Sharp.

Long-term monitoring of natural temperate grasslands in the Australian Capital Territory, Mark Dunford and Sarah Sharp.

Friends of Grasslands: printing and dissemination of a brochure on Friends of Grasslands and the preparation of a booklet on the conservation of grasslands.

Preparation of a poster on invertebrates in lowland native grasslands, CSIRO Division of Entomology.

Partial funding for a national field guide for plants in lowland native grasslands.

Copies of the reports are available at a cost of $5 ($15 for Sharp and Rehwinkel 1995). Contact: Sarah Sharp, Grasslands Project Officer, ACT Parks and Conservation Service, P.O. Box 1119, Tuggeranong, Australian Capital Territory, Australia 2901.

The Northern Hairy-nosed Wombat recovery program: trials and triumphs

Alan Horsup[1]

The Northern Hairy-nosed Wombat *Lasiorhinus krefftii*, Australia's rarest mammal species, survives in one small population of approximately 65 individuals on Epping Forest National Park in central Queensland. The species has been the subject of a major recovery program since 1991 which aims to establish at least one other geographically disjunct population within 10 years. There are many difficulties with the recovery program. *L. krefftii* is almost totally nocturnal and very difficult to trap. Wombat numbers have not increased in the last decade and the population is dominated by old individuals and skewed towards males. There has been a major drought in central Queensland during most of the 1990s. Epping Forest National Park is a remote, unstaffed park, which complicates the recovery program. On the plus side, *L. krefftii* is a long-lived and sedentary species whose current known range is totally protected on the one national park. Other aims of the recovery program are to improve our ability to accurately census the wombats and to reduce the stress caused by trapping. Both have been achieved through the development of a DNA fingerprinting technique which identifies and sexes individuals from hairs collected remotely at burrow entrances. Other projects include population viability analysis, a study of the feeding ecology of *L. krefftii* and pasture improvement trials on Epping Forest National Park. A study of the reproductive biology of the closely related Southern Hairy-nosed Wombat *L. latifrons* is developing artificial reproductive techniques. The establishment of a captive population is now underway so that these reproductive techniques can be applied to *L. krefftii*. This process will also meet the major objective of the recovery plan.

INTRODUCTION

THE Northern Hairy-nosed Wombat *Lasiorhinus krefftii* is Australia's rarest mammal species. It is listed as "Endangered" (ANZECC 1991; *Nature Conservation Act, Queensland* 1992) and is restricted to a single population of approximately 65 individuals on Epping Forest National Park in central Queensland. The history and estimated distribution of the species at the time of European arrival in Australia are summarized in Horsup (in press).

In the last two decades there has been a series of research projects aimed at better understanding the biology of *L. krefftii* and its habitat (Gordon *et al.* 1985; Johnson 1991a,b; Johnson and Crossman 1991a,b; Crossman *et al.* 1994; Taylor *et al.* 1994; Taylor *et al.*, in press; Hoyle *et al.* 1995; Hoyle *et al.*, in press). Three trapping studies have been carried out (in 1985–87, 1988–89 and 1993) to monitor population dynamics, wombat body condition and reproductive activity, and to attach radio-collars to determine home range and movement patterns (Crossman 1988; Johnson 1991a; Horsup 1994, respectively). These research projects culminated in the writing of a recovery plan which lists the actions necessary to bring about a recovery of the species (Davidson 1991; Horsup and Davidson 1994).

This paper summarizes the efforts to conserve *L. krefftii* and details the reasons for the species' decline, the steps taken to conserve it, the objectives and actions of the recovery plan, the difficulties faced by the recovery team in bringing about a reversal of current trends, and the progress to date.

REASONS FOR DECLINE

Competition with domestic cattle, especially during droughts, appears to have been the major reason for the decline of *L. krefftii* in the Epping Forest area (Crossman 1988). Considering the conservative feeding range of *L. krefftii* (Johnson 1991b) and the large dietary overlap that occurs between wombats and cattle (Crossman 1988), severe competition for food would probably have occurred during droughts. Evidence of major reductions in the species' range in the early 1920s, late 1930s, and mid-1960s correspond with periods of below-average rainfall in the Epping Forest area (Crossman 1988). The present colony underwent several contractions during droughts, the most recent in the mid-1960s (Crossman 1988). Declines during droughts were apparently not reversed during better rainfall years because of continued grazing pressure by cattle (Crossman *et al.* 1994). It seems likely from these trends in central Queensland that similar problems must have led to

[1]Queensland Department of Environment, P.O. Box 3130, Rockhampton Shopping Fair, Queensland, Australia 4701.

Pages 150–54 *in* BACK FROM THE BRINK: REFINING THE THREATENED SPECIES RECOVERY PROCESS ed by S. Stephens and S. Maxwell. Surrey Beatty & Sons, 1996.

the extinction early this century of the other two confirmed populations of *L. krefftii* in south-east Queensland and southern New South Wales.

CONSERVATION MEASURES

In recognition of the plight of *L. krefftii*, Epping Forest National Park (3 300 ha) was purchased from existing cattle properties with Commonwealth funding and gazetted in 1971. Funding was obtained from the then named World Wildlife Fund to fence the park to exclude cattle in 1981. It is now a National Park (Scientific), which under the *Nature Conservation Act 1992* prohibits public visitation. The park and particularly the area occupied by the wombats is protected from wildfire by an intensive system of firebreaks. Dingo *Canis lupus dingo* and Eastern Grey Kangaroo *Macropus giganteus* numbers on the park are regularly monitored and dingoes are controlled whenever their density exceeds a set level.

THE RECOVERY PLAN

The first recovery plan for *L. krefftii* was written in 1991 (Davidson 1991) and this was later revised (Horsup and Davidson 1994). The major funding for the recovery program comes from the Australian Nature Conservation Agency (ANCA), Queensland Department of Environment (DoE), and Australian Research Council (ARC). The cost to date of the recovery program is approximately $770,000, of which approximately $350,000 has been provided by ANCA. The recovery plan is implemented by a recovery team comprising researchers and managers from five universities and DoE, representatives of funding agencies, park neighbours, and non-government conservation organizations. The recovery team meets twice-yearly to assess progress on the recovery plan and to set future priorities. The recovery plan will run for an initial period of five years from 1992 to 1996. A major review will be undertaken in 1996 to assess the progress on individual actions of the recovery plan and to determine their contribution to the recovery of the Northern Hairy-nosed Wombat.

When writing the recovery plan, the recovery team had to differentiate between what *could* be achieved and what was desirable but was either too difficult, costly or risky to attempt. It was recognized that we could not expect a quick improvement in the situation as could be expected with small fecund mammals such as the Eastern Barred Bandicoot *Perameles gunnii* when taken into captivity (e.g., Backhouse *et al.* 1994). The actions of the recovery plan were therefore chosen according to their feasibility, cost and potential benefit, the risks involved. For example, a pure behavioural study was considered desirable but very difficult because of the species' secretive and nocturnal nature and was not given a high priority.

A recommendation to establish a captive population was made by the recovery team in early 1995. The issue of captive management had been avoided until better data on the population were obtained and because of the poor breeding record of wombats in captivity. The decision to recommend the establishment of a captive population was based on the results of the 1993 trapping study which indicated that there had been no increase in numbers since 1985 and that during this period the sex ratio had significantly skewed towards males (Horsup 1994). In addition, population viability analysis had indicated that the population could sustain removal of several individuals with little detriment to the source population (Hoyle *et al.*, in press). A captive population is currently being established.

Objectives of the Northern Hairy-Nosed Wombat Recovery Plan

The objectives of the Northern Hairy-nosed Wombat Recovery Plan are that within 10 years:

1. The population has doubled in size from its present level of approximately 65 individuals; and that
2. A second geographically disjunct and viable population has been established.

It is planned that these objectives will be achieved by: (a) developing new research and monitoring techniques; (b) improving the quality of available forage on Epping Forest National Park; (c) controlling potential predators and competitors; (d) using the closely related Southern Hairy-nosed Wombat *L. latifrons* to develop artificial breeding techniques; and (e) establishing new populations. The actions (which are detailed in Horsup, in press) are:

1. Appointment of a full-time recovery manager
2. Study of population ecology
3. Study of feeding ecology
4. Study of field energetics and water turnover
5. Study of conservation genetics
6. Study of reproductive biology
7. Monitor predators and competitors
8. Habitat management
9. Supplementary feeding
10. Translocation
11. Produce a GIS map of Epping Forest National Park
12. Establishment of a captive population

TRIALS

There are a number of factors which have complicated our efforts to bring about the recovery of the Northern Hairy-nosed Wombat. Epping Forest National Park is an isolated and unstaffed park, 120 km of dirt road from the nearest town and 250 km from the nearest DoE office. This factor complicates field work and may have contributed to the failure of supplementary feeding trials because of the irregular provision of fresh feed. Unfortunately the recovery program has coincided with a major drought in central Queensland which has seen Belyando Shire, in which the population occurs, drought-listed since 1 September 1992 (50 continuous months at the time of writing).

L. krefftii is almost totally nocturnal and so secretive that, despite spending a total of 10 months on Epping Forest National Park in the last five years (and apart from those wombats which I have trapped), I have seen only one Northern Hairy-nosed Wombat by day and five by night. The species is also extremely difficult to trap, with a trap success rate of only 2.6 captures per 100 trap nights during four months of intensive trapping in 1993. Trapping is very time consuming and involves fencing burrows so that the only exit for the wombats is through a trap. More importantly, trapping is extremely disruptive to the wombats. No wombat attractant has been discovered and trapping therefore essentially consists of "starving the wombats out". The mean time between setting traps and capture of a wombat is 5.5 days. Consequently, the wombats lose a significant amount of weight between captures (Hoyle *et al.* 1995).

The intensive trapping program over all active burrows in 1993 indicated that there had been no increase in the size of the Epping Forest population in the last eight years (Hoyle *et al.* 1995) and that the population continued to be dominated by old-age individuals. Trapping in 1993 also indicated that the adult sex ratio had skewed significantly towards males after being at parity in the other two trapping studies in 1985–87 and 1988–89. The trend may be partly explained by females remaining in their burrows longer than males before entering set traps (6.04 versus 4.93 days, respectively), suggesting that females are more disturbed by trapping than males. A skewed sex ratio, however, could greatly accelerate the rate of detrimental genetic changes (inbreeding and loss of variation between individuals). An excess of males in the population could also be a sign of inbreeding (W. Sherwin, pers. comm.).

Because of the difficulties involved in trapping and observing Northern Hairy-nosed Wombats we have few data on reproduction, recruitment and survivorship from which to make projections about possible population growth rates, etc. In the first two trapping studies (1985–87 and 1988–89), 45% and 40% of females, respectively, carried a pouch young at any one time (Crossman 1988; Johnson 1991a). In droughts such as that which continues to grip central Queensland, reproduction appears to be severely depressed. In the 1993 trapping study, only 10% of females carried a pouch young (Horsup 1994). It was during this period that the provision of a range of supplementary feeds was trialled, although none were eaten by wombats. Even when taken into captivity, Southern Hairy-nosed Wombats may initially refuse to eat for several weeks (V. Steele, pers. comm.)

The small size of the Epping Forest population and the possibility that wombat numbers may have been as low as 35 individuals in the early 1980s (Johnson 1991a), has created genetic problems. Analyses of Northern Hairy-nosed Wombat blood collected from 1988 onwards indicate that, relative to similar-sized populations of Southern Hairy-nosed Wombats, the Epping Forest population contains only 41% of the genetic variability (Taylor *et al.* 1994).

One of the biggest difficulties with recovering a species like *L. krefftii* is that while it is clear that there are major problems with inbreeding and the risks associated with small population size, it is not entirely clear why the species has not responded positively to its protection on Epping Forest National Park. This protection includes the exclusion of cattle, the wombat's main perceived external threat, the control of potential predators (dingoes), and the exclusion of the public. In most endangered species programs it is usually clear what the cause of the problem is, for example, predation, habitat destruction or disease. In the case of *L. krefftii* we are unsure whether drought (and the consequent lower quality of pastures) is the cause (although if it were, there should have been increases in population size in the late 1980s before the drought struck), whether there is some reproductive problem (sperm counts have been low in the six males at three times of the year in which semen has been collected), or whether there is another causative agent which we have not yet considered.

If *L. krefftii* was a small multiparous mammal it would probably have long been the subject of a captive breeding program. However, the other two species of wombat do not breed well in captivity and when the low numbers of *L. krefftii* and the risks involved in removing individuals from the population were taken into account, the establishment of a captive population was considered too risky by past managers of the wombat. When the population trend became clearer, it became obvious to the recovery team that the establishment of a captive population of Northern Hairy-nosed Wombats was a priority. This decision was taken for three main reasons: (1) that if population trends further deteriorate, the removal of individuals in the future may be potentially more detrimental to the population than it would now; (2) that captive reproductive techniques are now available which were not previously; and (3) that the current level of commitment (funding, time, and expertise) may not continue and should be taken advantage of now.

TRIUMPHS

At this stage we do not have many triumphs to report for the Northern Hairy-nosed Wombat recovery program. However, there are a few aspects of the biology of the species that are in our favour and there have been some promising developments during the recovery program.

L. krefftii is a long-lived species (one individual was at least three years old when captured and survived for another 27 years in captivity; J. Dennis, pers. comm.) and each individual is potentially capable of reproducing over many years or holding off its reproductive effort during drought years until conditions improve.

The entire current range of the species is protected on a national park from which cattle and the public are excluded. DoE has complete control over all management and research which occurs there.

There is a strong funding and moral commitment from DoE and ANCA to bring about the recovery of *L. krefftii*. The species has a high profile in Queensland and attracted $65,000 worth of funding during a public appeal in 1992.

The Northern Hairy-nosed Wombat Recovery Team comprises a multi-skilled group of people drawn from DoE, five universities, properties adjacent to Epping Forest National Park, funding agencies, and public conservation groups. Team members have backgrounds in endangered species research and management, including wildlife ecology, population genetics, reproductive biology, and park management.

The conservation genetics study is being carried out by a team led by Dr Bill Sherwin at the University of New South Wales. The study has developed a DNA fingerprinting technique which can identify individual wombats from single hairs collected on double-sided sticky tapes set at burrow entrances. Because all burrows can be hair-taped simultaneously, this technique will provide much more accurate census and sex ratio data than is currently possible by conventional trapping. Importantly, it will vastly reduce the disturbance caused by trapping and it is a technique which can be applied to other endangered species recovery programs.

Applied reproductive research on the closely related Southern Hairy-nosed Wombat is under way at Monash University by a team led by Dr Peter Temple-Smith. A successful pouch young cross-fostering pilot study was carried out in late 1995 on eight wild female *L. latifrons* and their pouch young. Upon release of the females (15 hours after the transfer of pouch young), all young were still attached to their surrogate mother's nipples (P. Temple-Smith, pers. comm.). Methods of semen collection from *L. latifrons* have been successfully developed and have recently been applied to *L. krefftii*. Artificial insemination is now another possible artificial reproductive technique which is being explored in *L. latifrons*.

A study of the feeding ecology of *L. krefftii* at James Cook University is nearing completion and will provide valuable insights into seasonal variation in the diet, activity patterns and movements of wombats relative to pasture conditions. The study has been successful in adapting monitoring techniques originally developed for other animal groups: bio-impedance metering monitors wombat body condition (a technique developed for humans), and alkane analysis monitors wombat dietary intake (a technique developed for domestic animals; A. Woolnough, pers. comm.).

Population viability analysis of Northern Hairy-nosed Wombat population parameters has indicated that the growth rate of the Epping Forest National Park population is most sensitive to adult female mortality. Creation of new wombat colonies has been simulated. The optimum number to translocate is 12 individuals, at which level the translocated population had less than a 1% chance of failing due to modelled stochastic processes, and all the modelled source populations survived (Hoyle *et al.*, in press).

CONCLUSIONS

Recovery of the Northern Hairy-nosed Wombat will require a long-term commitment from all parties involved: the Queensland and Commonwealth Governments, universities and research institutions, non-government conservation organizations and the general public. To date, there has been no improvement in the "on-ground" situation (i.e., there has been no apparent increase in numbers in the wild). However, the establishment of a captive population is underway, our knowledge of the wombat's biology has improved dramatically, and our ability to remotely monitor its numbers and behaviour has been optimized. The stage is now set for some major advances — the improvement of pastures on Epping Forest National Park and the application of reproductive techniques developed on the Southern Hairy-nosed Wombat. And surely the drought will end soon . . .

ACKNOWLEDGEMENTS

Funding and assistance for the Northern Hairy-nosed Wombat Recovery Plan has been received from the Australian Nature Conservation Agency, Queensland Department of Environment, Australian Research Council, Blair Athol Coal, National Geographic Society, Australian Geographic Society, Gordonstone Coal Management, Mt Isa Mines, QANTAS Airways, Dynamic Satellite Surveys, and a public appeal.

REFERENCES

ANZECC, 1991. List of endangered vertebrate fauna in Australia. Australian and New Zealand Conservation Council. Australian Nature Conservation Agency: Canberra.

Backhouse, G. N., Clark, T. W. and Reading, R. P., 1994. Reintroductions for recovery of the Eastern Barred Bandicoot *Perameles gunnii* in Victoria, Australia. Pp. 209–18 in Reintroduction biology of Australian and New Zealand fauna ed by M. Serena. Surrey Beatty & Sons: Chipping Norton.

Crossman, D. C., 1988. Population ecology and diet of the Northern Hairy-nosed Wombat, (*Lasiorhinus krefftii* (Owen)). Report to World Wildlife Fund Australia.

Crossman, D. C., Johnson, C. N. and Horsup, A., 1994. Trends in the population of the Northern Hairy-nosed Wombat *Lasiorhinus krefftii* in Epping Forest National Park, central Queensland. *Pac. Conserv. Biol.* **1**: 141–49.

Davidson, C., 1991. Recovery plan for the Northern Hairy-nosed Wombat, *Lasiorhinus krefftii*. Report to Australian Nature Conservation Agency. Endangered Species Program (Project 191). Queensland Department of Environment and Heritage.

Gordon, G., Riney, T., Toop, J., Lawrie, B. C. and Godwin, M. D., 1985. Observations on the Queensland Hairy-nosed Wombat. *Lasiorhinus krefftii* (Owen). *Biol. Conserv.* **33**: 165–95.

Horsup, A., 1994. Annual report on the Northern Hairy-nosed Wombat recovery program. Internal report to the Australian Nature Conservation Agency.

Horsup, A. (in press). The Northern Hairy-nosed Wombat: Finding a way out of the hole. *In* Wombats ed by R. Wells. Surrey Beatty & Sons: Chipping Norton.

Horsup, A. and Davidson, C., 1994. Recovery plan for the Northern Hairy-nosed Wombat, *Lasiorhinus krefftii*. Report to Australian Nature Conservation Agency Endangered Species Program (Project 191). Queensland Department of Environment and Heritage.

Hoyle, S. D., Horsup, A. B., Johnson, C. N., Crossman, D. G. and McCallum, H., 1995. Live trapping of the Northern Hairy-nosed Wombat *(Lasiorhinus krefftii)*: Population-size estimates and effects on individuals. *Aust. Wildl. Res.* **22**: 741–55.

Hoyle, S. D., Alpers, D. and Sherwin, W. B. (in press). Simulation of the population dynamics of the Northern Hairy-nosed Wombat, *(L. krefftii)* and possible translocation methods. *In* Wombats ed by R. Wells. Surrey Beatty & Sons: Chipping Norton.

Johnson, C. N., 1991a. Behaviour and Ecology of the Northern Hairy-nosed Wombat *Lasiorhinus krefftii.* Internal Report, Queensland National Parks and Wildlife Service, February 1991.

Johnson, C. N., 1991b. Utilization of habitat by the Northern Hairy-nosed Wombat, *Lasiorhinus krefftii. J. Zool. Lond.* **225**: 495–507.

Johnson, C. N. and Crossman, D. C., 1991a. Dispersal and social organization of the Northern Hairy-nosed Wombat, *Lasiorhinus krefftii. J. Zool. Lond.* **225**: 605–13.

Johnson, C. N. and Crossman, D. C., 1991b. Sexual dimorphism in the Northern Hairy-nosed Wombat, *Lasiorhinus krefftii.* (Marsupialia: Vombatidae). *Aust. Mammal.* **14**: 145–46.

Nature Conservation Act, 1992. Queensland Government legislation.

Taylor, A. C., Sherwin, W. B. and Wayne, R. K., 1994. Genetic variation of microsatellite loci in bottlenecked species: the Northern Hairy-nosed Wombat *Lasiorhinus krefftii. Molec. Biol.* **3**: 277–90.

Taylor, A. C., Alpers, D. and Sherwin, W. B. (in press) Remote censusing of Northern Hairy-nosed Wombats via genetic typing of hairs collected in the field. *In* Wombats ed by R. Wells. Surrey Beatty & Sons: Chipping Norton.

Recovery of the Mala *Lagorchestes hirsutus*: a 30-year unfinished journey

Kenneth A. Johnson,[1] David F. Gibson,[1] Donald G. Langford[1] and Jeffrey R. Cole[1]

Discovery of Mala *Lagorchestes hirsutus* in a remote part of the Tanami Desert in the early 1960s set in place a recovery program that persists more than 30 years on. The program has survived through substantial organizational and staffing changes, major shifts in tenure and conservation status of the land, and has required a partnership of Aboriginal and western cultures with clearly different conceptual views of the landscape. That the project has survived these challenges is testament to the accommodation of varying needs and perspectives of the participating interest groups.

The project has played a major role in alerting Australian conservationists to the importance of fire management and feral predators in the recovery of critical weight range desert mammals. It has also been responsible for preventing the last mainland representation of the species from becoming extinct.

There has been an evolution in the practical and theoretical approach of the project resulting from the many successes and failures experienced during the three decades of its existence. These and other experiences are discussed in the hope that they will be instructive to others embarking on recovery programs.

Key words: Mala, Aborigine, fire, feral predators, captive breeding, recovery program, desert.

INTRODUCTION

THE mainland form of the Rufous Hare-wallaby or Mala weighs approximately 1.3 kg and has a shaggy brown coat that was once used by Aborigines for making hair string. By day it shelters in a squat under a spinifex hummock, shrub or similar vegetation and may at times excavate a short burrow for added protection. It has an important mythology well known to the older desert Aborigines who once hunted it for food (Johnson and Burbidge 1995).

It was once widespread and common in the spinifex sandy deserts and fringing shrubland of the Northern Territory, South Australia and Western Australia. Finlayson (1935) regarded it as reasonably common during the time of his collecting expeditions through central Australia in the early 1930s, but it had become exceedingly rare just 20 years later when he made further visits in the mid-1950s (Finlayson 1961). The timing of this collapse is consistent with observations in the more westerly parts of its distribution in the Gibson and Great Sandy Deserts where it persisted until the 1950s (Burbidge *et al.* 1988; Burbidge and Johnson 1995). Two remnant populations persisted in the Tanami Desert of the Northern Territory until 1987 when one was extinguished by a fox and the influence of drought, and 1991 when the second disappeared following a wildfire that removed most of its spinifex habitat.

Recovery of the Mala began more than 30 years ago when Alan Newsome stepped from his Land Rover on the Tanami Track (now Tanami Highway) and strolled across the sandplain to investigate some low sandhills. Alan was studying Red Kangaroos in central Australia at the time and his broad ecological interests took him widely through the Northern Territory including the Tanami Desert.

He came across some odd looking macropod tracks in the red sand and later an Aboriginal man with him was quick to identify them as belonging to the Mala. There was little more to go on other than tracks until David Lindner and David Howe, both technical assistants in the fledgling wildlife research group at Alice Springs, revisited the area in 1964. They happened on the remains of a Mala from which a section of skull provided positive identification. This was the first hard evidence of the species' persistence and provided the confidence and impetus to commence a program to find out more about the species — a program that continues.

The project has revealed much about the Mala (Lundie-Jenkins 1993; Lundie-Jenkins *et al.* 1993a; Lundie-Jenkins *et al.* 1993b) and provided a basis for examination of causes of the species' decline (Johnson *et al.* 1989). It has also served as a general catalyst for discovery and investigation of other desert species, and functioning of the desert itself. These studies

[1]Parks and Wildlife Commission of the Northern Territory, P.O. Box 1046, Alice Springs, Northern Territory, Australia 0870.
Pages 155–61 *in* BACK FROM THE BRINK: REFINING THE THREATENED SPECIES RECOVERY PROCESS ed by S. Stephens and S. Maxwell. Surrey Beatty & Sons, 1996.

include investigations of the Greater Bilby *Macrotis lagotis*, Mulgara *Dasycercus cristicauda*, fire behaviour, species richness and distribution in palaeodrainage systems, and broad ecological survey of the Tanami Desert (Gibson 1986). There have also been many facets to the Mala project itself including fluctuating tenure conditions for the study site (Wildlife Sanctuary on vacant Crown Land to Aboriginal Freehold Title), strong Aboriginal participation in field activities, captive breeding, study of the species' ecology, translocation of captive bred animals, co-ordination between jurisdictions, and management of fire and feral animals in wild and reintroduced populations. It has been very much an exercise in action learning and it has involved a large number of people.

Few endangered species programs in Australia have endured like the Mala project and none (that we are aware of) have been conducted for long periods in such a remote and superficially inhospitable region. The experiment has been blessed with many exciting discoveries and periods of wide-eyed optimism as projects appeared to be on the brink of success. It has also suffered from periods of crushing disappointment as some of these emerging successes collapsed.

In this paper we will share some of these experiences, discuss their impact on the project, and consider their general relevance to species recovery programs in Australia: particularly in the desert environment.

THE STUDY AREA

The Tanami Desert is about 150 000 km^2 of relatively flat sandplain lying across the Northern Territory-Western Australian border about 600 km to the north-west of Alice Springs. It consists of a sea of sand broken occasionally by rocky outcrops and isolated dunes and dune fields. The rock outcrops form the basis of two lucrative gold mining operations but the land is generally unsuitable for grazing and there are only two pastoral stations in the region. One of these stations is now held under Aboriginal freehold title.

Several rivers flow into the Tanami but none flow out. Instead they carry their very occasional flows out into the desert to percolate down into the sand beds. Foxes and rabbits are rare but feral cats are common and widely distributed in all habitats. Feral camels are uncommon but appear to be increasing in numbers and feral cattle, donkeys and horses are absent in all but the most peripheral parts.

A salt lake system dominates the southwestern part of the desert and the general trend of flow is towards Lake Mackay on the Northern Territory-Western Australian border. The saline flats adjoining the lakes and channels appear to be a vital component of Mala habitat and the last known wild populations survived in association with these habitats until recent times.

Mala habitat is dominated by various species of spinifex (*Plectrachne* and *Triodia* spp.) which form a carpet of hummock grassland mixed with low Acacia shrubs. Termite mounds 3–4 m high are the major above-ground feature in low-lying areas of the Tanami.

Fire and highly variable rainfall are conspicuous environmental features associated with this desert region. Lightning strikes are responsible for much of the fire pattern in the area, and monsoonal remnants from the Kimberley and Top End are a major influence on rainfall patterns. Frosts are rare to non-existent and daily summer maxima can top 40°C for many days on end.

LAND TENURE

When Mala were first discovered, the Tanami Desert was a vast expanse of vacant Crown land with one road that was little more than a two-wheel track. There was a struggling gold prospect managed by a single family, and a pastoralist desperately trying to persuade the Northern Territory Administration that a tract of 4 200 km^2 should be leased to him as a cattle station — later to be called Mongrel Downs in a sarcastic reference to the administration's damning assessment of the block's pastoral value.

Largely as a result of Alan Newsome's efforts, the Tanami Wildlife Sanctuary comprising some 37 000 km^2 (more than half the area of Tasmania) was gazetted in 1964 (Fig. 1).

Following establishment of the *Aboriginal Land Rights (NT) Act* in 1976, a claim was lodged over much of the Tanami including the Sanctuary. The Land Commissioner, Justice Toohey, heard the claim in 1978 and found in favour of the Aboriginal claimants. He also recognized the conservation importance of the Tanami and included this in his report to the Minister for Aboriginal Affairs. The Minister accepted the recommendation that the land be granted and that the traditional Aboriginal owners negotiate with the then Territory Parks and Wildlife Commission, through the Central Land Council, for an appropriate arrangement for nature conservation.

Title was vested in an Aboriginal Land Trust in 1982. A Head of Agreement was established in principle with the Central Land Council in 1989, and agreement was reached for the Council to take the document to traditional Aboriginal owners for discussion. This is yet to happen but the Mala project has nevertheless continued through these substantial shifts in status, ownership and administrative processes associated with the land.

FRAMEWORK FOR THE MALA PROJECT

For more than 12 years the Tanami Desert Mala were known from just a single colony. In this period however a group of scientists in Alice Springs began to analyse the habitat characteristics of this colony with a view to surveying surrounding areas for additional populations. They concluded that Mala required sand dune areas associated with nearby caliche communities superimposed by a patchy fire pattern. The fire pattern produced a vegetation complex in varying stages of recovery, and included areas of long unburned and mature spinifex which provided cover for the Mala.

Field survey areas were selected using aerial photographs to map the saline systems, sand dunes and fire history associations. Two extensive surveys conducted

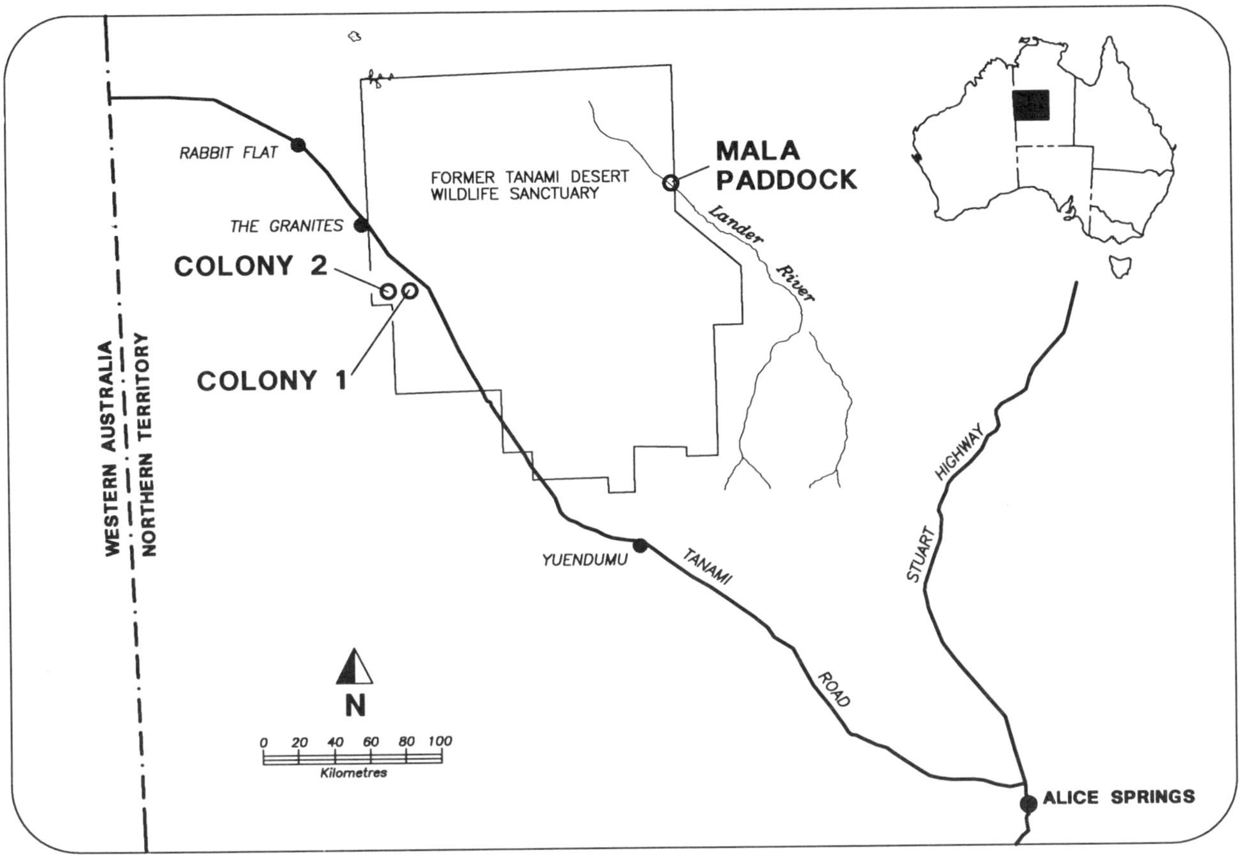

Fig. 1. Locations of the three mainland Mala populations in the Tanami Desert.

in 1976 revealed just one more population, later known as Colony 2, no more than 15 km to the west of Colony 1 (Fig.1).

The process of analysing habitat used by Mala in the two colonies led Bolton and Latz (1978) to propose the theory that fire was a major determinant in the survival of the species. Their watershed paper considered the traditional patchy burning practised by Aboriginal people before the arrival of Europeans as the chief requirement for maintaining suitable habitat for the Mala. They considered the very large wildfires that had become common since European settlement to have had a homogenizing effect that left Mala without the required mix of shelter and food resources. They further indicated that the unusual combination of salt channels and caliche vegetation which imposed natural fire breaks, and the more recent fire history imposed by European stockmen taking cattle to market were the key elements in the persistence of Mala at Colonies 1 and 2. If Mala were to survive in the Tanami Desert they concluded that fire would need to be properly managed to create the right mix of habitats.

This theory was to drive the Mala recovery project for the next 10 years after which predators, especially feral cats, were revealed as a further critical threat to survival of the species.

Following the Bolton and Latz survey there appeared to be a diminishingly small chance of Mala populations persisting anywhere else. Nevertheless, after Aboriginal people in local communities had been shown the two colonies they developed a firm belief Mala were still widely distributed in the more remote desert areas and that one just had to go out looking and they would be found. It should be remembered however that the people had mostly left their country at about the time of the Second World War and had subsequently moved to cattle stations, missions or communities set up specifically for the welfare of Aborigines. Their knowledge of the distribution of animals was therefore dependent on their memory of the situation some 30 to 40 years previously.

A survey program was initiated in 1979 and over the subsequent eight years this effort was intensified and extended to cover the whole Tanami region insofar as this is possible for an area of more than 150 000 km^2 (about two-thirds the area of Victoria). No additional populations were found and the older Aboriginal men accompanying the field teams were initially surprised at this finding and subsequently saddened by the situation.

The scientific dilemma in 1979, at the beginning of the second stage of the project, involved a series of choices on how best to manage recovery of the species.

There were two populations about 15 km apart. Fire was suspected to be a major force behind their decline and in fact a wildfire had destroyed a large portion of the Colony 2 habitat earlier that year. Numbers of animals in each population were known to be exceedingly small and in the order of 6 to 10 for Colony 1 and perhaps a few more for Colony 2. The spinifex hummock grassland in the area was exceedingly flammable and big fuel loads had built up around Colony 2. There was some cynicism about the project among the non-scientist members of the agency because scats and tracks were the only tangible evidence of Mala. Few people had actually seen a Mala and there was no photographic evidence. The entire wildlife research team at the time comprised a scientist and two technicians with responsibility for the southern part of the Northern Territory or about 800 000 km^2 of country (about the area of New South Wales). A lot of other work was waiting to be done and priorities had to be assigned.

The plight of the Mala seemed important so it was decided to undertake the following actions.

1. Undertake fire management in and around Colony 2.
2. Leave Colony 1 as a control so that if we got it wrong it was likely that one population would remain.
3. Refrain from removing individuals for captive breeding because of the very low numbers of wild individuals.
4. Acquire photographic evidence of Mala to silence the sceptics.

REINTRODUCTION

There was a shift in our philosophy on taking prisoners when a healthy extension to Colony 2 was found during fire management activities. The extended population was found by an elderly Aborigine, Darby Jampijinpa, when investigating options for locating fire breaks around the colony. The find allowed the population estimate to soar to a figure of 50 to 100 individuals; an order of magnitude greater than the previous assessment but still a perilously low number.

No reliable techniques for catching Mala were known at the time, but fortunately the first successful attempt to photograph a wild Mala showed that the use of a spotlight and a light-footed approach allowed wallabies to be approached closely enough for capture with a net. After a solid night's work in August 1981 the first five Mala were on their way back to Alice Springs.

We also secured a series of photographs which together with the live animals satisfied even the most committed sceptic in the organization.

The captive population grew rapidly and by mid-1984 there were sufficient animals with which to attempt a translocation, test the theory that fire was the key determinant, and if successful, provide improved security for the species through establishment of a third wild population.

A potentially suitable area near Lake Surprise (Yinapaka) had been found by Gibson (1986) during the Tanami surveys and this area was investigated again in May 1984. Warlpiri Aborigines from the nearby Willowra community accompanied us to the area, which involved about 60 km of travel over untracked spinifex. A site with a suitable mix of habitats and fire history was jointly selected, and later that year the Aboriginal owners helped to erect a 50 m by 50 m enclosure which we planned to use as a temporary holding pen.

On 1 September 1984 the first 12 Mala were released from the pen after 10 days of acclimatization. It is believed some were subsequently taken by dingoes and tracks interpreted by Aborigines showed that at least one was killed by a feral cat. This was regarded as acceptable because Colony 1 was known to be persisting with very few individuals that were also subject to predation. Mala were still present in the area when a second lot of 13 were released in August of 1985. A follow-up survey conducted in December of that year failed to locate any Mala left at the site.

There had been a continuing hot and dry spell of weather for much of the year and vegetation was dry and very sparse. Aborigines visiting the site with us commented about this lack of food and the probability that the Mala had moved off in search of green pick created by the scattered storms in the area. Whether the Mala had been killed, died of starvation, or had dispersed, they were effectively lost to the study and the program was in need of a substantial rethink.

The Aboriginal viewpoint was that we were using the wrong sort of Mala in the release program. Putting it in their context they felt that the captive Alice Springs bred Mala were like domestic dogs on the Willowra Aboriginal settlement. They were "Settlement Mala" and we needed wild Mala like the wild dogs or dingoes that were "living out bush" and capable of looking after themselves. After lengthy discussion and further examination of the area we agreed that we could not get the proper wild Mala but we might achieve the next best option. This would be to build a large enclosure in a remote part of the desert to hold wallabies for a lengthy period during which time they would develop a strong site fidelity and produce wild offspring that were not conditioned to humans.

A large enclosure of about 1 km^2 was built with considerable help from the Willowra people and without the aid of machinery to clear the fence line. To fit within budget the fence was an electrified wire design which we expected to contain the Mala and exclude predators.

The first release of 12 Mala into what had become known as the Mala Paddock by the Willowra people took place on 16 December 1986 and the first Mala escapes took place on the same night. Wires carrying 6 200 volts were not enough to contain Mala.

A plastic mesh approximately 20 mm high was fitted to the bottom of the fence and this was sufficient to stop the Mala escaping. It did nothing to stop dingoes which we subsequently discovered inside among the remains of many Mala. Excited by wallabies visible through the fence, the momentum of their attack probably carried them between the wires before the high voltage could act as a deterrent.

Additional wires were installed but a dingo trying to leap through the more narrowly spaced wires became entangled and short circuited the energizer. This dropped the voltage to zero and allowed other dingoes to get through the unelectrified wires.

By examination of tracks around the outside it was clear that feral cats were gaining access through and over the fence. With expert assistance from Willowra Aborigines we were able to track down and remove these enclosed cats. The dingo and cat incursions necessitated the addition of a black solar-resistant plastic netting down the body of the fence to stop animals leaping between the wires. A 1 m outrigger extension added to the top of the fence at 45° prevented cats jumping over. Despite almost 12 months of disappointment and frustration the fence finally evolved to meet project requirements and it has now operated satisfactorily since October 1987.

In the interim, disaster had struck Colony 2. The colony had been the subject of ecological studies (Lundie-Jenkins 1989) and although several dry years had caused the number of Mala to decline, it was evidently still in a healthy condition. After a few months in which no site visits were made, an inspection was carried out in November 1987 and fox tracks were found widely spread throughout the colony. Evidence where Mala had been chased was clear among the tracks in the sand and the chewed remains of carcases were found in depressing abundance. By 17 December of that year it was clear that the relatively more populous Colony 2 had been extinguished and Colony 1 was all that remained of the truly wild population.

Since the Colony 2 area was known to be suitable habitat for the Mala it was decided to attempt a parallel reintroduction. It had of course been intended to focus all reintroduction resources in the Mala Paddock area, where there had already been a substantial commitment, and to maintain Colony 2 with existing management operations. This new situation required a substantial rethink in servicing the Mala Paddock project, but after resolving the critical fencing problems, further Mala were released from the fenced Mala Paddock with radio transmitters attached. Animals survived up to 20 months after release and 18 survived longer than 12 months. They showed strong site fidelity and young were born and became independently established. This progress had initially given some cause for cautious optimism but every animal eventually disappeared from the area or was found dead.

At least 64% of deaths in these free ranging Mala were directly attributable to feral cats. The cause of death could not be determined for a further 29% of cases and 7% died as a result of trap injury. Cat predation seemed to come in waves. Mala would survive well for many months until hit by a wave of cat predation that might occur over a few short weeks. This caused yet another major rethink of the reintroduction process. Mala had plainly survived in comfortable numbers alongside cats for at least 50 years before their numbers crashed, so it seemed reasonable that released animals would not be seriously challenged. Our project demonstrated that the role of cats was critical.

Various trapping and baiting techniques were attempted in an effort to control cats but no method gave a satisfactory long-term solution. It has been necessary, therefore, to abandon any further release attempts pending the development of suitable techniques for feral predator control.

FIRE

When the Mala project began there was little known about fire behaviour and its use in conservation management of spinifex communities. Indeed the role of fire as a wildlife management tool was in its infancy and many Australian wildlife scientists at the time considered that fire should be excluded from wildlife habitat.

The fire management program began, therefore, with some trepidation and anxiety. A conservative approach was adopted whereby a 30 km fire break was graded around the perimeter and at a reasonable distance from the location of the Mala Colony 2. The break was intended to stop summer wildfires and to provide a baseline for progressively fragmenting the spinifex inside using a series of smaller winter prescribed burns.

The program was nearly cancelled in the initial stages when a fire escaped the fire break and swept into the colony area. In time however, some level of confidence returned and the project continued with even greater caution. Aboriginal people from Yuendumu played a major role in advising on the behaviour of fire and the procedures for achieving the desired result of spinifex in various stages of recovery from fire.

In time we had the courage to begin burning within the areas actually occupied by Mala and to gradually extend the pattern of fires through the colony area. We had learned that the large hummock spinifex vegetation occupied by Mala would not burn until it was about eight to 10 years old. At this time the individual hummocks had spread to a size whereby the spaces between the clumps closed sufficiently to provide a more continuous fuel load that allowed a fire to carry. Burning carried out under the relatively cool and still wind conditions of winter kept fires from spreading rapidly and allowed for considerable control.

By 1991 the vegetation in Colony 1 had developed to a stage where we were beginning to get concerned about the probability of wildfire sweeping the entire area. An effort was made that winter to burn breaks through the colony as a hedge against such an event but under the prevailing conditions it was not possible to get a fire to carry. This led to some complacency in the belief that the colony would be safe until at least the following year.

Weather conditions have a substantial effect on fire behaviour and late in 1991 a fire, almost certainly started by a passing motorist, broke out alongside the Tanami road. Under the influence of high winds and high temperature this fire devastated the colony area and intensive searches in the following days failed to locate a single surviving Mala. We were forced to conclude, therefore, that the last truly wild population had been extinguished. It was a devastating and depressing blow to the program.

ABORIGINAL PATRICIPATION

Apart from the captive breeding program in Alice Springs, the entire Mala project has been carried out on Aboriginal freehold land since 1982 and approval of traditional owners has therefore been an essential step in the process.

Our first involvement with Aboriginal people on the project was in 1980 when we invited a group of key Aboriginal men to visit the site. At this time it was stepping into the unknown because we did not know how they would respond to contact with Mala which were traditionally a highly sought food item having strong medicinal powers for old people. Many, including the principal owner of the Mala dreaming, took some persuasion to make the 200 km trip because he believed Mala to be "all finished up". They were pleased to find the converse and one very knowledgable old man who owned the Mala dreaming was able to share his vast knowledge of the species in finding wangku (or squats), and explaining the activity of Mala from tracks in the red sand. He and the others proved to be as concerned for the Mala as we were, and the possibility of a food hunt was not discussed.

Since then there has been considerable support for the project by Aboriginal people and they have played a significant and enduring role in the project. Their skills and knowledge have filled major gaps and their assistance with jobs like fencing has been of considerable value. More than 100 men, women and children have participated in the Mala Paddock project and some have assisted on numerous occasions since 1984. It has been a joining of Aboriginal and Western conservation technologies and we would judge it to have been a fairly successful and mutually beneficial exercise.

When the reintroduction project commenced 15 years ago there were many people with first-hand knowledge of Mala and a strong traditional interest. Their participation in the project required considerable flexibility in field procedures because, of course, they had agendas that went beyond pure scientific endeavour. Their particular interests included visits to sites that had not been seen for many years, and this sometimes entailed much cross country driving as particular areas were searched. Young people were also taken to country and taught aspects of culture that could not be readily presented at the communities.

Many of the older people who originally took part in the project have now either passed on or are unable to travel in the field due to age and ill-health. The younger people now taking their place have differing views and needs and there must be a constant evolution of a project to accommodate these changing requirements. Wildlife researchers and managers need to appreciate the rapidly evolving situation and be prepared to meet it with a considerable level of flexibility.

RESULTS OF THE MALA PROJECT

After more than 15 years of work in the current phase of the Mala project a satisfactory solution to reintroduction has not been found, but we consider we have learned much that will benefit conservation of the species. These outcomes include the following.

1. Husbandry techniques for captive Mala have been resolved — they breed prolifically.
2. A considerable amount of ecological data has been amassed.
3. Effective release techniques have been developed for reintroduction programs.
4. A cheap, functional and reliable electric fence has been developed that is suitable for remote localities.
5. Feral cats have been exposed as a major threat to the recovery of desert mammals.
6. Fire has been exposed as a major variable to be managed in any desert recovery program.
7. A valuable understanding and working relationship has been developed with traditional Aboriginal owners of the Tanami area.
8. Mala are *not* extinct!
9. There have been the following valuable lessons that have implications for future reintroduction programs, especially those scheduled for desert environments.
   a. Be prepared for the long haul. A clear organizational commitment is necessary as projects may span decades and involve staff changes that will affect project continuity.
   b. Know the distribution and status of the focus species so that educated decisions can be made about the allocation of resources. Taxonomic analysis may be necessary in poorly understood species such as has been the case for *Dasycercus* and *Notoryctes*.
   c. Establish a captive population as security against potential disasters.
   d. Be prepared to spend considerable time on technical issues — especially if a fence is involved.
   e. Set aside a substantial block of time for negotiation with Aborigines, sufficient to find a match between Aboriginal needs and western conservation objectives.

      Aboriginal social systems appear to be in rapid evolution and any program will need scientists sensitive to these changes and prepared to accommodate a shifting playing field.

      Scientists will need to invest an appreciable amount of field time in meeting Aboriginal needs, including visits to areas away from the study site.
   f. Feral cats are probably the main predator concern, especially in areas north of the fox line.

The Mala recovery project is now at a cross-roads. Mala are persisting at the Colony 2 release site but the programmed release of 10 to 20 animals at a time from the Mala Paddock have not succeeded due to the impact of feral cats. There is some indication from the analyses conducted by Tony Sinclair and Roger Pech (pers. comm. 1996) that the Mala may have withstood predation had they been released in larger numbers in the order of 50 individuals. Such numbers are available from the Mala Paddock and such an experiment would be possible.

To provide additional security for the species, investigations are currently under way to assess the possibility of establishing a population on an offshore island. This would provide a temporary solution while awaiting the long-term answer of developing effective and cost-efficient feral cat control measures to the extent that they are already available for foxes.

ACKNOWLEDGEMENTS

Recovery of the Mala has been something of a marathon (as indicated by the title of this paper) and it has involved many scientists, technicians, landholders and interested bystanders for a long time. It developed more or less in two phases. From the early 1960s until 1978 when the Parks and Wildlife Commission was formed, the project primarily involved Alan Newsome, Bill Bolton, Dave Lindner, and Dave Howe. Peter Latz had a large involvement in this phase as well as the second phase from 1978 until now. A very large number of Aboriginal people have also been involved but Darby Jampijinpa, Toby Japangarti, Sandy Japangarti, Larry Tracker Japiljarri and Salty Willy Japanangka have made particularly significant contributions. Geoff Lundie-Jenkins, Peter Jarman, Keith Bellchambers and Andrew Foster have also been major participants in the project and we gratefully acknowledge the respective contributions. Tony Robinson and Keith Morris made valuable comments on the paper.

REFERENCES

Bolton, B. L. and Latz, P. K., 1978. The Western Hare-wallaby *Lagorchestes hirsutus* (Gould) (Macropodidae) in the Tanami Desert. *Aust. Wildl. Res.* **5**: 285–93.

Burbidge, A. A., Johnson, K. A., Fuller, P. J. and Southgate, R. I., 1988. Aboriginal knowledge of the mammals of the central deserts of Australia. *Aust. Wildl. Res.* **15**: 9–39.

Finlayson, H. H., 1935. The Red Centre man and beast in the heart of Australia. Angus and Robertson: Sydney.

Finlayson, H. H., 1961. On central Australian mammals. Part IV — the distribution and status of central Australian species. *Rec. SA Mus.* **14**(1): 1–191.

Gibson, D. F., 1986. A biological survey of the Tanami Desert in the Northern Territory. Conservation Commission of the Northern Territory Technical Report No. 30.

Johnson, K. A., Burbidge, A. A. and McKenzie, N. L., 1989. Australian Macropodoidae: Status, causes of decline and future research and management. Pp. 641–57 *in* Kangaroos, Wallabies and Rat-kangaroos ed by Gordon Grigg, Peter Jarman and Ian Hume. Surrey Beatty & Sons: Chipping Norton.

Johnson, K. A. and Burbidge, A. A., 1995. Rufous Hare-wallaby *Lagorchestes hirsutus* Gould, 1841. Pp. 316–18 *in* The Mammals of Australia ed by R. Strahan. Reed: Chatswood.

Lundie-Jenkins, G. W., 1989. The ecology and management of the Rufous Hare-wallaby *Lagorchestes hirsutus* in the Tanami Desert. Unpublished M. Res.Sc. Thesis, University of New England: Armidale, New South Wales.

Lundie-Jenkins, G. W., 1993. Ecology of the Rufous Hare-wallaby, *Lagorchestes hirsutus* Gould (Marsupialia: Macropodidae), in the Tanami Desert, Northern Territory. I. Patterns of habitat use. *Wildl. Res.* **23**: 457–76.

Lundie-Jenkins, G. W., Phillips, C. M. and Jarman, P. J., 1993a. Ecology of the Rufous Hare-wallaby, *Lagorchestes hirsutus* Gould (Marsupialia, Macropodidae) in the Tanami Desert, Northern Territory II. Diet and feeding strategy. *Wildl. Res.* **20**: 477–94.

Lundie-Jenkins, G. W., Corbett, L. K. and Phillips, C. M., 1993b. Ecology of the Rufous Hare-wallaby, *Lagorchestes hirsutus* Gould (Marsupialia, Macropodidae) in the Tanami Desert, Northern Territory. III. Interactions with introduced mammal species. *Wildl. Res.* **20**: 495–511.

Developing and implementing a recovery plan: the Mala recovery program

Geoff Lundie-Jenkins[1,2]

The Mala was once one of the most abundant and widespread macropods of central Australia but has declined to the point where it is now one of the rarest and most limited in distribution. Efforts to recover and conserve the Mala have been considerable and predate the national *Endangered Species Protection Act 1992* which provides a framework for preparing and implementing recovery plans for threatened species and communities. The initial recovery plan for the Mala was developed in 1992 and focused on research and management actions related to the central Australian form of Mala. Implementation of this plan commenced in 1992 with ANCA funding to support major actions related to management and translocation of mainland populations. The formation of a recovery team in 1993 was the initial stage in the more formal review of the program. This review examined both progress towards implementing the plan and any need to refine existing actions and define new ones. Several program highlights were identified including establishment of captive colonies, implementation of translocation programs, co-operative programs with Aboriginal landholders and an enhanced understanding of the ecology of the Mala. A revised plan was finalized during 1995 providing a more national focus to the recovery effort, the critically endangered status of the Mala ensures that this effort will be ongoing. The development, implementation and revision of the Mala Recovery Plan has provided a useful insight into the potential achievements and shortcomings of the recovery process for conservation of threatened species.

INTRODUCTION

SINCE April 1993 management of Australia's threatened species has been guided by the *Endangered Species Protection Act 1992* which provides a framework for the preparation and implementation of recovery plans for listed endangered and vulnerable species and endangered ecological communities. Within the Act are provisions for the Commonwealth Minister to adopt recovery plans prepared by State and Territory agencies, and provide assistance for the implementation of these plans. Currently administration of this process is the responsibility of the Endangered Species Unit within the Australian Nature Conservation Agency (ANCA). At the time of writing, a total of 260 recovery plans had been prepared covering a range of plant and animal taxa and of these 199 were actively being implemented (ESU 1995). Given the infancy of this approach to the conservation of threatened species and communities in Australia, the process of developing and implementing these plans has evolved greatly since enactment of the legislation in 1993.

This paper examines the processes involved in the development, implementation and revision of a recovery plan for the Rufous Hare-wallaby or Mala *Lagorchestes hirsutus* and provides a summary of the achievements and shortcomings of the process in relation to this species to date. The Mala was once one of the most abundant and widespread macropods of central Australia but has declined to the point where it is now one of the rarest and most limited in distribution (Johnson *et al.*, this volume).

THE MALA RECOVERY PROGRAM

Efforts to recover and conserve the Mala have been considerable and predate the formation of the Endangered Species Program (ESP) and the protocol for developing formal recovery plans and teams (Johnson *et al.*, this volume). Such efforts commenced soon after rediscovery of the species in the Tanami Desert in the early 1970s (Bolton and Latz 1978). Direct conservation measures initiated by the then Conservation Commission of the Northern Territory (CCNT) to improve the status of the mainland populations of Mala included fire management (Bolton and Latz 1978), population and habitat monitoring (Lundie-Jenkins 1993), ecological research (Lundie-Jenkins *et al.* 1993a,b), establishment of captive populations and experimental reintroductions (Gibson *et al.* 1995; Lundie-Jenkins and Bellchambers 1995). Whilst a co-ordinated approach had been taken to implementing these measures no formal management/recovery plan for the species had been prepared prior to enactment of the Commonwealth *Endangered Species Protection Act 1992.*

[1]Parks and Wildlife Commission of the Northern Territory, P.O. Box 1046, Alice Springs, Northern Territory, Australia 0870.
[2]Present address: Queensland Department of Environment, PO Box 731, Toowoomba, Queensland, Australia 4350.

Pages 162–69 *in* BACK FROM THE BRINK: REFINING THE THREATENED SPECIES RECOVERY PROCESS ed by S. Stephens and S. Maxwell. Surrey Beatty & Sons, 1996.

Table 1. Summary of actions in original recovery plan for the Mala (from Johnson and Lundie-Jenkins 1992).

Research Actions

1.1 Assess the ecological processes by which translocated Mala establish new self-sustaining populations.
  1.1.1 Habitat and site fidelity of age/sex/reproductive classes upon release.
  1.1.2 Dispersal and establishment of F2 and later generations.
  1.1.3 Mortality and recruitment patterns.

1.2 Determine cost effective methods for stress mitigation to optimize establishment of reintroduced Mala.
  1.2.1 Delivery and use of artificial food and water supplements.

1.3 Determine habitat management requirements in the post-establishment phase.
  1.3.1 Movement patterns, survival and dispersal in response to fire.

1.4 Clarification of the status of the three recognized subspecies and establish genetic relationships among individuals in the captive breeding colony.
  1.4.1 Clarification of subspecific status through DNA analysis of island and mainland Mala.
  1.4.2 DNA fingerprinting of captive and released populations.

Management Actions

1.1 Maintenance of naturally occurring mainland and island populations and those reintroduced to Sangsters Bore and the Lander River.
  1.1.1 Monitoring of island populations.
  1.1.2 Management and monitoring of mainland populations.

1.2 Maintenance of captive breeding colony in Alice Springs.

1.3 Negotiation of conservation areas in the Tanami Desert with Aboriginal traditional owners.

1.4 Reintroduction of Mala to two major central Australian national parks.
  1.4.1 Negotiation with Aboriginal traditional owners and relevant park agency on agreed procedures for executing a reintroduction program.
  1.4.2 Preparation of a management prescription for cost-effective reintroduction and subsequent management.
  1.4.3 Implementation of prescription at two areas coincidently to maximize efficiency of resources.

The recovery plan

The first version of the Mala Recovery Plan was developed in 1992 (Johnson and Lundie-Jenkins 1992) and focused on research and management actions related to the central Australian populations of Mala. Consistent with the original format proposed by ANCA, the plan was structured to contain separate sections detailing research and management actions. A total of four primary research actions and four primary management actions were identified, these are indicated in Table 1. The criteria listed in the recovery plan for measuring progress towards recovery of the Mala were:

1. Maintenance of mainland and island Mala populations.
2. Establishment of self-sustaining populations at two sites in the Tanami Desert.
3. Maintenance of genetically diverse, self-sustaining captive populations in a number of zoological institutions.
4. Self-sustaining populations at a number of other mainland sites.

No formal consultation process was initiated in the development of the original recovery plan and hence ownership of the document was essentially limited to the Northern Territory. Whilst some actions relating to Western Australian island populations of Mala were incorporated in this plan, no accurate estimates of the resources, costs or commitment for these actions were included since no direct input from the Western Australian Department of Conservation and Land Management (WA CALM) was sought in developing these actions. Similarly, maintenance and establishment of captive populations was an important management action in the original recovery plan, but no consultation with the zoo community was undertaken to develop this action. With the wisdom of hindsight it is now apparent that whilst the plan as a document indicated an array of recovery actions, the potential to implement many of these would have been greatly enhanced by broader stakeholder involvement in its development.

Implementation of the plan

Formal implementation of the initial Mala Recovery Plan commenced in 1992 with ANCA funding to support two major actions, namely research action 1.1.2 and management action 1.1.2 (refer Table 1). Funding for research action 1.4.1 was received in 1993. Details of the specific programs initiated and their progress is summarized by Johnson *et al.* (this volume). The major achievements of this recovery program to date have been:

- Establishment of a genetically diverse captive population of Mala through the establishment of colonies in Alice Springs, Western Plains Zoo (Dubbo) and Monarto Zoological Park in South Australia.
- Successful translocation of captive bred Mala to remote sites in the Tanami Desert and increased understanding of the factors influencing the establishment of translocated populations, including introduced predators and fire.
- Development of equipment and techniques applicable to the translocation and conservation management of Mala and other arid zone species.
- Co-operative conservation of an endangered species on Aboriginal land.
- Improved understanding of the biology/ecology of the Mala and the arid ecosystems it previously inhabited.

The prime setback of the recovery program has undoubtably been the loss of the only known "naturally occurring" mainland population of Mala in 1991 when wildfire destroyed approximately 90% of its habitat (Johnson *et al.*, this volume). Clearly without planning to establish captive populations and initiate translocation/reintroduction experiments, this single event may have resulted in the extinction of the mainland form of the Mala. Whilst these actions may well have been initiated without the development of a formal recovery plan, the situation highlights the necessity for a broad approach to recovery of a species.

The recovery team

The inaugural meeting of a national Mala Recovery Team was convened in April 1993 in conjunction with a Reintroduction Biology Conference in Healesville. This action was initiated by officers of the then CCNT as the lead agency involved in Mala research and conservation activities. Invitations were circulated to 16 different organizations seeking their involvement in the recovery planning process. To ensure a balanced approach was made to the process, equal representation was sought from all the former range states of the Mala. Table 2 lists the organizations invited to participate on the Mala Recovery Team.

Table 2. Organizations invited to participate on the Mala Recovery Team.

Organization/Individual
Australian Nature Conservation Agency
Conservation Commission of the Northern Territory
Western Australian Department of Conservation and Land Management
South Australian Department of Natural Resources
World Wide Fund for Nature (Australia)
CSIRO Division of Wildlife and Ecology (Perth)
National Threatened Species Network (Arid Zone and Northern Territory)
Western Plains Zoo (New South Wales)
Monarto Zoological Park (South Australia)
Perth Zoo (Western Australia)
Central Land Council (Northern Territory)
National Farmers' Federation (Northern Territory)
Chamber of Mines (Northern Territory)
University of New England (New South Wales)
Independent Publicist

The initial recovery team meeting endorsed a proposal to establish a small core group of people actively involved in research or conservation of the species and a larger outer group who would be invited to be corresponding members of the team. The Australia-wide spread of recovery team members has made frequent meetings difficult and hence only one other full recovery team meeting has been held, in November 1994 in conjunction with the National Rock-wallaby Conference. Although there have been few meetings, formation of the recovery team has acted to formalize the contacts between organizations with interests in the Mala and has greatly improved the exchange of information on the species.

REVIEW AND REVISION

Since the commencement of active conservation measures for the Mala in 1978, the program has undergone considerable changes and internal review. The more formal process of reviewing the content and implementation of the Mala Recovery Plan commenced with the formation of the recovery team in 1993. The broad representation on the recovery team, including agencies from other former range States and non-government organizations, provided the impetus for this review and development of a more national approach to recovery of the Mala. This initial internal review by the recovery team consisted of the following discrete steps:

- matching actual recovery effort against specific actions within the written plan and hence identifying progress towards implementation of the plan;
- identification of changes in the species' status or information base which might require refinement of existing actions or the definition of new actions;
- identification of the relative roles of organizations represented on the recovery team in relation to each of the defined actions; and
- revision of recovery plan objectives, criteria and costings.

The points below summarize the outcomes of this review process in relation to implementation of the actions prescribed in the original recovery plan.

Research phase

1.1 Assess the ecological processes by which translocated Mala establish new self-sustaining populations.

1.1.1 Habitat and site fidelity of age/sex/reproductive classes upon release.

1.1.2 Dispersal and establishment of F2 and later generations.

1.1.3 Mortality and recruitment patterns.

Two experiments in reintroducing the Mala have been implemented: one at the site of the colony that went extinct in 1987 at Sangsters Bore and the other experiment in a freshwater paleodrainage line (Lander River) that traditional Aboriginal owners selected on the basis of habitat suitability and use by Mala in living memory. An interim report has been prepared covering the Sangsters Bore site (Lundie-Jenkins 1995), with a final report due to be finalized by July 1997. A draft report on the Lander River program has been completed (Lundie-Jenkins and Bellchambers 1995) and circulated internally for review. This report examines techniques, survival and fecundity, movements and dispersal, population estimates, diet, Aboriginal involvement and an assessment of the relative successes and failures of the program.

A small free-living population still exists at the Sangsters Bore site, however no further releases are currently planned for the Lander River site due to the loss of Mala to feral cats following each release to date. Monitoring of both populations is continuing. The enclosure-held population appears to be maintaining a *status quo* with recruitment closely matching losses.

The findings from both reintroduction sites will be combined in the preparation of articles for publication and a Management Plan for Mala in the Northern Territory, to be prepared following guidelines set out in the Conservation Commission's "Draft Strategy for Conservation of Endangered Species and Ecological Communities in the Northern Territory".

1.2 Determine cost effective methods for stress mitigation to optimize establishment of reintroduced Mala.

1.2.1 Delivery and use of artificial food and water supplements.

Information pertaining to this action including dietary analyses and isotope turnover studies have been derived from the Sangsters Bore reintroduction. The seasonal trends revealed by both investigations have important implications in terms of the timing and long-term supply of supplements to reintroduced populations of Mala.

1.3 Determine habitat management requirements in the post-establishment phase.

1.3.1 Movement patterns, survival and dispersal in response to fire.

This action has not been implemented subject to a more thorough revision of the recovery plan. It is recognized as an important management question but no action is recommended until reintroduced populations are established and island populations are resurveyed.

1.4 Clarification of the status of the three recognized subspecies and establish genetic relationships among individuals in the captive breeding colony.

1.4.1 Clarification of subspecific status through DNA analysis of island and mainland Mala.

1.4.2 DNA fingerprinting of captive and released populations.

The genetic component of the recovery plan has been a co-operative project involving the CCNT, WA CALM and Curtin University. Completion of this action was delayed due to the problems in extracting sufficient quality DNA from Dorre Island samples obtained from WA CALM. Further blood samples were collected by Dr Tony Friend of WA CALM during a trip to Dorre Island in August 1994; extraction of DNA and analysis of these samples is currently underway. Preliminary results indicate significant differences between the island and mainland forms of Mala similar to those revealed by the morphometric and allozyme studies of Courtenay (1993). In addition they reveal an encouraging level of heterozygosity within the captive population. Any differences between Bernier and Dorre Island populations should be revealed by analysis of the remaining Dorre Island samples. A final report and publication will be prepared for wider circulation.

Management phase

1.1 Maintenance of naturally occurring mainland and island populations and those reintroduced to Sangsters Bore and the Lander River.

1.1.1 Monitoring of island populations.

Trips to Bernier and Dorre Islands by WA CALM and CSIRO Division of Wildlife and Ecology personnel have been undertaken in relation to other research and management work in that area, but no detailed survey work has been undertaken since Short and Turner (1992). Correspondence received from Dr Tony Friend of WA CALM suggests a possible decrease in the Mala population on Dorre Island and highlights the need for a systematic resurvey of the island Mala populations.

1.1.2 Management and monitoring of mainland populations.

Monitoring of both mainland reintroduced populations has continued with regular field trips to each site. Controlled burning operations were increased in response to large wildfires which had burnt in the vicinity of both sites during the 1994–1995 summer seasons. NOAA satellite imagery was used to assist in these operations and the value of this approach has ensured that it will be incorporated in all future fire management.

A small free-living population of reintroduced Mala still exists at the Sangsters Bore site and trapping results indicate encouraging long-term survival by some released individuals and moderate recent recruitment of independent young to the population. Dingo activity at the site has increased over the past 12 months and there has been a coincident decrease in both fox and feral cat activity. A small number of Mala have fallen prey to dingoes and there has been some evidence of dingoes interfering with supplementary water points. Monitoring of the enclosure held Mala population at the Lander River site has continued, as has routine maintenance of the electric fence and feed and water supplements. The most recent estimates of the Mala population indicate no significant change in the size over the past 12 months. Significant recruitment of independent young to the population has occurred but this appears to be matched with a significant mortality of previously marked animals.

1.2 Maintenance of captive breeding colony in Alice Springs.

The captive breeding program is currently divided between three institutions following the establishment of a small population at the Monarto Zoological Park in October 1994. All breeding and management of the population is being co-ordinated from Alice Springs using the SPARKS database with assistance from Western Plains Zoo. There are no plans to greatly expand the current captive holdings of Mala.

Preliminary trials into artificial breeding of Mala were conducted at Western Plains Zoo involving Queensland University. Semen was collected from two males transferred from Alice Springs which had not successfully sired offspring. Future work will investigate storage and reclamation of semen samples and oestrus detection in females.

1.3 Negotiation of conservation areas in the Tanami Desert with Aboriginal traditional owners.

Negotiations between CCNT and the Central Land Council have continued for establishment of conservation reserves in the Tanami, as required by the findings of the successful 1978 land claim. Minimal progress towards this end has been achieved and no timetables have been set in relation to this process.

1.4 Reintroduction of Mala to two major central Australian national parks.

1.4.1 Negotiation with Aboriginal traditional owners and relevant park agency on agreed procedures for executing a reintroduction program.

1.4.2 Preparation of a management prescription for cost-effective reintroduction and subsequent management.

1.4.3 Implementation of prescription at two areas coincidently to maximize efficiency of resources.

These actions appear likely to be excluded from the revised recovery plan in favour of other translocation options covered in a discussion paper circulated to recovery team members (Lundie-Jenkins, pers. comm.). Alternative sites both on offshore islands and in other regions of the mainland were considered to offer enhanced chances of successful establishment of Mala.

The revised plan

The revised version of the Mala Recovery Plan was completed in 1995 (Lundie-Jenkins 1995) after significant consultation with the Mala Recovery Team and representatives of the member State agencies. This revised plan better reflected the national priorities for the Mala and incorporated results derived from implementation of the original plan. In addition the plan was modified to account for the change in the species' status as influenced by the extinction of the only known "naturally occurring" mainland population of Mala. The revised recovery plan has identified a number of potential new approaches in an attempt to secure the species' future. A copy of the summary page of the revised plan is shown in Appendix 1. The most significant amendments to the recovery plan are listed below:

1. Translocation of the mainland form of Mala to a predator-free off shore island.
2. Improved monitoring of the Bernier and Dorre Island Mala populations.
3. Translocation to secure mainland sites, including several sites in southwestern Western Australia as part of WA CALM's Western Shield Project (Keith Morris, pers. comm.).

In addition to changes to the actions within the revised plan, the recovery objectives and criteria have similarly been modified to reflect changes in the species' status and the direction of the recovery effort. Within the present document these criteria are listed only as single dot points and it would seem appropriate that future development of the Mala recovery plan will include expansion of these points into more measurable yardsticks. In this way the recovery team will be better able to evaluate the progress of the recovery plan towards achieving its stated objectives.

LESSONS, IMPROVEMENTS AND FUTURE DIRECTIONS

Clearly the processes of developing and implementing a recovery program have underdone significant evolutionary changes since ANCA initiated its Endangered Species Program in 1989. A considerable amount has also been learnt from actual involvement in the process as well as from the experiences of other recovery teams. The paragraphs below summarize some of the important operational questions which have been answered during the first five years of the Mala Program.

What comes first, the plan or the team?

Many of the early recovery plans developed under the ESP, including the initial version of the Mala Recovery Plan, were prepared in advance of any moves to establish an actual recovery team. As a consequence, this initial document represented only the views of the authors and not of a more broadly based recovery team. Since authorship of the Mala Recovery Plan rested with a single agency, only limited information was included in relation to resourcing, priority setting and existing recovery effort for other agencies. Subsequent revision of the recovery plan has benefited greatly from the diverse input of the recovery team. In addition, this team involvement has served to delegate responsibilities to team members and the agencies they represent. This increased "ownership" of the recovery plan should improve actual implementation of the plan. The clear lesson is that the process of developing a recovery plan should ideally be an interactive process involving all major stakeholders. Logistics and the location of the extant populations will however frequently dictate that the major input to preparing a recovery plan will be borne by a single individual or small group.

Who should be on the team?

During the initial formation of the Mala recovery team, significant concern was expressed with regard to the size and composition of the team. The basis for selecting representatives for the team clearly had implications for the development and implementation of the recovery plan. An initial criterion for determining membership of a recovery team would be based on the historic distribution of the species involved. Clearly however, for species with extensive continental distributions like the Mala, this could lead to an unworkably large team. Whilst it is probably unreasonable to exclude organizations or individuals from involvement on the team, it is appropriate that they should be able to justify their position/involvement on the team. Appointments based on purely political reasons or as a form of "tokenism" do little to progress the recovery of a species and may in fact hinder it. It is important that the recovery team is focused and proactive and not simply content to indulge in the process.

For species like the Mala where recovery teams cover several States, it may be appropriate to establish small individual State working groups to co-ordinate State actions and input into the planning and recovery process. The logistics of a central co-ordinator or broadly dispersed recovery team overseeing actual implementation of all actions make the formation of such working groups an important component of the recovery program.

What role do team members play?

Potentially the greatest strength of a recovery team is the diverse range of skills and experiences it draws together. The task of realizing this potential is not always so easily achieved. Development of a recovery plan and formation of a recovery team serves to some extent to identify the responsibilities of organizations and individuals in relation to recovery actions. However these actions alone do not specify the roles or duties of recovery team members, nor do they identify the skills these individuals bring to the team. It is clear that some attempt at defining roles and identifying team and individual skills would improve both the development and subsequent implementation of a recovery plan. Involvement of professional zoo-based staff and an independent publicist on the Mala recovery team has lead to significant improvements in the captive breeding and education/promotional aspects of the plan respectively.

How often should recovery teams meet?

It is apparent that the frequency of meetings will change considerably between recovery teams, principally as a function of both the geographic spread of its members and the types of actions being implemented. In the case of the Mala recovery team at least, annual "whole team" meetings appear to be sufficient to report on implementation and revision of national actions. Smaller State based working groups would benefit by more frequent meetings to discuss the finer technical aspects of local actions. As indicated in a previous section, teams must avoid indulging in the process of holding meetings without actually progressing species recovery. The majority of the people involved on the Mala recovery team have significant work loads in addition to membership of the team. The sole criteria for determining the frequency of meetings should be the need to progress implementation and reporting of the recovery program. In some instances it will be appropriate to include costings within the recovery plan to cover travel expenses for key team members to attend meetings. This should only occur where their personal or organizational resources are insufficient to meet these costs and their attendance is critical to advancing the recovery process.

How do we make sure it all goes according to "plan"?

Given that significant funding has been provided to the Mala recovery program on the basis of the original recovery plan it would seem, with the benefit of hindsight, that some form of external review of the plan prior to funding/implementation may have been appropriate. Such a review process could ensure that proposed actions are appropriate, scientifically valid and realistically costed. Whilst a recovery team will generally contain those people with the greatest experience and expertise in relation to a particular species, some form of external review ensures also that broader perspectives and alternative approaches are not overlooked. Models for similar reviews have been variously developed for other forms of project evaluation such as Australian Research Council grants.

The review process undertaken by the Mala Recovery Team has clearly shown the importance of the recovery plan being viewed as a dynamic document and not a static contract. To be effective, management of ecological systems and the species they contain must realistically be adaptable to changes in technology, species' status, priorities and economics. Clearly, actions within a recovery plan which on review appear unlikely to return the desired benefits towards recovery of a species should be suspended or modified. The enormous expense of recovery programs, particularly for mammal species, dictates that an "adaptive management" approach be taken to species recovery. The most important elements of the recovery plan to facilitate this approach are clear, measurable success/recovery criteria and regular internal review.

The Mala recovery program, as with other ANCA-supported recovery programs initiated in 1991, is scheduled for a major review during 1996. It is important that such reviews not be seen as adversarial, but as an opportunity to develop and refine the recovery program to the ultimate benefit of the species. As with the proposed initial review of the recovery plan, this review should involve people external to the recovery team. As an open and expected component of the recovery process, it should lead to improvements in the development and implementation of recovery programs for the growing list of endangered Australian plants, animals and communities.

CONCLUSIONS

Whilst some ground has been lost with the extinction of the only known population of the mainland form of the Mala, our knowledge of the species and the means of managing its recovery have improved substantially.

The critically endangered status of the Mala (using IUCN 1994 criteria) ensures that there will be an ongoing recovery effort for the species. The next 12 months will involve a period of consolidation and review of the existing recovery programs. A wealth of management related information has been derived which will be used to formulate management plans and to determine future directions. The final decision on these and other options for the species will rest with the recovery team. Clearly the prime objective in the short term will be to improve the status of the mainland form of the Mala and hence provide more opportunity to investigate higher risk options for re-establishing the species. The current critically endangered status for the species, with numerically small and geographically limited populations, dictates that our current approach be more conservative.

ACKNOWLEDGEMENTS

The recovery program for the Mala owes its origins and impetus to work initiated by Dr Ken Johnson in the Tanami Desert over 15 years ago. I am extremely grateful to the members of the Mala Recovery Team and specifically my co-workers at the Arid Zone Research Institute for their support and commitment to the conservation of this endearing arid zone wallaby. Implementation of the Mala Recovery Plan has been generously supported by the Australian Nature Conservation Agency, World Wide Fund for Nature (Australia), National Estate Grants-in-Aid Program, Pacific Asia Travel Association and the Central Australian Tourism Industry Association. Tony Robinson and Keith Morris provided valuable comments on earlier drafts of this manuscript.

REFERENCES

Bolton, B. L. and Latz, P. K., 1978. The Western Hare-wallaby *Lagorchestes hirsutus* (Gould) (Macropodidae) in the Tanami Desert. *Aust. Wildl. Res.* **5**: 285–93.

Courtenay, J., 1993. The systematics of the Rufous Hare-wallaby *Lagorchestes hirsutus* Gould 1844. International Theriological Society Congress: Sydney.

Endangered Species Unit, 1995. Status of recovery plans for species listed as endangered or vulnerable under the *Endangered Species Protection Act* 1992. Australian Nature Conservation Agency: Canberra.

Gibson, D. F., Johnson, K. A., Langford, D. G., Cole, J. R., Clarke, D. E. and the Willowra Community, 1995. The Rufous Hare-wallaby *Lagorchestes hirsutus*: a history of experimental reintroduction in the Tanami Desert, Northern Territory. Pp. 171–76 *in* Reintroduction Biology of Australian and New Zealand Fauna ed by M. Serena. Surrey Beatty & Sons: Chipping Norton.

IUCN, 1994. The International Union for the Conservation of Nature Red List of Threatened Animals. Gland.

Johnson, K. A. and Lundie-Jenkins, G., 1992. Recovery Plan for the Mala. A report submitted to the Australian Nature Conservation Agency: Canberra.

Lundie-Jenkins, G., 1993. Ecology of the Rufous Hare-wallaby, *Lagorchestes hirsutus* Gould (Marsupialia: Macropodidae), in the Tanami Desert, Northern Territory. I. Patterns of habitat use. *Wildl. Res.* **20**: 457–76.

Lundie-Jenkins, G., Corbett, L. K. and Phillips, C. M., 1993. Ecology of the Rufous Hare-wallaby, *Lagorchestes hirsutus* Gould (Marsupialia: Macropodidae), in the Tanami Desert, Northern Territory. III. Interactions with introduced mammal species. *Wildl. Res.* **20**: 495–511.

Lundie-Jenkins, G., Phillips, C. M. and Jarman, P. J., 1993. Ecology of the Rufous Hare-wallaby, *Lagorchestes hirsutus* Gould (Marsupialia: Macropodidae), in the Tanami Desert, Northern Territory. II. Diet and feeding strategy. *Wildl. Res.* **20**: 477–94.

Lundie-Jenkins, G., 1994. Recovery Plan for the Mala. Revised Edition. A report submitted to the Australian Nature Conservation Agency: Canberra.

Lundie-Jenkins, G., 1995. Conservation of the Rufous Hare-Wallaby, Northern Territory (Project No. 153) An interim report to the World Wide Fund for Nature (Australia): Sydney, Australia.

Lundie-Jenkins, G. and Bellchambers, K., 1995. Reintroduction of the Rufous Hare-wallaby into Aboriginal land in the Lander River region of the Tanami Desert, Northern Territory. Technical Report to the Conservation Commission of the Northern Territory: Alice Springs.

Short, J. and Turner, B., 1992. The distribution and abundance of the banded and rufous hare-wallabies, *Lagostrophus fasciatus* and *Lagorchestes hirsutus*. *Biol. Conserv.* **60**: 157–66.

APPENDIX 1

Summary page from the Mala Recovery Plan (Lundie-Jenkins 1995)

Current Species Status:

Endangered (ANZECC, 1991), Specially Protected (Regulations of the Territory Parks and Wildlife Conservation Act), Appendix 1 of CITES. Formerly distributed throughout spinifex deserts of central and western Australia. Mainland subspecies now presumed extinct in the wild and reduced to captive colonies and two experimental reintroduction programs. Bernier and Dorre Island subspecies (Western Australia) apparently stable.

Habitat Requirements and Limiting Factors:

Formerly in low shrubland of eastern Wheatbelt (Western Australia) but mainly in spinifex hummock grasslands of central deserts (Northern Territory, Western Australia and South Australia). Tanami colonies formerly associated with saline paleodrainage system, sanddunes and tight fire patterns. Large areas of spinifex desert appear suitable providing exotic predators (feral cats and foxes) and rabbits are at low densities or controlled and fire is properly managed.

Recovery Plan Objectives:

Secure and maintain existing natural populations and effect reintroduction of Mala into secure habitats.

Development of effective management and translocation prescriptions for the species.

Recovery Criteria:

1. Two self-sustaining populations of mainland sub species.
2. Clear management prescriptions for island and mainland populations.
3. Successful negotiation of conservation areas in the Tanami Desert.
4. Clarification of the species' genetics.

Actions Needed:

A recovery team comprising members from CCNT, Western Australia CALM, ANCA, CSIRO, Western Plains Zoo and other organizations, as appropriate, will be appointed to co-ordinate and supervise the following actions:

1. Monitoring of wild and reintroduced Mala populations and habitat.
2. Management of habitats in and adjacent to Mala populations.
3. Clarify the population genetics of the three recognized sub species of Mala.
4. Captive breeding to provide animals for translocations.
5. Translocation of mainland sub species to an offshore island.
6. Translocation of Mala to secure mainland habitats.
7. Recovery team administration.

Costs of Recovery: 1993 prices in $000s/year. Total cost (TC) and Endangered Species Program (ESP) funds required (= TC – other contribution).

	(1)		(2)		(3)		(4)		(5)		(6)		(7)		Total	
Action	TC	ESP	TC	ESP	TC	ESP	TC	ESP	TC	ESP	TC	ESP	TC	ESP	TC	ESP
1993	66	40			18	10									84	50
1994	66	40													66	40
1995	66	40							4	—					70	40
1996	112	75	11	4			38	17	10	1			11	3	182	100
1997	92	70	11	4			38	17	65	44			11	3	217	138
1998	92	70	11	4			38	17	10	—			11	3	109	41
1999	54	22	11	4			21	—	—	—	NB	NB	11	3	97	29
2000	39	17	11	4			21	—	—	—			11	3	82	24
2001	39	17	11	4			21	—	—	—			11	3	82	24
Total	573	338	66	24	18	10	177	51	89	45	NB	NB	66	18	989	486

NB = Not budgeted in current recovery plan.

Biodiversity Benefits:

Reintroduction, habitat management and exotic species control to benefit many C.W.R. species, two extant species and nine mammals and one bird that could be reintroduced.

Reviewing recovery programs for endangered species: some considerations and recommendations

Gary N. Backhouse,[1] Tim W. Clark[2,3] and Richard L. Wallace[3]

Implementing recovery programs for endangered species is one method commonly used for conserving biological diversity. These programs can be long-term and complex operations. Although many programs are now operating around the world, there are few signs of success, despite a steadily increasing number of endangered species. One important method for assessing outcomes of recovery efforts is undertaking program reviews. Reviews provide the opportunity to measure outcomes against recovery objectives, and can improve performance by early detection of problems and learning lessons from the review experience. Reviewing and evaluating some faltering recovery programs has been instrumental in improving their subsequent performance. A number of considerations for undertaking comprehensive reviews are presented, covering review types, integrating review into recovery planning, setting clear goals, stressing broad participation and involvement, using external reviewers, having all information available, undertaking a structured process, covering all aspects of recovery programs and documenting review outcomes. The application of a systematic approach can make the review process more comprehensive and aid other professionals in their own program reviews.

Key words: Endangered species, recovery programs, recovery planning, reviews.

INTRODUCTION

IMPLEMENTATION of recovery programs for endangered species is a high-profile aspect of biodiversity conservation. Scores of restoration efforts are underway world-wide, and the number is increasing. Some recovery programs have been running for many years; the Californian Condor *Gymnogyps californianus* program in California for about 50 years (Koford 1953; Snyder 1994), the Helmeted Honeyeater *Lichenostomus melanops cassidix* program for almost 30 years (Smales *et al.* 1995) and the Mala *Lagorchestes hirsutus* program for more than 15 years (Gibson *et al.* 1994).

Issues involved in recovery may be relatively simple, such as controlling predation in a single location and involving few agencies and individuals (e.g., Robertson *et al.* 1994). However, most often recovery programs are quite complex, covering a wide range of issues and involving individuals representing many agencies, groups and interests. For instance, the Florida Manatee *Trichenus manatus latirostris* Recovery Plan (FWS 1980; Wallace 1994) has 13 signatories representing major government agencies, industry and community interest groups. There may be many complex, vexing factors to simultaneously address in recovery, such as habitat destruction, disease, human consumption, political differentials among participants, special interests and poor planning impacting on program implementation, to mention only a few (e.g., Yaffee 1982, 1994; Alvarez 1993; Rabinowitz 1995).

The many recovery efforts under way, and those just starting or planned may eventually prove successful. Unfortunately, the record currently shows fewer endangered species recovery successes than might otherwise be expected given the relatively large number of programs and length of time and considerable effort involved, at least when measured by the number of fully recovered species (e.g., Tear *et al.* 1993). Given this tentative observation, it may simply be too soon to be drawing such conclusions about success rates, because final outcomes are not yet evident. However, part of the apparent low success rate in recovery of species may simply be the result of too few reviews of the actual effectiveness of many recovery programs. There may be more successes or partial success than is being reported currently.

Reviews should be an important and integral part of any policy process, especially biodiversity conservation efforts (Lasswell 1971; Brewer and deLeon 1983; Backhouse and Clark 1995). Whether called program reviews, evaluations, or appraisals, program-specific reviews should be undertaken "with the aim of determining how well a program is working,

[1]Department of Natural Resources and Environment, Flora and Fauna Branch, P.O. Box 41, East Melbourne, Victoria, Australia 3002.
[2]Northern Rockies Conservation Co-operative, Box 2705, Jackson, Wyoming 83001 USA.
[3]Yale University, School of Forestry and Environmental Studies, 205 Prospect Street, New Haven, Connecticut 06511 USA.

Pages 170–79 *in* BACK FROM THE BRINK: REFINING THE THREATENED SPECIES RECOVERY PROCESS ed by S. Stephens and S. Maxwell. Surrey Beatty & Sons, 1996.

and whether it needs to be improved or abandoned" (Leeuw *et al.* 1994:18). Reviews are the very basis for learning and "adaptive management" in conservation (Clark 1996). As Michael (1995:484) noted, "there are two kinds of learning: one for a stable world and one for a world of uncertainty", and endangered species conservation is part of the world of uncertainty. Learning in endangered species recovery must involve systematic, continuous internal and external review.

Various frameworks for reviewing endangered species programs exist (e.g., Clark *et al.* 1995a). Reviews by Reading *et al.* (1992) of the Eastern Barred Bandicoot *Perameles gunnii* program, Yaffee (1994) of the Spotted Owl *Strix occidentalis caurina* program in Washington, Oregon and northern California, and Clark (in press) of the Black-Footed Ferret *Mustela nigripes* program in the USA are examples of thorough programmatic reviews that are of interest to a broad audience and can aid species specific recovery efforts.

In this paper, we present a practice-based approach for reviewing endangered species recovery programs, based on our experience with recovery programs in Australia and the USA over the last 23 years. In the following three sections, we briefly examine some of the background, history, and benefits of reviews and list nine considerations to cover in reviews. This approach can aid other professionals in their own program reviews.

BACKGROUND AND REVIEWS IN ENDANGERED SPECIES RECOVERY

History of reviews

The concept of review is certainly not new in the field of biodiversity conservation. For example, scientific journals and publications use peer review prior to accepting papers for publication. However, moving the concept into the field and into ongoing endangered species recovery efforts in practical ways has not been easy for several reasons.

First, the review that does take place often concentrates on biological and technical aspects of recovery, and such self-constrained reviews dominate the published literature. Although many other important variables such as organizational and professional dimensions, economics, social issues, and politics can and do have profound effects on restoration efforts, few reviews address these factors explicitly or systematically (Reading *et al.* 1991). As a result, many reviews lack realism and completeness.

The literature seeking to overcome this deficiency is growing, especially in the USA, where there have been significant reviews covering many aspects of endangered species recovery programs (e.g., Yaffee 1982; GAO 1988; NRDC 1989; Tobin 1990; Kohm 1991; Tear *et al.* 1993). Several major, high profile recovery programs in the USA, including the Florida Panther *Felis concolor coryi,* Florida Manatee, Grizzly Bear *Ursus arctos horribilis,* Black-footed Ferret, Californian Condor and Red-cockaded Woodpecker *Picoides borealis* are reviewed in Clark *et al.* (1994). Collectively, these reviews identify and discuss many aspects of overall recovery efforts, including conservation science, conflict management, organizational, professional, and social issues, and the policy sciences; issues often neglected in conventional reviews.

Second, there is growing concern about the effectiveness of some recovery programs. For example, the Kakapo *Strigops habroptilis* recovery program in New Zealand was subject to external review when the New Zealand Government appointed specialists to review the program because of mounting public and sponsor concern over its directions and costs (Bryant 1995; Imboden *et al.* 1995). Reviews of the Sumatran Rhino *Dicerorhinus sumatrensis* program in Borneo (Rabinowitz 1995) and Leadbeater's Possum *Gymnobelideus leadbeateri* conservation efforts in Australia (Lindenmayer and Norton 1993) were also prompted by concern about the effectiveness of these recovery efforts. These concerns can be expected to grow in the future.

Third, there have been relatively few published systematic reviews of recovery programs in Australia to date. Short *et al.* (1992) recognized the need for program reviews, especially of those that did not work as planned, in their examination of the reasons for failures of translocations of threatened macropods in Australia. One program that is an exception and has been the subject of extensive review is the Eastern Barred Bandicoot recovery program in Victoria. This was spurred by the species decline to near extinction in the wild despite considerable management and research effort, and reviews have covered many aspects of that program (Reading *et al.* 1992; Backhouse *et al.* 1994; Clark *et al.* 1995b; Myroniuk 1995).

However, the trend is now toward increasing use of reviews throughout Australia. For example, aspects of conservation policy and several recovery programs were reviewed in Backhouse and Clark (1995). In Victoria, all major recovery programs are now subject to regular reviews, including several programs (Orange-bellied Parrot *Neophema chrysogaster*, Regent Honeyeater *Xanthomyza phrygia* and Spotted Tree Frog *Litoria spenceri*; K. Lowe, pers. comm.) that operate in association with other states and the Commonwealth. The Department of Conservation and Land Management in Western Australia now requires review of translocation proposals, and the review process is stated as part of its threatened flora and fauna translocation policies (CALM 1995). Clearly there are many positive benefits of reviews that are now being appreciated.

Benefits of reviews

Recovery programs are just one manifestation of nature conservation policy, and reviews are critical to assessing and improving the performance of any complex effort. Review and evaluation of recovery programs constitutes the point where the *problem* (a declining species or community threatened with extinction), the *solution* (legislation, regulation, recovery programs) and the *experience* (history and outcomes of recovery efforts) converge (Clark *et al.* 1995a). Reviews provide insight into how biodiversity conservation policy is actually being implemented in the field, in decision making processes, and it is the basis for determining whether legislative and administrative goals are actually being met or not (Backhouse and Clark 1995).

The goal of reviewing recovery programs is to encourage future improvements in program performance. Reviews offer the opportunity to evaluate a program's performance against stated goals, gain an understanding of how a program works, and facilitate change as circumstances demand. Reviews are critical to an organization's ability to learn from its actions and mistakes and as a basis for improving implementation. Reviews also provide a basis for identifying those elements that work in recovery programs for transfer to other programs. Frequent review is the key to successful recovery programs. Some reviews may lead to substantive changes in a program's structure or implementation strategy, whereas others may result only in the creation of "intellectual capital" or "better information, increased understanding of and/or better reporting on the program and its performance" (Leeuw *et al.* 1994:19).

Reviews may take many forms and be conducted with varying levels of rigour, leading to various types of results. In the field of conservation policy, as in all policy-related fields, "trade-offs are typically required as policymakers and managers try to obtain valid, timely, but not overly expensive evaluative information" (Wholey *et al.* 1994:1). While formal, large-scale reviews are important when a goal is to stimulate potentially broad programmatic changes, reviews of all types may result in information that is beneficial to implementing or reviewing programs in the future.

Though deciding upon the scale of a review is critical to what benefits it is desirable to achieve, the timing of reviews is also important. One important example of recovery program evaluation is the Eastern Barred Bandicoot recovery program (Reading *et al.* 1992). The review occurred at a time when the program was faltering, and proved essential in identifying weaknesses that led to termination of that recovery effort and commencement of a new broader-based program resulting in significant improvement in the species' status (Backhouse *et al.* 1994; Clark *et al.* 1995b). Improved review and analysis of data on captive breeding of Eastern Barred Bandicoots also led to rapid improvements after several years of limited success (Myroniuk 1995). The Helmeted Honeyeater recovery program in Victoria utilized rigorous evaluation that resulted in improvements to that program (Smales *et al.* 1995). These and other programs have benefited from reviews.

Program review should be constructive and identify ways to advance program implementation quickly and productively. This is most likely to occur in organizations that conduct evaluation internally in supportive environments. One benefit of institutionalizing reviews is to avoid the need for crisis-initiated reviews, which can be quite disruptive and stressful. Providing for timely reviews can alert managers to potential or developing problems so that they can be addressed early or avoided. Organizations that have developed internal review procedures are more likely to value the acquisition of knowledge that results from evaluation; that is, they are more likely to have an expectation and desire to learn from the review process and a desire to utilize what they learn to improve program implementation (Leeuw *et al.* 1994).

Reviews can be difficult, especially where information is sketchy or lacking, or if recovery efforts are too embryonic to evaluate performance or outcomes. In this situation evaluation is still important because the process can highlight uncertainty or the lack of critical information, issues that will need to be tackled for the program to move forward.

REVIEWING ENDANGERED SPECIES RECOVERY PROGRAMS

An explicit, systematic review process is required to ensure that all aspects of recovery are adequately covered. Too often reviews are inadequate and consist only of a superficial reporting of activities since the last meeting. Reviews should be rigorous, analytical, seek to determine if the recovery program is on track, how rapidly it is meeting its intended aims and what foreseeable problems loom ahead. Based on our experience with reviews of a number of programs in Australia and the USA we offer the following nine points for consideration.

1. Types of reviews

Reviews falls into two categories — internal and external (Leeuw *et al.* 1994). The two types are different, and both are useful for conducting successful programs. Most reviews (if they occur) are usually internal and informal. More external and formal reviews can significantly complement and aid endangered species recovery.

Internal reviews may occur amongst some or all participants, usually on a regular basis such as at meetings of recovery teams, working groups and operations groups as part of their routine functions. Internal reviews may take many forms. They may be unstructured discussions between two or more individuals from which little more than an understanding of what has occurred is achieved, or they may be formal, meticulous undertakings meant to provide agency chiefs with a program update or the rationale for program modification. Output may simply be recording information and decisions as part of the meeting minutes. Other internal reviews may be formal, either large or small scale, meant to monitor progress and to note changes that might be made to improve implementation. In Canada, for example, a government-wide policy created in 1977 directs agencies and departments to regularly review the performance of their programs in order to develop strategies to guide their improvement (Mayne 1994).

External review may also take numerous forms, and is usually a major structured process, involving people not directly associated with the recovery program (in some cases may even be conducted entirely by people not formally associated with the program). Formal reviews might be conducted by reviewing published and unpublished written materials, seeking written comments or conducting interviews, but the more usual form is in a workshop-type forum. The outcomes may be prepared in a formal written report, or papers on aspects of the recovery program prepared for publication along with recommendations for improving matters or a listing of lessons learned.

Such reviews might occur on an annual basis, or at the planned termination of a recovery program, where an extensive evaluation takes place as part of the transition to or preparation of a new program. Annual review is often convenient as it might coincide with the funding cycle for many organizations, and also be frequent enough to cope with rapidly changing situations in the field. There may also be changes in staffing involved in recovery efforts and membership of organizations and agencies represented on recovery teams, and annual reviews present the opportunity for new participants to quickly become knowledgeable of recovery issues. Major, external reviews may also be initiated on the basis of perceived problems with programs (e.g., Reading *et al.* 1992; Lindenmayer and Norton 1993; Imboden *et al.* 1995). Reviews may involve all direct participants and those who may have some peripheral involvement or who are affected by the program, such as line managers of participants. Regardless of the type or combination of reviews, they should be integral to the overall recovery effort.

2. Integrate reviews in recovery planning and practices

The review process needs to be fully integrated into recovery programs. Regular, internal reviews should be part of routine operations of key groups involved in recovery, such as recovery teams, advisory committees and working groups. Extensive, more formal review, either internal or external, could be held on a planned, regular basis, such as annually or bi-annually, and at the termination of a program. Reviews should be seen as a part of the recovery planning cycle — they simultaneously can be the end of one phase and the start of the next.

Although objective, external review is important in programs in which politics and other non-technical recovery-related factors influence decision making, internal evaluation is critical for in-house encouragement of program improvement. One substantial benefit of integrated internal reviews is that they can be managed by individuals who, after conducting or directing a review, may then be in a position to use the outcome to promote change from within their program or organization (Sonnichsen 1994). In short, internal evaluators may be designated in turn as both observer and advocate, dual roles that might save time, money and resources in programs operating on tight schedules and small budgets.

Reviews should be an action specified in all recovery plans. For instance, the recovery plan for the Eastern Barred Bandicoot specifies an annual program review (Backhouse and Slater 1995) and the Florida Manatee Recovery Plan mandates regular evaluations of recovery tasks and enforcement activities (FWS 1980). Recognizing the importance of reviews as part of recovery programs, all major recovery programs in Victoria now have a regular (usually annual) major review, as well as review on termination of the recovery plan. Action statements prepared under the Victorian *Flora and Fauna Guarantee Act 1988* also specify a review date (Mansergh *et al.* 1995; Wilson and Clark 1995). Reviews should be carried out seriously, not as a mere formality.

3. Setting clear goals

Setting clear goals, with measurable objectives, targets, timelines, and assigning responsibility and accountability are key aspects of successful implementation of recovery programs (Backhouse and Clark 1995). Ongoing feedback is essential for good programs and this depends on the program having clear goals against which performance can be measured (Clark *et al.* 1994). Without clear goals, tracking progress and movement toward recovery is impossible.

Each recovery component should have its own targets, timelines, and assigned responsibilities, which are derived from the overarching recovery program goals. For example, goals should be set for population targets, habitat, public involvement, decision criteria, and other essential components. With clear goals in place, effective reviews can quickly measure progress and detect developing problems before they become a crisis. Goals need to be specific, with quantitative targets, identified time lines and responsibilities for achievements, which then enables quick precise assessment as to whether the goal has been achieved or not and what contribution it has made (if any) to the recovery process.

The lack of clearly defined goals hindered both early progress and an informal, internal review of the Eastern Barred Bandicoot program (Reading *et al.* 1992). Subsequent to this review, explicit goals were set and they have significantly aided continuous and annual reviews over the last few years. Short, medium and long-term goals are now routinely set in recovery plans in Victoria. Goals should be set high enough to reflect real species recovery (Tear *et al.* 1993). However, goals should not be overly rigid as recovery programs are dynamic and highly uncertain and goals may need to be changed to reflect changes in the species, threatening processes and the program itself (Backhouse and Clark 1995).

4. Involve all participants

Participation and representation, as well as thorough deliberation, are essential features of good reviews. Whether the review is external or internal, involving all participants is essential for a successful review. Internal reviews might directly involve only members of the pertinent team or group. Major external or internal reviews utilizing a workshop format should involve as many participants as possible (certainly all the key participants) in the recovery program, including program managers, biologists, technicians, sponsors, and community representatives.

Each individual can contribute to the review, and broad inclusion is the best way of ensuring all aspects, issues, problems, and solutions are considered. Reviews are also about participants identifying and taking responsibility for their contribution to the recovery efforts. Few participants may possess a comprehensive, detailed overview of the whole recovery effort, especially if they are working on a small section of the program or in a single location. Participating in reviews may be the only way for some individuals working on part of a program to fully understand and appreciate the entire program. All participants should be assured that they have something to offer and that they are obliged to contribute to ensure the best outcomes. Reviews rely on honesty and full participation.

It is also helpful to include other people who can contribute to the review, such as managers and experienced participants in other recovery programs or external experts in conducting program evaluations. Such individuals can often provide invaluable insights and comments and different perspectives on recovery programs or program implementation, even if they have not been involved in the specifics of the particular program under review.

5. Using external reviewers

External, professional review by experienced reviewers is very useful in appraisal and may open up insights that would otherwise be missed by the team working closely with a species. An external reviewer is someone who is not directly involved in the program but may be familiar with it or aspects of it and other endangered species programs. The person may be a manager of an agency involved in the recovery effort, someone from another recovery team, a professional facilitator, or a specialist in organizational behaviour or program implementation and review. The individual should at least be knowledgeable with recovery processes for endangered species, and will need to be fully briefed and have access to all appropriate people and information.

An external reviewer facilitating a recovery workshop can enable all participants of recovery programs to fully participate in the review, rather than have a single individual who is directly involved in the program running the workshop as chair or facilitator and also trying to contribute as well. In Victoria, external reviewers are usually invited by program managers as a part of regular program review and audit. Finding someone with experience can be a problem. However, the basic skill required is in running a productive meeting, following an agenda, running to time, keeping discussion on issues of relevance, coming to clear conclusions wherever possible and identifying actions and responsibilities.

In some instances, external review may be prompted by perceived problems with a program (e.g., Lindenmayer and Norton 1993; Rabinowitz 1995) or rarely might be imposed on a recovery program where a crisis or negative perceptions persist over program performance and outcomes (e.g., Imboden *et al.* 1995). These can be stressful and traumatic for participants. With several high-profile recovery programs in the USA perceived to be faltering, some authors have argued for impartial periodic review of the whole endangered species recovery process as well as each species recovery program (e.g., Reading and Miller 1994). Jackson (1994) calls for an independent review organization as one element of the entire national recovery effort in the USA. Program managers might do well to seek periodic external review to achieve the best results; previous examples (Reading *et al.* 1992; Imboden *et al.* 1995) well illustrate the value of external reviewers.

6. Use all available information

Information is the primary resource in decision making, and therefore in program review (Browne 1993). Creating a constructive climate and ensuring the use of all available information is essential — comprehensive reviews are only possible in an environment of full and freely available information which can enable more detailed problem definition, evaluation and solution preparation (Dery 1984; Weiss 1989). Information should be available to all recovery team participants to assess its reliability, rather than left to one person or a few individuals to determine its utility. Using all available information helps to reduce uncertainty in decision-making, and is essential in applying techniques such as decision analysis and risk analysis (Maguire 1986, 1990) to help in option selection and decision making in the complex environments of endangered species recovery efforts. Information on biology and ecology of the species or community and threatening processes is obviously important. In addition, information on less obvious aspects such as organization and management of recovery efforts, staffing, budgets, resources and decision-making processes are all relevant to reviews.

In some instances, information may be withheld from reviewers for various reasons. Some research may not be complete or results may not be made available until they have been formally published. Issues of confidentiality may be cited as reasons for withholding information. In some cases, information may be withheld by agencies responsible for conservation management for reasons of self-interest (Mattson and Craighead 1994), or to avoid agency or individual embarrassment or as a result of one individual's or agency's desire to control program processes by not sharing information. Poor data collection, collation and lack of regular summation hindered recovery efforts for the Eastern Barred Bandicoot in Victoria (Myroniuk 1995). Faulty interpretation and withholding information was a major cause in delay in effective recovery management of the Californian Condor (Snyder 1994), contributing to the animal's extinction in the wild.

This phenomenon is called *goal inversion* (or *displacement*) (Perrow 1986), where the initial conservation goal is subsumed by a new goal that is not directly related to recovery, and can have a chilling effect on recovery efforts. Fortunately though, in most cases, seldom is information withheld because other agendas take precedence over species recovery. However, program participants should be aware that the process unfortunately exists. The guiding principle should always be to make decisions using the best and most complete available information.

7. Have a structured procedure for review

Reviews may follow a variety of formats, largely determined by those managing the process. A workshop format or series of meetings often works best. A review in a workshop format is ultimately about good meeting management. An agenda should be set and the goals of the review clearly explained beforehand, so that all participants are aware of the format and purpose of the exercise. It is also important to ensure that each participant understands their responsibilities and contribution during the review. All participants need to be thinking about how their work contributes to overall goals of the recovery program and how to improve their own and others' performance.

At the outset the terms of reference for the review should be stated. The reasons for the recovery program also need to be clearly defined (e.g., the status of the species/community, causes of decline, goals and objectives in recovery to the extent these are known). This helps to define the problem setting clearly to sharpen the focus of attention for all participants so that they are working to a common, but evolving problem definition and gives them a means to locate themselves in the overall recovery effort. The common recovery problem will likely change with time and therefore will require updating on a continuous basis. Providing an overview of the recovery program, including a brief history to date, and outlining all major and minor components of recovery is useful and helps put the review in context.

Following the introduction ("setting the scene"), the review should identify and then examine in detail each of the main components, which could include goals/objectives (What is the program trying to achieve? Have these changed since they were last defined?), program organization and management, field manipulation, research, captive breeding, community education, staffing, funding and other resources, problem-solving approaches, decision-making process, leadership, team functioning, information generation and sharing and communication. Each component may have its own set of objectives related to the overall goals of the recovery program.

Each component should be analysed and presented by an individual or small team working on that area, and a brief written report prepared for the review group and for the record. A key issue is to get each presenter to provide their information in a consistent format. Reporting needs to be measured against set targets for that component. For instance, a presentation could include the following points: (1) what the work is and how it contributes to overall recovery, (2) goals and objectives for area of responsibility, (3) targets set and achieved, (4) targets not achieved, (5) issues, (6) future directions, and (7) targets for the next time period. These are certainly applicable for major reviews, and could be a useful framework for informal internal reviews.

Targets are derived from the goals of the recovery program, which should be specified in recovery or management plans. Once participants become familiar with preparing and discussing reports in terms of achieving targets, and identifying issues influencing target achievement, reviews (particularly regular informal reviews) can become highly efficient.

A key issue is the "net assessment" — concluding the review with a thorough discussion and a concise summary of the component reviews, to place all of the components in context and "tie" the overall review together. This should include a summary of the major achievements as well as major challenges and how they will be met. Net assessment is often a difficult task requiring very good skills in the important function of integrating and synthesizing a wide variety of information and summarizing issues and outcomes.

The review should finish with a shared sense of renewed purpose; advancement and achievement recognized without being overly positive; problems, failures and challenges remaining honestly acknowledged and a commitment to solve these without punishing underperformance; goals, responsibilities and timelines redefined if required; and a revitalized commitment to tackle and solve the recovery challenge.

8. Address all programmatic elements in reviews

Recovering endangered species is a complex, multi-dimensional task (Reading *et al.* 1992; Clark *et al.* 1994). Running recovery programs successfully is not simply or only about the biology of the species or community involved and the technical challenges of recovery. It is also about how organizations operate and how professionals behave, about community group involvement, about economics and intangibles like "politics". Therefore, reviews need to consider the entire recovery effort, not just the biological and technical elements. Professionals in charge of recovery programs are mostly biologists and may not be adequately skilled in organizational, social, economic, or political issues, and so may need to enlist help from outside professionals skilled in those areas.

An increasing number of papers graphically illustrate the profound impact that professional and organizational behaviour, economic, social and political variables can have on the conduct and outcomes of recovery programs (e.g., Warshall 1994; Crosthwaite 1995; Rabinowitz 1995), and reviews should examine all of these dimensions openly and honestly. There may be conflict in the way programs are operated (e.g., Lindenmayer and Norton 1993; Macfarlane and Loyn 1994; Rabinowitz 1995). For instance, Silviera (1995) argues against the case for expending resources on recovery of the Black-eared Miner *Manorina melanotis* in Australia, despite a recovery program involving three states and the Commonwealth (Backhouse *et al.* 1995). Warshall (1994) provides a telling study of the impact of vested interests and politics on efforts to conserve the Mount Graham Red Squirrel *Tamiasciuris hudsonicus grahamensis* in the USA.

Several papers in Clark *et al.* (1994) examine the broad range of issues involved in recovery programs, including conservation science (Minta and Kareiva 1994), conflict management (Wondollek *et al.* 1994), organizations involved in recovery efforts (Westrum 1994), professional behaviour (Clark and Reading 1994), sociological aspects (Kellert 1994) and policy sciences (Brewer and Clark 1994). These issues should be addressed in all reviews.

Among the important questions to consider are: (1) What is the organizational structure of the program and how does it function (e.g., who has power and authority and who does not? How are power and authority used in program implementation?) (2) How effective is communication — are all participants kept informed of progress? (3) Is the broad set of skills required for effective operation present — for example, group or team leadership, research, program management, community participation, economics, etc.? (4) Who are the leaders of the program and do their goals and skills for program implementation match the goals of species recovery? (5) How do individuals and groups within the program interact with each other? (6) How do groups within the program interact with individuals and groups outside the program? (7) How is new information integrated and used within the program? (8) What types of information are most influential in decision making, and why? (9) How do you proceed when things work well and when things are clearly not working well?

9. Document review outcomes

Publication of reviews adds to the review itself and enhances future analysis of recovery programs. Documentation of review outcomes helps to ensure a comprehensive process and identify the actions required for any remedial attention. As such, reviews directly contribute to the ongoing and continuous improvement of conservation processes. Where major reviews are undertaken, it is important to document them as a learning aid useful to program participants and to the larger conservation community.

Published literature on comprehensive reviews of recovery programs is still quite limited, but growing. Reasons for publication may vary; some reviews report on increasing successes, others may be a comprehensive assessment of novel programs such as new legislation (e.g., Wilson and Clark 1995). Reviews may also be published to express concerns about possibly inadequate conservation processes (e.g., Lindenmayer and Norton 1993; Silviera 1995) or in extreme cases to draw attention to corrupted processes (e.g., Jackson 1994; Warshall 1994; Rabinowitz 1995). However, the prime reason for reviews and their documentation is similar in all cases: to measure progress, rapidly identify existing or developing problems, and increase prospects for successful recovery outcomes.

Much of the review record includes internal and often unpublished government reports, a few journal papers (e.g., Lindenmayer and Norton 1993; Rabinowitz 1995), conference proceedings (e.g., Backhouse and Clark 1995) and books (e.g., Clark *et al.* 1994). Many reports still emphasize biological and technical aspects to the total or near exclusion of non-biological issues, but a growing number of reviews are beginning to take a look at how governments, agencies, organizations and individuals function in recovery.

Journals such as *Conservation Biology, Pacific Conservation Biology* and *Environmental Management* are suitable outlets for reviews as they cover a wide range of topics relating to biodiversity conservation, recognizing that effective recovery involves considerably more than just biology and ecology. They also reach an international audience, an important consideration in sharing experiences and learning lessons for program improvement.

CONCLUSIONS

Reviewing recovery efforts is essential in not only evaluating the progress of particular programs but, taken collectively, provides a measure of how biodiversity conservation policy is actually being implemented. The review of recovery programs included in this volume provides further insight into how policy aims are being implemented, and whether they are meeting the goal of legislation, strategies and plans. Reviews also provide a basis to identify and rapidly tackle problems, learn from experience, identify lessons and to continually improve performance. All recovery efforts can benefit from their own reviews, and the outcomes and lessons from other programs.

Reviewing performance is crucial for learning from the accumulated experiences. It is also the only rational way to move forward effectively and efficiently in restoring endangered species and for conserving biodiversity (Clark 1993). Review of threatened species cases represents a step in this learning process — a step that should become a regular, institutionalized and ongoing process of sharing and comparing experiences and seeking constructive ways to improve successes.

Many issues are usually involved in recovery and all need to be reviewed in a systematic fashion. There are too many examples of limited or no reviews and faltering or failing programs. Even though thorough reviews should be part of all endangered species recovery programs, they often are not. What little review that does take place in species conservation work is often insufficiently systematic and rigorous to practically aid recovery programs. As Tear *et al.* (1995:194) note for the USA ". . . recovery plans all too often manage toward extinction rather than away from it". Clearly there is a pressing need for comprehensive program reviews.

Reviewing complex endeavours such as endangered species programs is not an easy process. However, application of a systematic approach, including the elements we describe here, can help make the process more comprehensive. Certainly as the literature grows the process will become more systematic, and hopefully become a regular institutionalized part of recovery efforts. Successful implementation in the field is what ultimately matters in endangered species conservation, and review and evaluation can be important to support the goal of successful biodiversity conservation.

This paper contributes to this aim and this volume represents an important step in taking a national approach to reviewing the many variations on the recovery process. Together they encourage what will hopefully become a regular, systematic and comprehensive process of review and learning in the challenge to improve recovery of endangered species.

ACKNOWLEDGEMENTS

We thank Dr Kim Lowe and Mr Peter Menkhorst (Department of Natural Resources and Environment, Victoria) and Dr Alan Saunders (Department of Conservation, New Zealand) for their comments on the manuscript. K. Lowe also provided information on the operation of recovery teams in Victoria for inclusion in this paper.

REFERENCES

Alvarez, K., 1993. Twilight of the Panther: Biology, Bureaucracy and Failure in an Endangered Species Program. Myakka River Publishing: Sarasota.

Backhouse, G. N., Clark, T. W. and Reading, R. P., 1994. The Australian Eastern Barred Bandicoot recovery program: Evaluation and reorganization. Pp. 251–71 *in* Endangered Species Recovery: Finding the Lessons, Improving the Process ed by T. W. Clark, R. P. Reading and A. L. Clarke. Island Press: Washington DC.

Backhouse, G. N. and Clark, T. W. (eds), 1995. Case Studies and Policy Initiatives in Endangered Species Recovery in Australia. Pp. 1–116 *in* People and Nature Conservation: Perspectives on Private Land Use and Endangered Species Recovery. Transaction of the Royal Zoological Society. Surrey Beatty & Sons: Chipping Norton.

Backhouse, G. N., Bennet, S. and McLaughlin, J., 1995. Recovery Plan for the Black-eared Miner *Manorina melanotis*. Department of Conservation and Natural Resources: Melbourne.

Backhouse, G. N. and Slater, G., 1995. Recovery Plan for the Eastern Barred Bandicoot *Perameles gunnii* 1995–1996. Department of Conservation and Natural Resources and Zoological Board of Victoria: Melbourne.

Brewer, G. D. and Clark, T. W., 1994. A policy sciences perspective-improving implementation. Pp. 391–413 *in* Endangered Species Recovery: Finding the Lessons, Improving the Process ed by T. W. Clark, R. P. Reading and A. L. Clarke. Island Press: Washington DC.

Brewer, G. D. and deLeon, P., 1983. The Foundations of Policy Analysis. Dorsey Press: Homewood.

Browne, M., 1993. Organizational Decision Making and Information. Ablex Publishing Corporation: Norwood.

Bryant, S., 1995. Te Papa Atawhai. *Wingspan* 5(3): 12–14.

CALM, 1995. Translocation of Threatened Fauna Policy Statement No. 29 (as revised). Department of Conservation and Land Management: Perth.

Clark, T. W., 1993. Creating and using knowledge for species and ecosystem conservation: Science, organizations and policy. *Perspectives in Biology and Medicine* 36: 497–525.

Clark, T. W., 1996. Learning as a strategy for improving endangered species conservation. *Endangered Species UPDATE* 13(1&2): 5–6, 22–23.

Clark, T. W. (in press). Averting Extinction: Reconstructing the Endangered Species Recovery Process. Yale University Press: New Haven.

Clark, T. W. and Reading, R. P., 1994. A professional perspective: Improving problem solving, communication, and effectiveness. Pp. 351–70 *in* Endangered Species Recovery: Finding the Lessons, Improving the Process ed by T. W. Clark, R. P. Reading and A. L. Clarke. Island Press: Washington DC.

Clark, T. W., Reading, R. P. and Clarke, A. L. (eds), 1994. Endangered Species Recovery: Finding the Lessons, Improving the Process. Island Press: Washington DC.

Clark, T. W., Wallace, R. L. and Kellert, S. R., 1995a. Endangered species protection: A framework for analysing and improving performance. Pp. 104–09 *in* People and Nature Conservation: Perspectives on Private Land Use and Endangered Species Recovery ed by A. Bennett, G. Backhouse and T. Clark. Transactions of the Royal Zoological Society. Surrey Beatty & Sons: Chipping Norton.

Clark, T. W., Backhouse, G. N. and Reading, R. P., 1995b. Prototyping in endangered species recovery programs: The Eastern Barred Bandicoot experience. Pp. 50–62 *in* People and Nature Conservation: Perspectives on Private Land Use and Endangered Species Recovery ed by A. Bennett, G. Backhouse and T. Clark. Transactions of the Royal Zoological Society. Surrey Beatty & Sons: Chipping Norton.

Crosthwaite, J., 1995. Identifying and addressing social and economic issues in recovery planning: Experience with the Victorian *Flora and Fauna Guarantee Act* 1988. Pp. 78–87 *in* People and Nature Conservation: Perspectives on Private Land Use and Endangered Species Recovery ed by A. Bennett, G. Backhouse and T. Clark. Transactions of the Royal Zoological Society. Surrey Beatty & Sons: Chipping Norton.

Dery, D., 1984. Problem Definition in Policy Analysis. University of Kansas Press: Lawrence.

FWS, 1980. Florida Manatee Recovery Plan. US Fish and Wildlife Service, Southeast Region: Atlanta.

GAO, 1988. Endangered Species: Management Improvements Could Enhance Recovery Programs. General Accounting Office: Washington DC.

Gibson, D. F., Johnson, K. A., Langford, D. G., Cole, J. R., Clarke, D. E. and Willowra Community, 1994. The Rufous Hare wallaby *Lagorchestes hirsutus*: a history of experimental reintroduction in the Northern Territory. Pp. 171–76 *in* Reintroduction Biology of Australian and New Zealand Fauna ed by M. Serena. Surrey Beatty and Sons: Chipping Norton.

Imboden, C., Jones, P. and Atkinson, I., 1995. Kakapo Recovery Program Review. Report to the Department of Conservation: Wellington.

Jackson, J. A., 1994. The Red-cockaded Woodpecker recovery program: Professional obstacles to co-operation. Pp. 157–82 *in* Endangered Species Recovery: Finding the Lessons, Improving the Process ed by T. W. Clark, R. P. Reading and A. L. Clarke. Island Press: Washington DC.

Kellert, S. R., 1994. A sociological perspective: Valuational, socioeconomic, and organizational factors. Pp. 371–90 *in* Endangered Species Recovery: Finding the Lessons, Improving the Process ed by T. W. Clark, R. P. Reading and A. L. Clarke. Island Press: Washington DC.

Koford, C. B., 1953. The California Condor. National Audubon Society Research Report No. 4. National Audubon Society: New York.

Kohm, K. A. (ed), 1991. Balancing on the Brink of Extinction: The Endangered Species Act and Lessons for the Future. Island Press: Washington DC.

Lasswell, H. D., 1971. A Pre-view of Policy Sciences. American Elsevier Publishing Company: New York.

Leeuw, F. L., Rist, R. C. and Sonnichsen, R. C. (eds), 1994. Can Governments Learn? Comparative Perspectives on Evaluation and Organizational Learning. Transaction Publishers: New Brunswick.

Lindenmayer, D. A. and Norton, T. W., 1993. The conservation of Leadbeater's Possum in southeastern Australia and the Northern Spotted Owl in the Pacific north-west of the USA: management issues, strategies and lessons. *Pac. Conserv. Biol.* **1**(1): 13–18.

Macfarlane, M. A. and Loyn, R. H., 1994. Management for the conservation of Leadbeater's Possum, (*Gymnobeledius leadbeateri*) — a reply. *Pac. Conserv. Biol.* **1**(2): 84–86.

Maguire, L. A., 1986. Using decision analysis to manage endangered species populations. *J. Environ. Manage.* **22**: 245–60.

Maguire, L. A., 1990. Risk analysis for conservation biology. *Conserv. Biol.* **5**: 123–25.

Mansergh, I., Jelinek, A. and Clunie, P., 1995. A review of the Action Statement process under the Victorian *Flora and Fauna Guarantee Act* 1988. Pp. 68–77 *in* People and Nature Conservation: Perspectives on Private Land Use and Endangered Species Recovery ed by A. Bennett, G. Backhouse and T. Clark. Transactions of the Royal Zoological Society. Surrey Beatty & Sons: Chipping Norton.

Mattson, D. J. and Craighead, J. J., 1994. The Yellowstone Grizzly Bear Recovery Program: uncertain information, uncertain policy. Pp. 101–29 *in* Endangered Species Recovery: Finding the Lessons, Improving the Process ed by T. W. Clark, R. P. Reading and A. L. Clarke. Island Press: Washington DC.

Mayne, J., 1994. Utilizing evaluation in organizations: The balancing act. Pp. 17–43 *in* Can Governments Learn? Comparative Perspectives on Evaluation and Organizational earning ed by F. L. Leeuw, R. C. Rist and R. C. Sonnichsen. Transaction Publishers: New Brunswick.

Michael, D. N., 1995. Barriers and bridges to learning in a turbulent human ecology. Pp. 461–88 *in* Barriers and Bridges to the Renewal of Ecosystems and Institutions ed by L. H. Gunderson, C. S. Holling and S. S. Light. Columbia University Press: New York.

Minta, S. C. and Kareiva, P. M., 1994. A conservation science perspective: conceptual and experimental improvements. Pp. 275–304 *in* Endangered Species Recovery: Finding the Lessons, Improving the Process ed by T. W. Clark, R. P. Reading and A. L. Clarke. Island Press: Washington DC.

Myroniuk, P., 1995. Captive management of the threatened Eastern Barred Bandicoot: Zoos and co-operative conservation. Pp. 63–67 *in* People and Nature Conservation: Perspectives on Private Land Use and Endangered Species Recovery ed by A. Bennett, G. Backhouse and T. Clark. Transaction of the Royal Zoological Society. Surrey Beatty & Sons: Chipping Norton.

NRDC, 1989. Extinction in Paradise: Protecting our Hawaiian Species. Natural Resources Defence Council: Honolulu.

Perrow, C., 1986. Complex Organizations (3rd edition). McGraw Hill: New York.

Rabinowitz, A., 1995. Helping a species go extinct: the Sumatran Rhino in Borneo. *Conserv. Biol.* 9(3): 482–88.

Reading, R. P., Clark, T. W. and Kellert, S. R., 1991. Towards an endangered species paradigm. *Endangered Species UPDATE* **8**(11): 1–4.

Reading, R. P., Clark, T. W., Arnold, A., Fisher, J., Myroniuk, P. and Seebeck, J. H., 1992. Analysis of a Threatened Species Recovery Program: the Eastern Barred Bandicoot *Perameles gunnii* in Victoria, Australia. Report to Department of Conservation and Natural Resources: Melbourne.

Reading, R. P. and Miller, B. J., 1994. The Black-footed Ferret recovery program: Unmasking professional and organizational weaknesses. Pp. 73–100 *in* Endangered Species Recovery: Finding the Lessons, Improving the Process ed by T. W. Clark, R. P. Reading and A. L. Clarke. Island Press: Washington DC.

Robertson, H. A., Hay, J. D., Saul, E. K. and McCormack, G. V., 1994. Recovery of the Kakerori: an endangered forest bird of the Cook Islands. *Conserv. Biol.* **8**(4): 1078–86.

Short, J., Bradshaw, S. D., Giles, J., Prince, R. I. T. and Wilson, G. R., 1992. Reintroduction of macropods (Marsupialia: Macropodoidea) in Australia — A review. *Biol. Conserv.* **62**: 189–204.

Silviera, C. E., 1995. The Black-eared Miner. *Aust. Bird Watch.* **16**: 96–109.

Smales, I., Menkhorst, P. and Horrocks, G., 1995. The Helmeted Honeyeater recovery programme: A view of its organization and operation. Pp. 35–44 *in* People and Nature Conservation: Perspectives on Private Land Use and Endangered Species Recovery ed by A. Bennett, G. Backhouse and T. Clark. Transactions of the Royal Zoological Society. Surrey Beatty & Sons: Chipping Norton.

Snyder, N. R. F., 1994. The California Condor recovery program: Problems in organization and execution. Pp. 183–204 *in* Endangered Species Recovery: Finding the Lessons, Improving the Process ed by T. W. Clark, R. P. Reading and A. L. Clarke. Island Press: Washington DC.

Sonnichsen, R. C., 1994. Effective internal evaluation: An approach to organizational learning. Pp. 125–41 *in* Can Governments Learn? Comparative Perspectives on Evaluation and Organizational Learning ed by F. L. Leeuw, R. C. Rist and R. C. Sonnichsen. Transaction Publishers: New Brunswick.

Tear, T. H., Scott, J. M., Hayward, P. H. and Griffith, B., 1993. Status and prospects for success of the Endangered Species Act: A look at recovery plans. *Science* **262**: 976–77.

Tobin, R. J., 1990. The Expendable Future: U.S. Politics and the Protection of Biological Diversity. Duke University Press: Durham.

Wallace, R. L., 1994. The Florida Manatee recovery program: Organizational learning and a model for improving recovery programs. Pp. 131–56 *in* Endangered Species Recovery: Finding the Lessons, Improving the Process ed by T. W. Clark, R. P. Reading and A. L. Clarke. Island Press: Washington DC.

Warshall, P., 1994 The biopolitics of the Mt Graham Red Squirrel *(Tamiasciuris hudsonicus grahamensis). Conserv. Biol.* **8**(4): 977–88.

Weiss, J. A., 1989. The powers of problem definition: The case of government paperwork. *Policy Sciences* **22**: 92–121.

Westrum, R., 1994. An organizational perspective: Designing recovery teams from the inside out. Pp. 327–50 *in* Endangered Species Recovery: Finding the Lessons, Improving the Process ed by T. W. Clark, R. P. Reading and A. L. Clarke. Island Press: Washington DC.

Wholey, J. S., Hatry, H. P. and Newcomer, K. E. (eds), 1994. Handbook of Practical Program Evaluation. Jossey-Bass Publishers: San Francisco.

Wilson, B. A. and Clark, T. W., 1995. The Victorian *Flora and Fauna Guarantee Act* 1988: A five-year review of its implementation. Pp. 87–103 *in* People and Nature Conservation: Perspectives on Private Land Use and Endangered Species Recovery ed by A. Bennett, G. Backhouse and T. Clark. Transactions of the Royal Zoological Society. Surrey Beatty & Sons: Chipping Norton.

Wondollek, J. M., Yaffee, S. L. and Crowfoot, J. E., 1994. A conflict management perspective: applying the principle of alternative dispute resolution. Pp. 305–26 *in* Endangered Species Recovery: Finding the Lessons, Improving the Process ed by T. W. Clark, R. P. Reading and A. L. Clarke. Island Press: Washington DC.

Yaffee, S. L., 1982. Prohibitive Policy: Implementing the Endangered Species Act. MIT Press: Cambridge.

Yaffee, S. L., 1994. The Wisdom of the Spotted Owl: Policy Lessons for a New Century. Island Press: Washington DC.

FURTHER READING

Miller, B., Reading, R., Conway, C., Jackson, J. A., Hutchins, M., Snyder, N., Forrest, S. and Frazier, J., 1994. A model for improving endangered species recovery programs. *Environ. Manage.* **18**(5): 637–45.

Tear, T. H., Scott, J. M., Hayward, P. H. and Griffith, B., 1995. Recovery plans and the Endangered Species Act: Are criticisms supported by data? *Conserv. Biol.* **9**(1): 182–95.

Down the track: a critique of recovery programs in Tasmania

Sally L. Bryant[1] and Stephen Harris[1]

In Tasmania, 59 threatened species recovery projects were written and/or implemented between 1990 and 1995. Of these 22 are still active. Most recovery projects have been successful in implementing their recovery actions but few have achieved the broader goals of increasing populations in the wild or reversing population trends. Some of the problems encountered and benefits to the recovery process are identified. Over 70% of the funding received has been directed to fauna programs, most of which are species specific. Changing our emphasis to include a landscape or regional approach is inevitable. So too is the recognition that species recovery is co-operative and must be supported by a range of other processes.

Key words: Tasmania, recovery programs, endangered species, threatened flora, threatened fauna.

BACKGROUND

IN Tasmania the recovery plan process for threatened species and communities, administered by the Endangered Species Unit (ESU) of the Australian Nature Conservation Agency (ANCA), began in 1990 as part of the National Conservation Strategy. This meant that for the first time, threatened flora and fauna projects could be undertaken which were adequately funded and well focused on actions to improve species' status. Prior to 1990, assessment and survey of threatened species was already well under way but funding was sporadic, usually short-term and unable to provide the continuity of actions required either for long-term species recovery or community involvement. This nationally co-ordinated recovery plan process has made an enormous contribution to conserving biodiversity and raising the profile of threatened species Australia wide and internationally.

In May 1996, Tasmania had 72 species on Schedule 1 of the Commonwealth *Endangered Species Protection Act 1992,* 44 plants and 28 animals. Five of the six threatening processes listed on Schedule 3 impact on the Tasmanian ecosystems, namely root-rot fungus *Phytophthora cinnamomi*, feral cats, rabbits and goats and incidental bycatch from longline fishing. The sixth process relates to foxes which, although absent in Tasmania, still remain a potential threat. Despite Tasmania often being referred to as a stronghold and refuge for many nationally threatened species, it also has significant claims on the recovery process. For example, Tasmania formed the first national recovery team being for the Orange-bellied Parrot *Neophema chrysogaster*, has one of Australia's most threatened vertebrates, the Pedder Galaxias *Galaxias pedderensis*, the first national listing of an invertebrate *Astacopsis gouldi* and marine fish *Brachionichthys hirsutus,* and the first preparation of a multispecies management plan using a quasi functional group approach (Coates 1991).

Five years on it is timely to evaluate the effectiveness of the recovery process in Tasmania and determine whether project objectives and outcomes have translated into primary benefits for threatened species, such as an increase in population size, species security in the wild or downlisting.

RECOVERY PROJECTS 1990–1995

Between 1990 and 1995 in Tasmania 59 recovery projects were written and or implemented, of which 37 have been completed and 22 are still active. The breakdown of these projects is shown in Table 1. ANCA funding provided during that five year period exceeded $2.1 million, of which over 70% has been for single species fauna programs. The paucity of invertebrate and non-vascular plant projects reflects our comparatively poorer level of understanding of these groups and their national conservation status.

The positive outcomes of 16 well established flora and fauna recovery projects in Tasmania are complex trade-offs between improving habitat security, identifying more wild populations and artificially increasing numbers (Table 2). All but one, the Pedder Galaxias *Galaxias pedderensis* project, have been able to implement their recovery actions, which suggests that most recovery plans had well defined and realistic tasks. Only one fauna project for the Swan Galaxias *Galaxias fontanus* had increased its species population in the wild (by reintroduction) but five flora recovery projects had successfully increased wild populations mainly through propagation. The Tunbridge Buttercup *Ranunculus prasinus*, known only from four wetlands on private land, had been successfully introduced into appropriate habitat in a nature reserve.

[1]Wildlife Branch, Parks and Wildlife Service, Department of Environment and Land Management, GPO Box 44A, Hobart, Tasmania, Australia 7001.
Pages 180–84 *in* BACK FROM THE BRINK: REFINING THE THREATENED SPECIES RECOVERY PROCESS ed by S. Stephens and S. Maxwell.
Surrey Beatty & Sons, 1996.

Table 1. Threatened species recovery projects in Tasmania 1990–1995.

	Writing	†$ (ANCA)	Implem.*	$ (ANCA)	Survey	$ (ANCA)
Flora	18	36 900	10	415 235	3	75 248
vascular	18	36 900	10	415 235	3	75 248
non-vascular	—	—	—	—	—	—
Fauna	10	49 780	13	1 473 810	2	19 178
vertebrates	8	43 780	11	1 286 690	—	—
invertebrates	2	6 000	2	187 120	2	19 178
Threats	1	5 500	2	94 658	—	—
Totals	29	92 180	25	1 983 703	5	94 426

*Implementation includes research and/or management phase, some species may have both.
†The State government has made "in kind" contributions which are not shown here.

There are eight cases where the potential exists for downlisting species over the next five years, either through the discovery of more populations in the wild and/or by securing critical habitat (Table 2). For example, new colonies of the Forty-spotted Pardalote *Pardalotus quadragintus* have been discovered on Flinders Island and the largest contiguous breeding colony on private land was secured as a State Reserve in 1995 (Bryant 1991). Several populations of Curly Sedge *Carex tasmanica* have been discovered throughout the state, however, none occur on reserved land. Management prescriptions in the Forest Practices Code (Forestry Commission 1993) for the Wedge-tailed Eagle *Aquila audax fleayi* have ensured that critical habitat surrounding nest sites is identified and protected from disturbance during forestry operations. Similarly, other guidelines in the Code and in Jackson and Taylor (1995) identify important blue gum habitat for the Swift Parrot *Lathamus discolor* and management of grasslands for the Ptunarra Brown Butterfly *Oreixenica ptunarra*.

While all recovery projects are achieving positive results, important broader goals such as securing species in the wild or reversing current population trends are more difficult to obtain. Of the 16 recovery projects reviewed, only one had reversed the current population trend, while the remainder reported the species were either stable or in fact continuing to decline.

PROBLEMS ENCOUNTERED

A number of common problems were identified as impeding the recovery process in Tasmania. Firstly, the role and composition of some recovery teams were cause for concern. Most teams met only once a year and there was often a significant imbalance in member contribution. For example, some members did not contribute to the process at all while others, like a landholder with a species occurring solely on his property, have rejected recovery actions outright. Secondly, annual team meetings provided an overview of progress but did not foster the true team participation that is vital for overcoming problems or providing that extra input to boost a recovery action at a critical time. Hence, the composition and role of some recovery teams in Tasmania need review in the manner outlined in the new ANCA guidelines.

To date the co-ordination of recovery projects within the state has been difficult. As with all Commonwealth projects which are fully costed, host agencies are either explicitly or implicitly nominated to make in-kind contributions. While this is possible to some degree, the framework required to facilitate or encourage the recovery process, especially the interchange of skills and collective planning, has not existed. Only now are opportunities for synergistic benefits being realized. For example, the recovery teams for all threatened flora and fauna of grasslands and grassy woodlands are combining to unify their approach to shared problems. Similarly, one recovery team to oversee all threatened Galaxiid projects has now been recently established. The creation of a Tasmanian Threatened Species Unit in the Parks and Wildlife Service is timely and this unit should take on the role of co-ordinating projects of national and state listed species and ensuring the continuity needed for their long-term recovery.

Many of the basic biological questions on species' habitat requirements, population dynamics, threats and even distribution and population size remain unanswered, yet there is an urgent need for this information to feed into the conservation and reservation assessment process. A way to improve understanding will be to strengthen links between academic research and conservation managers. An agreed annual list of recommended projects, with shared supervision, would help foster a closer link between applied research and management and also raise the collective expertise in field skills. Including a list of priority research topics in recovery plans has already been adopted by the Threatened Species Unit in New Zealand.

ASSISTING THE RECOVERY PROCESS

A number of processes implemented during the 1990 to 1995 period have complemented and assisted recovery efforts in Tasmania. The compilation of State lists of threatened species has not only stimulated research but also improved the management of locally threatened species as well as those with a national perspective (Flora Advisory Committee 1994; Invertebrate Advisory Committee 1994; Vertebrate Advisory Committee 1994). For example, a spider *Plesiothele fentoni*, previously listed as extinct has been rediscovered, and by late October 1995, 42 known or suspected locations of rare or threatened invertebrates were reported from State forest and private land prior to logging (Taylor and Bryant, in press). These State lists were the basis for the

Table 2. Outcomes from 16 multi-year recovery programs in Tasmania.

	Increased population in the wild	Found new populations	Protected more habitat	Artificially increased population	Run an education program	Achieved recovery actions	Gained essential bio. info.	Current population trend	National status in 1991	Downlisting over next five years?
Fauna										
Orange-bellied Parrot	No	Yes	Yes	Yes	Yes	Partially	No	Stable	Endangered	No
Forty-spotted Pardalote	No	Yes	Yes	No	Yes	Yes	No	Stable	Endangered	Possibly
Wedge-tailed Eagle	No	No	Yes	No	Yes	Yes	No	Stable	Endangered	Possibly
Swift Parrot*	n/a	No	Yes	No	Yes	Yes	Yes	Declining	Vulnerable	No
Pedder Galaxias	No	No	No	Yes	No	No	No	Declining	Endangered	No
Swan Galaxias	Yes	Yes	No	Yes	No	Yes	No	Increasing	Endangered	Possibly
Eastern-barred Bandicoot*	n/a	Yes	No	No	Yes	Yes	Yes	? Unknown	Vulnerable	Possibly
Ptunarra Brown Butterfly	No	Yes	Yes	No	Yes	Yes	Yes	Stable	Endangered	Possibly
Flora										
Carex tasmanica	Yes	Yes	No	Yes	Yes	Yes	?	Stable	Vulnerable	Yes
Phebalium davesii	Yes	Yes	No	Yes	No	Partially	Yes	Declining	Endangered	No
Ranunculus prasinus	Yes	No	Yes	Yes	Yes	Yes	Yes	Stable	Endangered	No
Eucalyptus morrisbyi	Yes	No	No	Yes	Yes	Partially	Yes	Declining	Endangered	No
Epacris stuartii	No	No	No	Yes	Yes	Yes	Yes	Declining	Endangered	No
Prasophyllum concinum	No	Yes	Yes	No	Yes	Yes	No	Stable	Endangered	Yes
Euphrasia spp. (4 species)	No	Yes	No	Yes	No	Yes	Yes	Stable	End/Vuln.	For some
Barbarea australis	Yes	Yes	No	Yes	No	Yes	Yes	Declining	Endangered	No
Ratio Yes/No	6Y/8N	10Y/6N	7Y/9N	10Y/6N	11Y/5N	12Y/1N	10Y/6N			8Y?/8N

*Research phase programs not specifically aimed at increasing or securing population in the wild, i.e., not appropriate.

schedules for the *Threatened Species Protection Act 1995* and will ultimately feed into the process of assessing species for listing at a national level. Revision of the State lists using quantifiable criteria such as those in the IUCN Red List Categories (IUCN 1994) or Chalson and Keith (1995) will minimize any inconsistencies of species assessments in the future and the adoption of a ranking system could prioritize flora and fauna non-discriminately.

Nearly all endangered species programs are calling on the public to assist in recovery, and off-reserve conservation is an important alternative to reservation to conserve biodiversity (Bryant and Harris 1994; Wells 1996). Networking and community based groups such as Landcare, Understorey Network, Save the Bush, Whole Farm Planning, Greening Australia and the mobilization of specialist groups such as those by the Royal Australasian Ornithologists Union (RAOU) have enabled recovery teams to not only increase the scope of tasks but also to speed up the process. Incorporating education and community awareness in most recovery projects (Table 2) has laid a good foundation for the Tasmanian public taking increased responsibility for threatened taxa.

The proclamation of the Tasmanian *Threatened Species Protection Act 1995* provides a legal mandate to list species, identify threats and critical habitats and undertake financially assisted management agreements on private and Crown land. The Act also makes provision for the preparation of a "threatened species strategic plan" which should identify both the enormity of the conservation task and how it may best be tackled.

A comprehensive regional assessment (CRA) of Tasmania's forests is currently under way which involves many studies on forest biota and the massive compilation of species and community assemblage data. This work is being carried out under a one-off intergovernmental Regional Forest Agreement (RFA) for the purpose of combining the conservation of wildlife and heritage with planning for long-term ecologically sustainable management of forests. The results of this work will be quickly evident and used to more accurately rank species in terms of threat, provide or improve management guidelines and be considered in developing a comprehensive, adequate and representative reserve system. Much of the data will be readily available for preparing listing statements under the *Threatened Species Protection Act 1995,* for reassessing species' status and also importantly for identifying data gaps.

FUTURE DIRECTIONS

While the demands for threatened species protection are increasing, the funds available for recovery are not. Single species recovery programs are expensive and many may soon be incorporated into multispecies plans, functional groups or a landscape approach which integrates flora and fauna recovery. Habitat Conservation Plans (HCP) offer a realistic approach to integrating endangered species issues with production and development on non-federal land in the USA (Hansen 1995). HCP are multispecies and apply to all listed taxa that occur now and potentially in the future and include a legally binding implementation agreement.

Some HCP have attempted to be ecosystem based, protecting habitat across broad areas rather than limiting protection to known occupied sites (Raines 1995). For example, Bruny Island in Tasmania contains many national and State listed species (e.g., Forty-spotted Pardalote, Swift Parrot, Wedge-tailed Eagle, Hooded Plover, Eastern Quoll, Curly Sedge, Mt Mangana Stag Beetle, etc.) and yet there is no integration of recovery actions on the island. Shared skills and resources to tackle costly issues may well be a more practical and effective approach to recovery in the future. It is clear that co-ordination of various programs and mechanisms at the catchment, landscape and regional scale is required, however, there is a long way to go before maximum synergism is obtained. It is also important not to overlook that basic information like species distribution, threatening processes, and conservation and reservation status, is fundamental to any assessment or management plan being developed.

In Tasmania, the greatest threat to flora is vegetation clearing and conversion to agriculture (Leigh and Briggs 1992), while to fauna an even greater number of threats including land clearing have been identified (Bryant 1993; Bryant 1994). Measures to slow down these processes are need urgently if recovery programs are to succeed and the threatened species list shortened. The relatively new *Land Use Planning and Approvals Act 1993,* if coupled with a State native vegetation policy could be used to control land clearance. Rural and urban zoning, negotiating covenants with landholders and voluntary mechanisms such as "Land for Wildlife" will slow the cascade of species onto the rare and threatened species lists. Lessons and directions from the case studies presented in Bennett *et al.* (1995) indicate that a mix of legislation, agency support, private and corporate involvement, research and problem solving are required to facilitate the recovery process. However, an underlying fact remains and that is "prevention is better and cheaper than cure".

ACKNOWLEDGEMENTS

Thank you to David Rounsevell who compiled the information on ESU projects and who has been responsible for much of the co-ordination of recovery projects to date. Also to project officers for supplying information and the two external referees for their valuable comments.

REFERENCES

Bennett, A., Backhouse, G. and Clark, T. (eds), 1995. People and nature conservation: Perspectives on private land use and endangered species recovery. Transactions of the Royal Zoological Society of New South Wales. Surrey Beatty & Sons: Chipping Norton.

Bryant, S. L., 1991. Recovery Plan for the Forty-spotted Pardalote *Pardalotus quadragintus*: Management Phase. Parks, Wildlife and Heritage: Tasmania.

Bryant, S. L., 1993. Tasmania's threatened biodiversity. Pp. 94–101 *in* The benefits of sustaining biological diversity on the farm ed by R. Donaghey and K. McKay. Proceedings of the TCT-TFGA Nature Conservation Conference.

Bryant, S. L., 1994. Priorities for Conserving Tasmania's Flora and Fauna. Unpublished report to ANCA, Parks and Wildlife Service: Tasmania.

Bryant, S. L. and Harris. S., 1994. Evaluating Tasmania's rare and threatened species. *The Tasmanian Naturalist* **116**: 52–57.

Chalson, J. M. and Keith, D. A., 1995. A risk assessment scheme for vascular plants: pilot application to the flora of New South Wales. Project No. 450. Unpublished report to Australian Nature Conservation Agency.

Coates, F., 1991. Conservation ecology and management of five rare species in the Rhamnaceae family. Wildlife Scientific Report 91/3. Parks, Wildlife and Heritage: Tasmania.

Flora Advisory Committee, 1994. Native Higher Plant Taxa which are Rare or Threatened in Tasmania. Edition 1. Parks and Wildlife Service: Tasmania.

Forestry Commission, 1993. Forest Practices Code. Forestry Commission: Hobart, Tasmania.

Hansen, C., 1995. Multispecies plan for forest habitat. *Endangered Species Bulletin* **20**(6): 6–9.

Invertebrate Advisory Committee, 1994. Interim List of Native Invertebrates which are Rare or Threatened in Tasmania. Edition 1. Parks and Wildlife Service: Tasmania.

IUCN, 1994. IUCN Red List Categories. Gland, Switzerland: IUCN.

Jackson, J. and Taylor, R., 1995. Threatened Fauna Manual for Production Forests in Tasmania. Forest Practices Unit, Forestry: Tasmania.

Leigh, J. H. and Briggs, J. D., 1992. Threatened Australian Plants: overview and case studies. Australian National Parks and Wildlife Service for ANZECC.

Mace, G. and Stuart, S., 1994. Draft IUCN Red List Categories, Version 2.2. *Species* **21–22**: 13–24.

Raines, C., 1995. Negotiating for Conservation. *Endangered Species Bulletin* **20**(6): 22–23.

Taylor, R. J. and Bryant, S. L., in press. Compilation of a list of threatened invertebrates: the Tasmanian experience. *Mem. Mus. Vict.* **56**.

Vertebrate Advisory Committee, 1994. Native Vertebrates which are Rare or Threatened in Tasmania. Edition 1. Parks and Wildlife Service: Tasmania.

Wells, P., 1996. Off-reserve conservation in Tasmania: its role in the maintenance of native bushland on private property. Pp. 44–53 *in* From Conflict to Conservation: Native vegetation management in Australia. South Australian Department of Environment and Natural Resources: Adelaide.

Realism in response to the ecological performance of populations in recovery programs

Peter J. Jarman[1]

The shortage of completed species recovery programs means that we have little experience on which to judge the long-term progress of a reintroduced or protected population. Without established expectations, recovery teams may respond with either optimism or pessimism to changes in the target population. This paper reviews some ecological theory to try to bring realism into recovery teams' responses. Individual organisms must be expected to perform less well in the wild than in well-managed captivity, growing, maturing and reproducing more slowly, and surviving less well. Populations reintroduced or protected at low density may grow initially exponentially, consuming accumulated resources; but that will not last. Using up the standing crop of consumable resources, they will fall back to the level sustainable by the productivity of those resources. Community composition and habitat parameters will also change. Without realism, any of these might be seen as a symptom of the recovery program's impending failure. Properly understood, they may in fact signal just the opposite. A realistic view is essential if the recovery team is to justify correctly their needs for on-going funding for the recovery program.

INTRODUCTION

THIS contribution uses ecological theory as a basis for discussing what might happen in the ecology of some species during their recovery from low density or after reintroduction. No one has yet looked for and reported on the phenomena that I suggest may emerge, so I cannot give references to published examples of them. This is an exercise in consciousness raising. It is intended to help recovery teams and managers to think about the ecological performance of the endangered populations whose recovery they are working for. It ignores almost all non-ecological aspects of an endangered species' recovery except one: the need to justify on-going funding. I do not dismiss the importance of non-ecological aspects of the recovery process, but I see the need to broaden our view of the possible performances of recovering species.

Despite many publications in the past few years that report and discuss early stages of the recovery process, managers planning and implementing the recovery of endangered species of animals or plants still have a poor reserve of successful and long-term experience to call upon in setting expectations for their recovery programs past the first phase of "halting" the decline. For example, of the 22 single-species cases of reintroductions and translocations of endangered species discussed by contributors to *Reintroduction Biology of Australian and New Zealand Fauna* (Serena 1994), no case contained detailed data on the demographic performance of populations for several generations after establishment. There is a similar absence of such cases in journals, for example, *Biological Conservation* 1994–96. Yet, as current recovery programs mature, increasing numbers will reach a point at which decisions must be made about their success or failure, and continuation or termination. While many programs state their aims in terms of the species' recovery in the wild, few can set simple criteria for judging when re-establishment has occurred or when the remnant wild population has recovered to a point of self-sustainability (under a specified management regime). Those making decisions may respond unrealistically to their own expectations about what the target species' population "ought" to be doing, without having fully considered how they might realistically expect the species to perform. This paper suggests some possible performances of reintroduced populations.

Managers of recovery programs usually hope that the animals will live and breed, and the local densities or the distribution of their target species will steadily increase. They may see anything else, for example a fall in numbers or deaths among their individuals, as indicating failure of the program.

[1]Department of Ecosystem Management, University of New England, Armidale, New South Wales, Australia 2351.

Pages 185–89 *in* BACK FROM THE BRINK: REFINING THE THREATENED SPECIES RECOVERY PROCESS ed by S. Stephens and S. Maxwell.
Surrey Beatty & Sons, 1996

This pessimistic view may be as unreal as their optimism if their species at first breeds prolifically. Although we have little direct experience to guide our expectations, we have a wealth of theory. This paper reviews some of the relevant theory with the aim of setting some realistic expectations for the performance of individuals, populations and communities that are subject to species recovery programs.

This paper is written in terms of vertebrate species, although the principles apply just as well to invertebrates or plants. It also concerns itself with recovery past the first aim of an endangered species recovery plan (to halt and reverse decline), looking at the next phase, growth in the population towards a stage of self-sufficiency and stability.

THE ECOLOGICAL CONTEXT

A recovery program will be implemented only when a target species' population has shrunk drastically in density or distribution. The program's aims will include the halting and reversal of that shrinkage. That may involve:

1. removing an excessive mortality factor (such as predation by a novel predator, or the impact of a disease);
2. restoration of destroyed, degraded or fragmented habitat and resources; or
3. removal of competition.

If necessary, the program may involve restoring captive-bred or wild-caught animals to the wild in chosen places. By any of these actions, management should set up a population of the target species living at low density and in a situation of plentiful resources for each individual. Moreover, some of their consumable resources may have accumulated while the target species was locally extinct or at critically low density, or as a result of management's restoration of resources prior to release. These are, of course, the very circumstances in which a population can increase at its maximum intrinsic rate, r_m, as discussed below.

The situation created by the management program is intended to be dynamic: the recovery program aims for change. The easiest phase of that dynamic process to make predictions about is the start, for perhaps the first generation or two, when the protected or re-established population will either fail (for some unmanaged reason) or will grow relatively spectacularly. For example, six reptile species (five of them endemic) on Round Island, Mauritius increased population densities by values from 54% to >3 000% in three years following eradication of rabbits (North *et al.* 1994). Failure at this stage, however, will lead to pessimism. Yet all very small populations are vulnerable to stochastic events (e.g., one fire, one feral cat, or one storm, none of which could devastate a larger or better dispersed population). The danger in such failures is that the recovery team may falsely interpret them as indicating that the management is inappropriate and terminate the program prematurely.

If the population grows well for the first two generations following management (as many have), the prematurely optimistic manager may then be dismayed to find growth slowing, individuals failing to breed, and even to see numbers falling again. As explained below, ecological theory suggests that it is realistic to expect such changes. Moreover, despite the re-established or protected animals at first being in a resource-rich, low-density environment, they still experience challenges that they are buffered against in captivity. Managers should not be too alarmed if individuals perform less well in the wild than in captivity, although the causes still need to be sought and understood.

THE INDIVIDUAL'S PERFORMANCE

A manager who has formed expectations about an individual's growth, maturation, survival and reproduction from records taken from captive populations, is likely to be disappointed by the performance of monitored animals released to the wild. All these parameters are likely to be "worse" in the wild, or to get worse as the population's density rises during recovery. Compared with captivity, animals in the wild may grow more slowly and reach reproductive maturity later; they may also be socially restrained from breeding to an extent not shown in captivity. Once they do start breeding, inter-birth or inter-clutch intervals are likely to be longer; clutch or litter sizes may be smaller; and weaning or fledging rates may be lower. Above all, mortality rates may be higher, especially in pre-breeding stages. It is common for populations of wild mammals and birds to experience high mortality between weaning or fledging and the onset of breeding, followed by quite low and constant mortality thereafter. This means that few individuals that are born ever breed. This is exactly what the manager of a captive breeding colony tries hard to avoid; the success achieved by those managers in minimizing pre-breeding mortality may leave us with a false expectation of juvenile and sub-adult survival rates in the wild.

The manager of a captive colony that is breeding animals for release to the wild will also strive for even genetic representation of founder individuals or gene lines. Success in this aim, coupled with better survival of individuals to breeding age and removal of social constraints on breeding, leaves a record of low variance among captive individuals in their reproductive performance. However, this is contrary to what has been found out about the reproductive success of individuals in the wild, where it is common for a small proportion of adults of either sex to dominate the population's reproduction. For example, in a studied population of wild Eastern Grey Kangaroos *Macropus giganteus*, less than 30% of physiologically mature males achieved any matings each year, and among those that did, one male in five was likely to achieve over half the population's matings. Surprisingly, variance in lifetime reproductive success among female kangaroos was also high; a few females left as many as five adult offspring in a lifetime, while the great majority left no reproducing offspring (Jarman, unpubl. data) Such variance in a released or protected population — with some individuals not breeding at all — although disappointing, should not be seen as a sign that the recovery program is not working.

THE POPULATION'S PERFORMANCE

If the reintroduced or protected individuals survive, grow, mature and reproduce less well and less consistently than in captivity, the derived population parameters (fecundity, survivorship, generation time, etc.) will of course also be less apparently favourable. However, if the target population's size, density or distribution grow satisfactorily, the recovery program manager is unlikely to be too worried.

Ecological theory warns us, however, that such good initial growth is not going to continue. Even the simple logistic model of growth suggests that, as the population density climbs towards the "carrying capacity" of the environment, resources for each individual will diminish until each animal can only just sustain itself and can produce only one replacement offspring in its lifetime. Variance among individuals means in practice that some will breed while others die young. The effect is that the population density stabilizes about that level. This form of population growth may be seen when management induces slow restoration of habitat with which the growth of the population can keep pace. Following steps to restore their habitat on Cousin Island, the population of endemic Seychelles Warblers *Acrocephalus sechellensis* grew from 26 individuals in 1959 to a plateau of about 320 birds in the 1980s and 1990s, in a good example of logistic growth to stability at "carrying capacity" (Komdeur 1994).

However, we know that in practice the logistic model is a simplification that applies poorly to populations introduced to a pristine environment or one where preferred resources have accumulated. Yet this is effectively the circumstance enjoyed by a newly released population, or one newly protected from a limiting factor (such as excessive predation or disease) that had previously held down its density relative to its resources. Such a population at first enjoys not only a high production of resources per head of population, but also a standing crop of resources that have accumulated in the absence of consumer animals. The population can grow rapidly (at a rate near to its maximum intrinsic capacity) in those most favourable circumstances, and achieve abnormally high densities. But in doing so it consumes the accumulated resources, or alters the composition of the resource-producing community to its own disadvantage. If, by the time the accumulated resources are exhausted, its density has exceeded what the current production of the resource base can support, individuals will be left bereft of resources and will stop reproducing and die.

This pattern of performance of a population introduced into a pristine environment has been demonstrated repeatedly, overseas (e.g., Klein 1968; Caughley 1970) as well as in Australia: by domestic sheep *Ovis aries*, and wild European Rabbits *Oryctolagus cuniculus*, Brown Hares *Lepus capensis*, Red Foxes *Vulpes vulpes* (Jarman 1986a,b) and Cane Toads *Bufo marinus* (Eastel *et al.* 1985; Freeland 1986; Freeland *et al.* 1986). Each of these species entered a new district, exploded to high densities, and then fell back to lower densities once the standing crop of resources was consumed. Freeland *et al.* (1986) also showed that the resources being eaten by Cane Toads changed with time since arrival in an area. As explained above, we can expect a population of an endangered species reintroduced to a site in its former range, or released from a density-limiting factor, to respond in the same way. Such a performance was recorded for Beavers *Castor fiber* reintroduced to Sweden (Hartman 1994).

Ecological theory predicts three possible outcomes for such introduced populations and their resources. Either the productivity of resources is insufficient to support the introduced species, which goes extinct; or the species and its resources fluctuate between limited extremes; or in time, after a series of diminishing oscillations of both the resources and the animal density, the animals will tend towards a fairly stable density relative to the productivity of their resources. In all these cited examples the last seems to be what has happened or is happening.

Sheep graziers tended to see the peak of sheep density achieved in the first surge of population growth as being what they should expect from that land, and they were invariably disappointed when such densities could not be sustained. In New South Wales, sheep densities stabilized at about half that initial peak density; and theory suggests that is a reasonable expectation. Managers of endangered species recovery programs must guard against being as disappointed as the graziers. Their target species, even if they are never as numerous as sheep, will still explode as they consume the accumulated capital of resources and must inevitably fall back to a lower density, but one that reflects the immediate productivity of the resource base. Such falling back is not a failure of the program. Nevertheless, the target population will for a time revert to a smaller size or lower density, with some of the attendant risks of vulnerability to stochastic events.

One aspect of population performance that cannot easily be predicted from captive behaviour is the species' probability of dispersing when reintroduced to, or protected in, the wild. We understand the process of natural dispersal of individuals in an established population quite poorly, and dispersal of individuals in a spatially expanding population not at all. Perhaps the only generalization to be made is that the individuals first dispersing into uncolonized areas may not be a cross-section of the population. Indeed, they are likely to over-represent subadults, females that have not yet bred, or young adult males. Under any circumstances, dispersing animals run risks that established residents do not. Managers should not be dismayed, therefore, to detect lower survivorship among dispersers. Conversely, once established in their new area, the dispersers may breed at a younger age and more prolifically than animals in the longer-colonized areas.

THE COMMUNITY'S PERFORMANCE

The above account of population growth in a pristine or long-unoccupied environment described the growing population consuming accumulated resources. In doing so, the animals obviously change the environment. "Accumulation" may include the growth of (say) plant species that are sensitive to being eaten by the target animal but flourish in its absence, or the increase of a prey species that is vulnerable to predation by the target animal. Consumption of the accumulated resources would, in those cases, involve the reintroduced or protected target species reducing the population density of the sensitive plant species or vulnerable prey species and allowing less vulnerable species to flourish in their stead. Copley (1994), reviewing translocations (in these cases not reintroductions) of native vertebrates in South Australia, noted the "major impacts" of Tammar Wallabies *Macropus eugenii*, Western Grey Kangaroos *M. fuliginosus*, and Koalas *Phascolarctos cinereus* upon their novel environments.

The effect is to change the species composition of the trophic resource base. That in turn may lead to subsequent changes in other species that consume or compete with those species that are released by the first induced changes. And there may be consequences of those changes; and so on. In any interacting community, changes to the population density of one species could trigger a chain reaction of changes until the community either reaches a new quasi-stable state, or settles to long-term cycles in which the relative abundance of its component species fluctuate out of phase with each other between some limits; it is most unlikely that the community would go extinct, although some species might do so. Whatever new state emerges, some species will be at a lower density than previously, and some might even be locally extinct. The recovery program manager must be aware that these are possible outcomes, and consider such possibilities in planning the recovery program.

Similar readjustments can be expected when a whole community is relieved of a widely threatening factor, such as predation by foxes, or the impact of too-frequent fire, or habitat fragmentation. Elimination of any of these would favour some species (including the targets of recovery programs) but disfavour others. Moreover, if a set of species had been held to low densities by any of these eliminated factors, they will all begin to increase in density. At some point they can be expected to start to have effects upon each other that were not apparent at lower densities. To take a hypothetical example: if foxes were greatly reduced or eradicated in a Western Australian woodland, having previously held down the densities of Woylies *Bettongia penicillata*, Tammar Wallabies *Macropus eugenii* and Common Brush-tailed Possums *Trichosurus vulpecula*, all three herbivores would begin to increase in density. At first there would be so few of each that their foraging would not interfere with that of any other species; but as each population's density rose, one species' removal of a resource would be increasingly likely to affect, not just members of its own species, but also members of any other species that also used that resource. As each species' density approached being limited by the availability of food resources, so each would become increasingly sensitive to competition from other species for those resources. Species that, while limited by fox predation, were effectively not in competition might become each other's major limiting factors. The importance of this insight is that the recovery program manager should expect the factors limiting a target species to change through time.

DISCUSSION

There may come a time in an endangered species recovery program when a manager finds the first corpse of the target species, or sees thin animals, or traps adult females that are not breeding when they "should" be. This paper aims to help recovery teams to rationalize such dismaying observations, suggesting that there are times when they are to be realistically expected.

Endangered species recovery plans aim firstly to halt or reverse decline in distribution or abundance of the target species, and to do so in a specified and usually short time, often less than two or three generations. Success in that first aim can often be judged unambiguously: within the specified time the target population has either persisted or become extinct, and the program (to that point) has therefore either succeeded or failed. This paper is concerned, however, with judgments made about recovery programs after that stage, when teams have to decide whether the recovered population is making good progress in withdrawing from the brink of extinction. This is a crucial stage, because at this point government Ministers will wish to claim success for their actions, managers will be under pressure to divert funding to other endangered species, and sponsors may wish to scale down their support.

However, any reasonably cautious manager will be loathe to terminate a recovery program without substantial reassurance that everything is on track. Yet within more than a few generations of re-establishment or protection, a target population may show many signs (listed above) of apparent trouble: individuals may be growing and maturing more slowly, breeding less prolifically, and surviving poorly; the population's growth will slow and reverse, numbers possibly falling very fast to a density well below the peak; and the species may be having obviously detrimental effects upon other species and upon resources in its environment. If it has not been possible to monitor the population by sampling at short intervals, some of these changes may appear as a sudden decline in population parameters, with dead, starving or non-reproducing animals suddenly being seen.

This is no time for thoughtless optimism, for using the theory I have cited as an excuse for enquiring no more deeply. The manager would, above all, need to make sure that the new poor performance does not signal the return of whatever threatening process drove the population to the brink in the first place. If a population is meant to be recovering from too high a rate of predation by foxes, then finding several fox-killed carcases need not simply mean that the population has peaked once more; nor need it mean that management has not "cured" the fox problem. The carcases alone only show that foxes are present and killing the target species. The crucial question remains: does recruitment to the population still exceed mortality? The effectiveness of management actions must be monitored directly as well as indirectly.

Every population is going to be limited by something. The ecologically most convenient populations to manage are those (exemplified by many vertebrates) limited by a simple and detectable density-dependent relationship with a measurable resource; the least convenient are some small vertebrates, invertebrates and plants whose limitation occurs through periodic or random crashes after intervening stages of rapid growth. The former are convenient because managers have a good chance of monitoring the limiting relationship; that will be difficult with the latter. It may take a substantial change of mind for a manager to recognize that the local population of a formerly endangered species has stopped growing and has, through deaths and reproductive failures, fallen in density, but to a sustainable level, and that the recovery program has therefore achieved its second target: a sustainable population in the wild. Once that success is accepted, the third phase, long-term monitoring, can then begin, to ensure that new threats or other changes in the community do not once again endanger the species.

ACKNOWLEDGEMENTS

I would like to thank an anonymous referee, Peter Copley and Margaret Brock for constructive comments on earlier drafts of this paper.

REFERENCES

Caughley, G., 1970. Eruption of ungulate populations, with special emphasis on Himalayan thar in New Zealand. *Ecology* **51**: 53–72.

Copley, P. B., 1994. Translocations of native vertebrates in South Australia: a review. Pp. 35–42 *in* Reintroduction Biology of Australian and New Zealand Fauna ed by M. Serena. Surrey Beatty & Sons: Chipping Norton.

Eastel, S., van Beurden, E. K., Floyd, R. B. and Sabath, M. D., 1985. Continuing geographical spread of *Bufo marinus* in Australia: range expansion between 1974 and 1980. *J. Herpet.* **19**: 185–88.

Freeland, W. J., 1986. Populations of cane toad, *Bufo marinus*, in relation to time since colonization. *Aust. Wildl. Res.* **13**: 321–29.

Freeland, W. J., Delvinqueir, B. L. J. and Bonnin, B., 1986. Decline of cane toad, *Bufo marinus*, populations: status of urban toads. *Aust. Wildl. Res.* **13**: 597–601.

Hartman, G., 1994. Long-term population development of a reintroduced beaver *Castor fiber* population in Sweden. *Conserv. Biol.* **8**: 713–17.

Jarman, P. J., 1986a. The red fox: an exotic, large predator. Pp. 45–61 *in* The Ecology of Exotic Animals and Plants. Some Australian Case Histories ed by R. L. Kitching. John Wiley and Sons: Brisbane.

Jarman, P. J., 1986b. The brown hare: a herbivorous mammal in a new ecosystem. Pp. 62–76 *in* The Ecology of Exotic Animals and Plants. Some Australian Case Histories ed by R. L. Kitching. John Wiley and Sons: Brisbane.

Klein, D. R., 1968. The introduction, increase and crash of reindeer on St. Matthew Island. *J. Wildl. Manage.* **32**: 350–357.

Komdeur, J., 1994. Conserving the Seychelles Warbler *Acrocephalus sechellensis* by translocation from Cousin Island to the islands of Aride and Cousine. *Biol. Conserv.* **67**: 143–52.

North, S. G., Bullock, D. J. and Dulloo, M. E., 1994. Changes in the vegetation and reptile populations on Round Island, Mauritius, following eradication of rabbits. *Biol. Conserv.* **67**: 21–28.

Serena, M., 1994. Reintroduction Biology of Australian and New Zealand Fauna. Surrey Beatty & Sons: Chipping Norton.

Workshop Reports

The *Back from the Brink: Refining the Threatened Species Recovery Process* conference included workshops on four issues vital to the recovery process. Each workshop ran twice with different participants, with about 25 people in each session. Two leaders ran both sessions of each workshop, to ensure continuity.

The leaders then consolidated the outcomes of both sessions of their workshops and presented these at a full forum of the conference.

The workshop reports presented here provide a summary record of the workshops, based on the notes taken by the scribes at the workshops and incorporating comments from the participants.

The recommendations arising from the workshops are also included here. These recommendations do not necessarily indicate consensus of the whole conference forum. They do, however, provide a basis for further discussion and refinement of the recovery process. Some recommendations are already being incorporated into the recovery process.

The recommendations arising from the workshops, together with the papers and discussions at the conference, are expected to form the basis of revised guidelines for practitioners implementing the recovery process. This will replace the existing Endangered Species Program *Recovery Plan Guidelines for Endangered and Vulnerable Species and Ecological Communities.*

WORKSHOPS

Workshop 1: Recovery Plan Implementation

Leaders: Peter Copley and Stephen Garnett
Scribes: Liz Dovey and Katrina Jensz

SUMMARY

Several issues concerning the current recovery process were discussed and problems identified.

As well as focusing on the protection and recovery of species and ecological communities, it is essential to address threat mitigation, including broad-scale processes.

Some problems were identified with having both research plans and recovery plans. It was concluded that recovery plans should incorporate research actions as required, removing the need for separate research plans.

Opportunities should be provided to ensure the involvement of key stakeholders in the recovery process. An information gathering workshop could be held and as far as possible all stakeholders should be represented. Community involvement and "ownership" is essential for all recovery programs.

Planned actions should be reviewed at the beginning of the recovery process and regularly (at least annually) thereafter. Thresholds for failure should be identified in recovery plans so that appropriate changes can be made. An appropriate time frame for recovery plans is five years, although many recovery programs will need to operate for longer. Interim plans could be useful for species requiring urgent attention.

Effective communication is essential to the recovery process and may be assisted by the involvement of a trained or proven communicator. Recovery teams should meet regularly and maintain good records. Guidelines could help the recovery team to function effectively. Members' roles need to be clearly identified and team skills developed. Membership of teams should be reviewed annually. The core recovery team could be assisted by operational groups where required.

The recovery plan should identify the conditions to be met for completion and termination. Ongoing maintenance and monitoring may be required.

RECORD OF WORKSHOP AND RECOMMENDATIONS

What are the problems with the current recovery process?

The current process focuses on species rather than threatening processes:

- Need to prevent species from becoming endangered or vulnerable.
- Not enough threatening processes listed as yet — major ones still missing.
- Need to address more generic threatening processes, e.g., land clearing, weeds, grazing.

Other issues discussed were:

- General habitat protection — the community is becoming more aware and this should be encouraged at ground level.
- If a species is spread over a large area, it is difficult to get community members to participate. Travel should be budgeted for in the recovery plan.
- Community involvement has not been adequately recognized as a resource and used.
- A need for sufficient survey money to clarify "Rare" and "Poorly Known" status was recognized.
- Techniques for dealing with invertebrate and other "unfashionable" taxa need to be developed and encouraged.
- That the standard of recovery plans and the science that underpins them varies enormously between plans.

It was agreed that recovery plan implementation could be improved with State-based threatened species units and that these could assist with defining and addressing the problems. The Northern Territory and South Australia do not have such units. Such units could focus on issues, determine priorities, seek alternative funding sources, raise the profile of threatened species and co-ordinate the State management effort.

It was also suggested that there should be regional threatened species and communities co-ordinators to co-ordinate recovery actions in a specific geographic area. This may depend on the different ways that State agencies are managed.

There was discussion concerning mechanisms for co-ordination of actions across recovery plans. There should be a process to co-ordinate rather than compete for funds. Funding of different priorities within project proposals could be negotiated and may be of value across other projects.

Methods for improving the quality of recovery plans and research undertaken in that context could involve peer review of the project proposal which will be submitted for funding, either using a seminar or forum approach (as used in CSIRO and NPWS for scientific programs) or a training program of relevant people as undertaken in New Zealand. May need a separate avenue for involvement of interested people external to the management agency. External reviews could be tendered out at any stage of the project. Costs of reviews need to be anticipated in the budget.

It is sometimes possible to have people and equipment in common across recovery teams and this should be encouraged as it is more efficient to share expertise and equipment across projects and to encourage linkages between agencies and programs.

It was felt that a national threatened species and communtities priority ranking system is important to provide a clear basis for decisions on funding and to prevent an *ad hoc* process. New Zealand and Western Australia have single systems for priority setting and the possibility of developing a national standard approach across all taxa and communities needs to be addressed. Ranking of threatened communities may need to include assessment of reservation status as a factor as well.

In recognizing that many vulnerable (and little known species) may be on the way to becoming endangered, it was considered that regional recovery plans would be of benefit. At present, very few "projects" for species not listed as endangered are funded. An area-based approach and threat-based approach could assist in addressing this problem. The regional recovery plan approach could be a cost effective method, as has already been shown with several projects in Western Australia.

Recommendations:

1. That there is a need to address both species and threats.
2. The conference recognizes that the protection and recovery of threatened species and ecological communities is essential, but will not in itself address the ongoing problem of loss of biodiversity across the Australian landscape and recommends the establishment and support of a program (as a logical successor to the Feral Pests Program) aimed at dealing with threatening processes. The program must be aimed at developing strategic approaches to mitigating problems such as those caused by:
   — invasive species of plants, animals and pathogens,
   — overgrazing by domestic stock,
   — land clearing and habitat destruction,
   — unsustainable resource use, and
   — urbanization.
3. That community involvement be recognized as an important resource.
4. That there should be review of planned research and management actions at the beginning of the recovery process and regularly throughout the process.
5. That State agency regional threatened species co-ordination be encouraged.
6. That there needs to be a national ranking system across all threatened species and communities.
7. That the regional recovery plan approach be applied where possible.

Is there a need for research plans?

The issue of whether separate research plans were beneficial was discussed. It was generally agreed that dichotomy is a mistake. Research is needed in many cases but should be part of the recovery plan.

Problems perceived with research plans included:

- No recovery team.
- No management actions defined.
- Gives impression of lack of urgency.
- Often nothing happens.

It was felt that it was important to set up a recovery team and get management actions underway at the same time as doing researh. The recovery plan is focused and keeps things on track, including research actions important for recovery. In-depth research is often not necessary for recovery to succeed.

Recommendation:

8. That a separate research plan is not necessary. Research actions should be incorporated into recovery plans.

Who are "we"? The people involved in the recovery process

Key stakeholders include the people directly involved with the management of threatened species and habitat, the people who make decisions about the land on which the species/habitat occurs and the community who may be interested in the actions being proposed (i.e., community involvement).

A recovery plan may exist, but it may not be a product of all stakeholders. A recovery plan is often written by project officers who consult with relevant individuals. It was felt that there should be an opportunity for others, including community members, to contribute. This may also allow these people to be involved with the running of the project, which is important for long-term management.

There is a need to identify the range of people who have information and could contribute. There is then a need to communicate with those people. This process has to be appropriate to the circumstances.

An open forum at the beginning of the recovery process could be a first step to determine who should be involved. It was suggested that a workshop be held for all those interested, from which individuals could be chosen for the recovery team. It was taken as a given that the process is usually initiated by the State conservation agency.

Participants identified that the communication channels within the process were a problem. It was suggested that there was a need for someone in the recovery team who is a good communicator to act as facilitator.

It is important to use existing networks.

It would be useful to tick off the processes as they are done, e.g., different people's tasks.

Recommendations:

9. That there is a need to identify key stakeholders and to get them involved from the initial stages of the recovery process.
10. That a process be developed to assist in identifying the appropriate people to be involved in the recovery process, taking into account:
    — communication,
    — circumstances,
    — need for flexibility, and
    — need for assignment of tasks to appropriate people with appropriate timelines.

Recovery teams: who should be on them and how can they work effectively?

The following problems were identified:

- That it is difficult to get members together when the species is widespread.
- That there may be differing levels of involvement desired (e.g., planning/research versus operational) which can be exacerbated by a wide geographic distribution of the species/ecological community.
- That it can be difficult to find a person who represents a community (e.g., Aboriginal community) as it requires someone who is accepted by the community and who is able to represent that entire — often very diverse — community.
- That poor communication between members of recovery teams can interfere with the objectives of species/community recovery.

It was suggested that a recovery team could include someone who is capable of getting the message to the community. There was also discussion on the need for a person on the recovery team who is good at communication. Someone with the appropriate skills and qualifications should have the role of coordination and keeping effective communication going.

It was recognized that team composition may need to change over time, and the recovery team establishment process must allow for changes in membership as the needs of the project change. An annual review of membership was suggested. Member roles should be identified and written into the recovery plan.

It was agreed that regular meetings are vital.

Operational groups should be used so that the recovery team does not get too unwieldy. These should be represented on the main recovery team (e.g., there is a Victorian operations group for the Orange-bellied Parrot that is represented on the national recovery team).

There was concern that the recovery team may not always have all the necessary expertise where there are a range of issues involved. It was felt that there is a need for guidelines to help manage the process, develop skills and handle conflict.

It is important to identify what skills there are in the team and how team skills can be developed. It was suggested that guidelines for the operation of recovery teams could be established which should include:

How teams operate (many books written).

- Establish a set of ground rules.
- Assistance to develop people skills.
- Leadership and delegation are crucial roles.
- Agree on the rules at the beginning.
- Review membership.
- How to handle conflicts of interest.
- Identify problems.
- Code of conduct.
- There may be a role for ESU to train the team on some aspects.

There was discussion on whether membership of a recovery team should be voluntary or paid. Some members participate by virtue of their employment while others volunteer. Agencies should recognize that travel costs are committed when they agree that an officer is on a recovery team. Out of pocket costs may be paid for volunteer members (e.g., travel and accommodation) and costed into the recovery plan budget. It was agreed that sitting fees for attendance at meetings should not be paid.

Recommendations:

11. That guidelines be developed to help the recovery team work as an effective team.
12. That recovery teams should have regular meetings and keep minutes.
13. That there be annual reviews of recovery team membership.
14. That member roles should be identified and written into the recovery plan.
15. That the recovery team make up the core group and operational groups be used where required.

Time-frames and budgeting for recovery plans

It was suggested that interim recovery plans could be prepared for urgent species, as is currently done in New Zealand. These are not as detailed as draft recovery plans. An interim plan may secure the population from immediate threats but there may not be any actual recovery. This should be built into the recovery plan process.

The New Zealanders are writing 5–10 year plans. A 5-year plan provides a level of certainty for funding. There is an option to apply for an extension of money. A lot is learnt in the first five years. There are changes in priority for actions. The next five years is not such a steep learning curve.

It needs to be recognized that the content of a recovery plan will often lag behind actual actions, because of the dynamic nature of management.

Recovery plans prioritize actions and sometimes receive only partial funding for the essential actions. Should it be a case of all or nothing? Should a recovery plan only include essential actions necessary for recovery? Who decides what is critical or essential?

It does not matter who funds or does the work as long as the recovery team, having the right composition, co-ordinates the process. Although the legislative responsibility remains with the State agency, contracts with other organizations are certainly possible.

It is important that the criteria, objectives and timelines for recovery are realistic.

Surveys, or a review of status, are often top priority and may lead to change in perspective on actions required. The process needs to be flexible to move on to other actions, issues or even taxa depending on results of the surveys.

It was felt that there is a need to identify thresholds for failure where funding should cease or be re-allocated. We do not want to put good money after bad.

Recommendations:

16. That the recovery process must have flexibility to be able to respond to changing circumstances.
17. That funding for recovery plans undergo major review every five years.
18. That thresholds for failure of recovery actions should be recognized.
19. That there be annual reviews, including ability to ensure that modified objectives fit in with overall objectives.
20. That the use of interim recovery plans be considered for critically threatened taxa or populations.

Student projects

Problems associated with the use of students undertaking recovery research as part of their post-graduate studies were discussed:

- It is difficult to change research priorities half way through their study period.
- Students have an academic requirement to meet.
- Recovery plan funding could be terminated.
- Requires the supervisor to develop the study appropriately.

It was recognized that incorporating student projects into the recovery plan can be very cost effective.

There are cases where student research has contributed significantly to recovery actions.

Community involvement

It was felt that we should not focus on community *groups* but the community.

The Recovery Plan Guidelines flow chart for preparing recovery plans does not include community involvement. This is often an important action and should be written into all plans.

It was suggested that a consultation and advocacy (education) phase should be an essential component within each project. This is often a major weakness in projects and should be carefully identified and targeted. Trained extension/media people can be an important part of the programs.

Community involvement should not be a one-way process, but recognized as an integral part of the recovery process. "Ownership" of the process is a key to maintaining the community's commitment to the management of the species or community, particularly when the formal recovery process ends and long-term maintenance of appropriate management is required.

Community involvement should start during the writing of Recovery Plans.

Recommendations:

21. That community involvement is an essential action for all Recovery Plans.
22. That community involvement be incorporated into the Recovery Plan Guidelines flow chart.
23. That a consultation and advocacy (education) phase be incorporated into each plan.

Phases of recovery plan implementation

1. Initiation

It was suggested that an information gathering workshop could be held, taking into account technical, socio-economic and organizational issues. As far as possible all stakeholders should be represented.

It was agreed that there is a need for transparency and accountability. It is also important to define the problems and issues. Program goals need to be clearly identified.

There is a need for a logical stepwise plan for preparing recovery plans. Future changes in the plan need to be allowed for as it is recognized that the recovery process is dynamic.

Recommendations:

24. That an information gathering workshop be held to identify the issues, problems and people involved.
25. That the Recovery Plan Guidelines incorporate consultation processes.

2. Selection: choosing a solution

The following actions were identified:

- Goal setting.
- Team establishment.
- Define roles and responsibilities.
- Write recovery plan.
- Develop implementation schedule.
- Assess options.
- Justify costings — external peer review of draft plans.
- Make recovery plans more widely available.

Recovery plans should be reviewed regularly and it is important to ensure that modified objectives are compatible with the overall objectives of the plan. It was agreed that it is important to include in a plan that a review has been done or is to be done.

Recommendation:

26. That recovery plans be reviewed regularly and that the plans include details of review requirements.

3. Implementation

The important elements identified to assist in successful implementation were:

Recovery Teams:
— Training/education
— Leadership
— Delegation
— Teamwork

Resources:
— Human
— Skills
— Time
— Money

There is a need to identify possible sources of resources and information that may be of use, and examine their applicability e.g., zoos, genetics laboratories.

4. Termination

Stopping the effort or changing it to solve new problems:

- Review progress/achievements.
- Measure success and learn from experience.
- Maintenance (ongoing management and monitoring).

Recommendation:

27. That the scope of the recovery plan should be identified early in the planning process, particularly the conditions that must be met for completion and termination.

Workshop 2: Community Involvement

Leaders: Lynn Baker and David Baker-Gabb
Scribes: Fiona Mandelc and James Pook

SUMMARY

Recovery planning should be presented as a joint partnership between community and government. We need to clarify what is meant by "the community" and recognize that the whole community is part of the problem and the solution.

Before undertaking a recovery program, the lead agency could hold an independently facilitated public forum to discuss the issues. The decision to write a recovery plan should be publicly advertised.

When selecting members of a recovery team all key community stakeholders should be considered. Membership should reflect the political, social/cultural, scientific and economic components of the community. Members should have clearly identified roles, and there may be a need for sub-groups with specific responsibilities.

Community representatives can assist at the local government level and will be aware of local legislation, priorities and pressures. Existing opportunities (e.g., Landcare and Integrated Catchment Management groups) and constraints need to be identified and addressed in the recovery plan. Community groups can also facilitate wider community education.

Recovery teams and meetings need to be flexible to cater for the diverse and changing needs of the program and the people involved. In some circumstances, it may be useful to open particular recovery team meetings to the public. It may be useful to have a co-ordinator or liaison officer assisting in the operation of recovery teams. Procedures for dealing with conflicts or differing priorities should be clearly established within the recovery team.

Issues associated with volunteer involvement (such as insurance, OH&S, travel) in recovery plan implementation need to be considered in the development of the recovery plan. It is important to provide feedback to volunteers. Locals should also be considered for casual labour for survey, monitoring, feral animal and fire management, etc. Funding of community action needs to be flexible with the possibility of funds going directly to a community group, with the State agency being advised.

Once the main recovery program is completed, an ongoing monitoring and management plan that is accepted by the community should be developed.

Recovery Plans and proposals unfunded by ANCA could be circulated to organizations such as zoos and non-government organizations (NGOs), so that alternative funding can be sought. Community groups also have the potential to assist in raising funds.

RECORD OF WORKSHOP AND RECOMMENDATIONS

"Recovery planning should be an equal partnership between community and government that is open and participatory."

Refer to the draft *Conservation of Australian Species and Ecological Communities Threatened with Extinction: a National Strategy:*

> "Objective 6: Prepare and implement Recovery Plans for all endangered species and ecological communities . . . via:
> 6.1–6.6: . . .
> 6.7: Ensuring that landholders, industry, business, individuals and volunteers have access to information about endangered species and ecological communities and opportunities to participate in preparation and implementation of Recovery Plans."

Problem recognition

What is "community"? — political, economic, social, cultural, scientific. The community is part of the problem and part of the solution.

Identify community stakeholders.

The lead agency should identify publicly the need for the establishment of a recovery team and co-ordinate a public forum to discuss the issues. The lead agency should organize an independent facilitator to assist at that forum.

The public forum should include objectives to identify stakeholders, and should establish the structure of the operation or process, i.e., establish recovery team membership, identify the need for specific working groups, "Friends of" groups etc.

In selecting the members for a recovery team, consider all key community stakeholders and ensure that their views can be expressed and taken into consideration either by representation on the recovery team or other forums that can feed into the recovery team process. Community representatives must have effective communication lines with their constituents to ensure the flow of information.

It is important to have land owners and managers involved in the recovery team. This will help build confidence in other land owners that management actions are justified and appropriate.

The decision to write a recovery plan should be publicly advertised so that interested groups can identify themselves.

Community representatives should be involved in the scoping process of the recovery plan through public forums run by an informed independent facilitator. Draft recovery plans should be circulated within the community for comment. Finalized recovery plans should be readily available to the community and circulated to key groups.

For species that have a large range, regional community groups may need to be involved.

Recommendations:

1. That all key community stakeholders be considered when forming the recovery team.
2. That a trained communicator be included on the team to facilitate community support.
3. That the decision to write a recovery plan be publicly advertised.
4. That an independently facilitated public forum be held to discuss the issues, identify stakeholders and establish the recovery process structure.

Project initiation and design: operation and membership of recovery teams

Present recovery planning as a joint partnership between community and government to make the process "transparent" and "participatory". Transparency and equality are important if the full potential of all recovery team members is to be realized. Community members can feel threatened and less than equal.

Flexibility is important in membership, schedules, locations of meetings etc.

Membership should reflect "expertise" from the four areas of community operation: political, social/cultural, scientific and economic (i.e., not just scientific).

Recovery plans should state the community groups that have been consulted and involved in the preparation of the recovery plan. Community consultation needs to recognize the potential role of the community in the recovery process and be tailored to suit that species to avoid getting delayed unnecessarily in the consultation process.

Community ownership of projects is important and may need a good communicator.

Recovery teams and their subcommittees or working groups should be based on "equity" for membership representation — acknowledgment of viewpoint and to aim at building up trust (i.e., equity in every sense where possible including financial).

It should be recognized that representatives of key community stakeholders have a separate role from that of seeking broader community support for a project, which is most appropriately undertaken by people with specialist skills in this area. Should consider having a trained communicator on the recovery team to facilitate community support/education (e.g., newsletter).

It is important to ensure that all recovery team members have a clear role.

The recovery plan needs to clearly identify the role of the community in the recovery process (e.g., community education, management, survey, etc.).

Recovery teams should clearly identify the roles and responsibilities of volunteers and if actions need paid or unpaid volunteers. If paid participants are to be employed, the numbers and conditions need to be clearly identified.

Travel and accommodation costs should be built into recovery plan costs where applicable e.g., for non-government representatives. Where possible, in-kind contributions of NGOs, individuals or communities need to be costed, based on identified rates. ESU could consider the development of baseline rates which may be used nationally.

The issue of "public insurance" needs to be investigated and clarified particularly when using volunteers and across jurisdictions.

It is important to consider having a co-ordinator or liaison officer whose primary role would be to disseminate and feedback information throughout the community. Such a liaison person could be shared between a number of teams.

It is important that government agency representatives bring knowledge of political and bureaucratic/policy framework (i.e., from Commonwealth and State) to the recovery team — it is their responsibility to inform on these matters.

Recovery teams can be "political" entities — proactive towards recommending/directing government agency policy and towards community involvement and delivering on expectations (e.g., to take problem back to agency and talk through how to resolve politically, or landowner on recovery team meets with peers and explains).

Should recovery team meetings be open to the public?

Recommendations:

5. That transparency and equality be fundamental to the recovery team.
6. That flexibility be built into recovery team operation.

7. That recovery team membership reflect expertise from all components of the community.
8. That the role of volunteers be recognized and in-kind contributions be identified in the recovery plan.
9. That casual labour costs and conditions be clearly identified.
10. That draft recovery plans be readily available to the community for comment.

Survey, research, monitoring and rehabilitation

Community group involvement is important in recovery; however, there are costs involved. There is often scope for community involvement in survey. Although it is often easier for agencies to carry out actions themselves, there may be long-term benefits in involving the community.

It is important to provide timely feedback to volunteers who have helped collect the data in a form that will encourage their continued involvement.

Where employment of casual labour is necessary, locals should be considered.

Issues such as insurance, OH&S, expenses, etc. need to be considered in the development of the recovery plan.

Recommendations:

11. That the community be involved in appropriate project work, and that costs and conditions be considered in the recovery plan.
12. That feedback be prompt and continued community involvement be encouraged.

Education

Community groups can be involved in facilitating wider community education and involvement in the recovery process. Broader community education will often be more readily accepted if it comes from their peers.

Recommendation:

13. That the local community be involved in community education and involvement.

Policy

Government agencies can penetrate national/State maze.

Community can assist at local government/Shire level; locals will be familiar with local legislation.

There is a community expectation that government departments should be able to reconcile their differing priorities to work towards threatened species conservation. Procedures for dealing with conflicts or differing government priorities should be clearly established within the recovery team.

Existing opportunities (e.g., Landcare, Total Catchment Management/Integrated Catchment Management groups) and constraints need to be identified and addressed in recovery plan.

Recommendations:

14. That procedures for reconciling differing priorities be established within the recovery team.
15. That use be made of existing opportunities and structures (such as Landcare, Integrated Catchment Management) where appropriate.

Funding

Recovery Plans have been focused on obtaining dollars from ANCA. They need to take into account other issues and potential resources (e.g., other funding sources, community resources) that may allow recovery to proceed in the absence of, or in conjunction with, ANCA or State/Territory agency funding.

There is a need for increasing flexibility in funding community action — e.g., funds could go direct to the community group with State agency being kept informed.

How do State/Territory agencies gain funds to run public forums and employ independent facilitators without a recovery plan in place (as the recovery plan is the normal basis on which they seek funding for implementation)? Suggest: that ESU recommends to ESAC to seek a commitment from lead State/Territory agencies for this initial action.

The Endangered Species Unit needs to improve its feedback on draft recovery plans, project applications/annual reports.

Projects unfunded by ANCA/ESU: ESU could circulate project details, with priorities (e.g., to zoos, NGOs,) so that these, or sections of them, can be considered for funding from other sources (e.g., industry).

Community groups have the potential to assist in raising funds for the project. Recovery plans lend themselves to fund raising.

For "non-charismatic" species, a strategic approach is needed, identifying linkages with names, habitat, locality, "any opportunity" and potential sponsors.

Recommendations:

16. That funding sources other than ANCA be considered in recovery plans.
17. That flexible funding arrangements be considered for community action.
18. That the Endangered Species Unit recommends to ESAC to seek a commitment from lead State agencies to fund initial public forums.
19. That ESU improve feedback on project applications, draft recovery plans and annual reports.
20. That ANCA /ESU circulate unfunded project details to zoos, NGOs etc., to seek alternative funding.

Signing off

As part of the signing off process after the completion of the main recovery program, there should be the development of an ongoing monitoring and management plan that is accepted by the community, where it is expected that they will have a continuing involvement.

Recommendation:

21. That an ongoing management plan acceptable to the community be developed.

Intellectual copyright

Needs to be clarified for individual participants, recognizing that participants will change throughout the life of the team.

Mechanisms for providing feedback in appropriate format through appropriate channels — to the different providers of information.

Recommendations:

22. That intellectual copyright issues be clear.

23. That feedback to providers of information be ensured.

Workshop 3: Monitoring and Evaluation of Recovery Programs

Leaders: Sally Bryant and Bruce Male
Scribes: Anne Duncan and Lloyd Van der Wallen

SUMMARY

Monitoring

Monitoring should address three components; the scientific and biological factors (data, techniques and costs relating to species/habitat/threats); the social factors external to the recovery team (community attitudes and organizational/political commitment) and the social factors within the recovery team (team performance).

Performance indicators for all aspects of recovery, including management of the recovery process, should be explicit in recovery plans. The monitoring program should be cost-effective, well designed and built into actions.

Evaluation

Recovery plan proposals should be reviewed by a State panel with government and non-government agencies before submission to ANCA.

The recovery process should be evaluated at each recovery team meeting, with a formal review each year. Major external reviews should be conducted at the end of the funding cycle, and review panels should be arranged by ESAC. Such evaluation should be built into recovery plans (carried out against objectives/criteria and actions) and should be costed as are other actions. The ESP Guidelines should include advice on evaluation.

Recognition that a recovery plan is failing should initiate external review, with lessons learnt from the experience being disseminated to other practitioners.

Roles

As the recovery team is crucial to effective implementation and evaluation of a recovery plan, training or support for team members may be needed. More detailed guidelines on the roles and responsibilities of teams would assist members.

ANCA should take a lead or co-ordinating role in the development of national plans where species cross State borders. State/Territory plans should incorporate the national perspective.

The responsibility for ongoing management of species once a recovery plan has been successful needs to be identified and a plan for ongoing management written. Involvement of non-government groups or corporate sponsors in a team approach to management may still be appropriate.

RECORD OF WORKSHOP AND RECOMMENDATIONS

1 Monitoring

What should be monitored?

There was consensus that, in addition to the scientific and biological aspects of recovery (species/habitats/threats), there are a raft of social aspects which also need monitoring. These include the recovery process itself (e.g., team performance, leadership) as well as the agency and community context of recovery (attitudes and political support outside the recovery team). Costs and cost effectiveness also need to be addressed. It was suggested that progress towards objectives rather than the objectives themselves should be monitored.

Scientific aspects included:

— Information base and problem definition:

- Have in place a system for obtaining and evaluating existing and new information in the field on taxon, habitat integrity and threats, as well as analogue taxa, research, management and techniques;
- Monitor the defined problem against incoming information;
- Monitor data standards, data duration and data co-ordination.

— Tools (techniques)
- Set performance criteria for appropriateness and functioning of techniques and check against those criteria.

— Costing/budgeting
- Review costing/budgeting against predictions (include in kind values).

Social aspects included:

— Organizational commitment
- Have clear, explicit, agreed roles for team members.
- Define functions in achieving recovery plan goals and monitor against those expectations.
- Define resourcing requirements in recovery plan and monitor delivery against those predictions in recovery plan.
- Outline consequences of resourcing shortfalls.
- Monitor turnover of representatives on recovery team, checking for coverage and continuity.

— Political commitment
- Monitor letters to Minister, newspapers, etc. and Ministerial responses.
- Seek criteria for community "ownership" of recovery plan, so that it can be monitored, e.g., via attendance of community representatives at meetings, "Friends of" groups, etc.

— Public relations
- Establish criteria for monitoring effectiveness of publicity for maintaining community support, and political will, for the recovery plan and process.

Recommendation:

1. That monitoring address three components:
   — the scientific and biological factors (data, techniques and costs relating to species/habitat/threats);
   — the social factors external to the recovery team (community attitudes and organizational/political commitment); and
   — the social factors within the recovery team (team performance).

How should it be monitored?

Firstly, clear objectives and well-defined recovery criteria are necessary as a starting point for effective monitoring. Short-term checkpoints as well as long-term goals are also valuable for effective monitoring.

One objective of a recovery plan should relate to successful management of the recovery process (social factors) and recovery plans should contain explicit performance indicators for both social and scientific factors. An evaluation of team performance as part of annual reporting was suggested. However, the lack of experience within recovery teams for self-monitoring and in team leadership may be a problem and training in these aspects may be required. It was recognized that as a whole, self-monitoring may be time-consuming and expensive and so needed to be well-directed and designed.

Recommendations:

2. That performance indicators for all aspects of recovery, including management of the recovery process, be explicit in recovery plans.
3. That the monitoring program should be cost-effective, well designed and built in to actions in a continuous and ongoing way.

2. EVALUATION

When should evaluation take place, and by whom?

It was felt that there should be ongoing analysis built in to recovery team meetings and reporting, plus more formal review periodically and that one should feed into the other. There was discussion about whether the periodic review should be annual and whether it should have some external input. It was suggested that if a few meetings were scheduled together, people from one team could be involved in review of other recovery teams. It was also felt that there should be a longer term review at the end of the funding cycle (3–5 years) and that it was entirely appropriate that it be external.

In addition it was suggested that a technical audit may be required at any time during the project and that the recovery team, or any member of the team, should be able to request this.

It was suggested that a review (external and authoritative), be conducted both at the start and the finish of recovery plan implementation. An external review can provide a critical appraisal of procedural steps, objectives and timelines. States and Territories could put together a panel, including government and non-government agencies, to assess and prioritize recovery plan proposals. This may also give non-government agencies a better chance of running recovery programs.

Recommendations:

4. That a review be undertaken on three timescales:
   — an on-going internal review at every recovery team meeting,
   — a formal annual review, and
   — a major external review conducted at the end of the funding cycle (usually three to five years).
5. That it be possible for a technical audit to be instigated by the recovery team, or by a member of the team, when difficulties with the plan are not being resolved.
6. That recovery plan proposals be reviewed and prioritized by a State panel involving government and non-government agencies, before submission to ANCA.

Where should independent reviews come from? Who should pay?

It was suggested that review panels should contain managers and scientists and both include at least one member from outside the recovery team. However, it was also thought necessary that the recovery team leader should be part of, or provide advice to, any external review panel. The difficulties of finding someone truly independent who would be useful in a review were discussed. The concept of an "Australasian review group" was discussed but it was recognized that this would be expensive. Ultimately the membership of external review panels has to be acceptable to ESAC, and should be arranged by them. It was agreed that the review process should be costed into the plan and funded by ANCA.

Recommendations:

7. That review be costed as part of the recovery plan.
8. That major external review panels be arranged by ESAC.

How should evaluation be conducted?

Evaluation should be built into the format of the recovery plan (objectives/criteria/actions). A mechanism for stimulating review at every recovery team meeting could be to have every member report on their own past actions and future objectives in the framework of the recovery plan. It was suggested that some skill may be needed to facilitate meetings to ensure that review occurred. The need for ongoing review to satisfy the needs of all participants, not just the funding agencies, was recognized. If review was to be successful at every meeting, then there was a need for all recovery team members to have all information and copies of reports prior to meetings. If regular reviews and audits were prescribed and expected, they would be seen as a positive part of the process and not some sort of "policing". If a recovery plan is not in place, there should at least be a checklist for reporting achievement against prescribed tasks.

Recommendations:

9. That evaluation be built into recovery plans, (i.e., carried out against objectives/criteria and actions).
10. That procedures for evaluation be explicitly included in the ESP guidelines for the recovery process.

How and when do we make the hard decisions? (e.g., to change actions or abandon the recovery plan)

It may become appropriate for a recovery plan to go into "recess" for a period or to have a drastic change in approach or to make changes to the recovery team membership. In all cases it was agreed that people find it hard to let go, to make the hard decision to change or stop. It was suggested that success and failure (termination) criteria should be built in to recovery plans. Termination does not necessarily mean the end of the recovery effort. It may mean major review followed by a change in direction or approach as well as a change in the recovery team composition. It was agreed that the external review process should pick up the need for "termination". It was felt that better guidelines on team construction and running from ANCA would assist the problem of recovery team members not wanting to "let go".

Recommendations:

11. That success or failure criteria be built into recovery plans.
12. That ongoing internal review and periodic external reviews be an explicit part of the recovery process.

What should the approach be when a recovery plan is failing or problems seem insoluble?

It was agreed that there was a need to generally salvage what is tangible in physical terms (e.g., for zoo collections) as well as learning as much as possible from the experience. It was recognized as particularly important to disseminate information about lessons learnt to other practitioners.

A recognition by the recovery team that the recovery plan is failing should initiate an external review process, which should explicitly address:

- why the recovery criteria have not been met;
- requirements for revised research;
- requirements for revised program delivery; and
- requirements for changes to recovery team.

Recommendations:

13. That failures in recovery plan operation or implementation initiate external review.
14. That lessons learnt from the reviews be disseminated to other practitioners.

3. Roles

What is the role of the recovery team in monitoring and evaluation?

It was recognized that the recovery team itself has a crucial role in monitoring and evaluating recovery plan performance, setting directions, and in making recommendations for change. It was felt that the recovery team should be sufficiently independent and flexible to set direction and evaluate programs with the ability to invite specialists and community groups to participate as required.

It was also agreed that the effectiveness of the recovery team was significantly increased by more frequent meetings.

There was some general discussion about effectiveness of recovery teams. It was felt that some team members did not contribute much to the process, mainly because they did not know what their role was. The skills of the whole group needed to be recognized and utilized for the team to be most effective. It was recognized that recovery team leaders may not necessarily have the skills to manage the social side of the recovery process and may need training and support. It was suggested that perhaps in some situations a professional facilitator may be useful to improve recovery team performance.

It was suggested that the ANCA person on the recovery team should have the role of independent observer, be able to monitor group dynamics and recommend reviews. It was further suggested that the

ANCA role be formalized as independent observer and conduit for problems and that feedback on recovery team performance be provided by ANCA routinely. However, it was also recognized that roles of members should not be too prescriptive — whoever had the skills should do the job.

Recommendations:

15. That recovery team members may be given training or support necessary to carry out their responsibilities effectively.
16. That recovery team roles be clearly detailed in ANCA guidelines on the roles and responsibilities of teams and individuals.

How do team members incorporate regional needs into national plans?

It was recognized that where species cross State or Territory borders the recovery plans should be national and that ANCA should take a lead or co-ordinating role in the development of such national plans. Where there are State or Territory plans they should incorporate the national perspective as well as State priorities.

Recommendations:

17. That ANCA take a lead role in developing national plans for species across State or Territory borders.
18. That State or Territory plans incorporate the national perspective.

Who is responsible after a recovery plan reaches its objectives?

For successful recovery plans which reach their objectives, a species may become "conservation dependent" at which time ANCA may no longer be able to continue funding. The responsibility would then fall to the State or Territory agency or to a non-government group or corporate sponsor. However, the team approach to management should not necessarily be abandoned at this stage. Regardless of who takes responsibility, monitoring and evaluation of ongoing conservation action is necessary outside the recovery plan. Such ongoing actions need to be detailed in an ongoing management plan. Perhaps guidelines for "ending" recovery plans should include the recommendation to write a plan for ongoing management.

Recommendations:

19. That responsibility for ongoing management of species once a recovery plan has been successful be identified and a plan for ongoing management written.
20. That non-government groups, corporate sponsors and local communities be involved in a team approach to ongoing management as appropriate.

Workshop 4: Ecological Communities and Groups of Species

Leaders: John Blyth and Margaret Brock
Scribes: Elsa Dexter and Ann Jelinek

SUMMARY

The workshop initially was asked to consider the recovery process for ecological communities, groups of species, regions and habitats, and threatening processes. There was consensus that the recovery process for ecological communities should be considered the prime topic. Recovery of habitats, species groups and regions was recognized as important but it was felt that these areas could benefit from modification of the recovery process as developed for species, and the process that needs to be developed for threatened ecological communities.

Workshop aim: *To consider aspects of the process of recovery of threatened ecological communities.*

Plant communities or other recognisable units can assist in identifying ecological communities. Established classification systems, including floristic, species and abiotic information should be used if possible. Scale may be a problem in the definition of some communities, and boundaries may overlap, or may change.

The plant priority ranking scheme could be modified for communities, though threatened communities need not contain threatened species. It would be valuable to deal with some vulnerable communities before they become more threatened. Funding priority may not be the same as the ranking of degree of threat.

Recovery planning needs to accommodate the dynamic nature of a community and recognize that actions may not equally benefit all components of a community. Actions should ensure the maintenance of ecological processes and conserve the potential for successional change. Threat abatement may be very important. Research and management need to be integrated and well documented.

Some individual community issues will be similar to species issues in the same region, and similar approaches may be relevant. Threatened communities without listed species should not be neglected. Terrestrial, aquatic and marine communities may require different approaches.

Involvement of the human community will need education, and use of some flagship taxa and "charismatic" communities. Time frames for recovery may be extremely long, and will need sustainable and measurable indicators of success.

Threat abatement approaches may benefit groups of species within a region. A group or regional approach may benefit less threatened species as well as those that are critically endangered, as well as maximising use of available funds and expertise.

RECORD OF WORKSHOP AND RECOMMENDATIONS

Ecological communities

How do we identify threatened communities?

- Agreed that definition may be difficult and that we should use the criteria defined in the *Endangered Species Protection Act 1992* (ESP Act) to proceed on a case by case basis.
- Nominations will need detailed information and clearly identifiable boundaries wherever possible.
- Nominations will need to show reasonable confidence that all significant occurrences are known.
- Plant communities or other recognizable or quantifiable units can often be used as surrogates for identifying ecological communities.
- Scientific judgement on what each community is may be needed on a case by case basis. Wherever possible, references to published descriptions, recognized authorities or relevant scientists should be included.

Recommendations:

1. That nominations for listing contain detailed information, including information on community distribution and identifiable boundaries.
2. That plant communities or other surrogates be used to identify ecological communities.

What are the problems of identifying threatened communities?

- Abiotic conditions vary and species shift, so it is hard to identify the boundaries.
- Boundaries may overlap e.g., wet and dry sclerophyll forest.
- Ultimately, scientific judgement will be the way a community is assessed.
- To describe a community we should use an established classification system if possible and this should include floristic, species and abiotic information e.g., soils, which are less susceptible to change.
- There may be problems with scale e.g., old man saltbush community with a very large distribution.

Recommendation:

3. That the classification system used to describe a community include floristic and abiotic information.

How do we establish relative priorities?

- Recognize the need and purposes of ranking. Ranking for funding priority may not be the same as ranking "degree of threat" or "potential for recovery".
- Threatened communities do not need to contain threatened species to be nominated.
- Use "degree of threat" and consider "potential for recovery" and "uniqueness" separately.
- Look at current species ranking schemes and adapt them where necessary.
- The plant priority ranking scheme (discussed in more detail by Jensz, this volume) could be modified for communities.
- Is a common ranking scheme for species, communities and threatening processes possible, or do we need to consider funding for communities separately? Separate funding was thought more practical.
- Realize the value of tackling some vulnerable communities before they become endangered.

Recommendations:

4. That communities need not contain threatened species in order to be nominated.
5. That "degree of threat" be the main factor considered, with "potential for recovery" and "uniqueness" considered separately.
6. That the plant priority ranking scheme be considered for use, with modifications, for communities.

How do we deal with differing requirements or threats to components of the ecological community?

- There is a whole range of components of ecological communities which may or may not benefit from individual recovery actions.
- Communities are dynamic in space and time and recovery planning needs to recognize this.
- Could be dealt with in a recovery plan on a case by case basis.

Recommendation:

7. That recovery planning recognize the dynamic nature of communities.

How will the recovery process differ for communities?

- An ecological community is dynamic in space and time and all elements may not be physically represented at any one time.
- The breakdown of ecological processes and interactions in the community can be important.
- Threat abatement plans may be particularly important.
- Use the present guidelines and modify and adapt as necessary.
- The selection of recovery teams will be particularly important and we may need to modify the species model.
- Indicators of recovery may need to include "sufficient", "sustainable" and "representative" type indicators as species and population parameters will not be relevant.
- We must recognize the need to conserve the potential for expression of the various successional stages of a community.
- Approaches will be more complex than species approaches.
- Case-specific approaches will be needed, based on the best available scientific and management information.
- We need to recognize the role of experimental manipulation and management in community recovery.
- Document the recovery process for ecological communities well so we can learn from successes and failures.
- Incorporate research actions within recovery plans for communities. Research actions are important and will be complex and we can not afford to wait for research plans to be completed before recovery plans start. Adaptive management will be needed to take account of the results of research.
- We need to make sure that the threatened communities without listed species are not biased against in priorities and ranking.
- Many individual community issues will be similar to species issues in the same region, and similar approaches may be relevant.
- Marine, inland water and terrestrial community approaches may differ.

Recommendations:

8. That threat abatement plans be considered as methods for conserving communities.
9. That the current recovery plan guidelines be used and adapted as necessary.
10. That indicators of recovery be of the "sufficient", "sustainable" or "representative" type, as species and population parameters are not applicable.
11. That care be taken to conserve the potential for succession in a community.
12. That adaptive management take into account the results of research.

How will threatened ecological communities be monitored?

- Both biotic and abiotic parameters will be relevant.
- Sustainable indicators need to be identified for each case.
- Time frames for recovery may be an order of magnitude longer, say 50–100 years.
- Need measurable and achievable indicators.

Recommendation:

13. That monitoring cover biotic and abiotic parameters and include achievable and measurable indicators.

How easy will it be to get people (e.g., local community) involved?

- This may be more difficult for ecological communities than for species, ownership of the problem will not come as easily.
- Education, awareness and flagship taxa may be useful.
- Charismatic mega-communities may be easier than the less charismatic.

Recommendation:

14. That community involvement and education be encouraged through flagship taxa or ecological communities.

How do we promote the recovery of ecological communities and obtain funding commitments?

- Be proactive.
- Initiation of more funding.
- Links with related programs.
- Links with current programs.
- Rename the ESP "The Endangered Species and Communities Program".

Recommendation:

15. That links be made with related programs.

GROUPS OF SPECIES AND REGIONS

Although the main workshop aim was ecological community recovery, in the second session of the workshop some attention was given to the recovery of groups of species and regions. Some of the issues considered included:

- The variety of group approaches and the need to define what is being considered in each case as approaches to recovery will differ, e.g., regional approaches and phylogenetic group approaches to species recovery.
- Will resources spread further in group approaches than for individual species?
- A threatening process-based approach may be relevant for groups of species in a region. This may work well for groups faced with a common threat e.g., foxes, land clearance, *Phytophthora*.

- Could have advantages in economies of scale, funding, expertise, etc.
- A group or regional approach may allow vulnerable, endangered and data deficient (or poorly known) species to be tackled at the same time as critically endangered species.

Recommendation:

16. That the potential advantages of group or regional approaches be taken into account when considering recovery.

Threatening processes and threat abatement

This workshop identified the need to consider the process of threat abatement alongside the recovery planning process.

- The need for prioritising threats was recognized.
- The potential to solve many species, group and ecological community issues through threat abatement was recognized.

Recommendation:

17. That threat abatement be identified as an integral part of the recovery process.

List of Participants

Name	Organization Represented
Dr Tony Auld	NSW National Parks and Wildlife Service
Mr Gary Backhouse	Department of Conservation and Natural Resources[1], Victoria
Ms Lynn Baker	Consultant
Dr David Baker-Gabb	Royal Australasian Ornithologists' Union
Mr John Benson	Royal Botanic Gardens, Sydney
Mr Lindsay Best	Department of Environment and Natural Resources, South Australia
Mr John Blyth	Department of Conservation and Land Management, Western Australia
Ms Kathy Breen	NSW National Parks and Wildlife Service
Mr John Briggs	NSW National Parks and Wildlife Service
Dr Margaret Brock	Endangered Species Advisory Committee
Dr Andrew Brown	Department of Conservation and Land Management, Western Australia
Dr Mick Brown	Forestry Tasmania
Mr Peter Brown	Department of Environment and Land Management, Tasmania
Dr Sally Bryant	Department of Environment and Land Management, Tasmania
Dr Andrew Burbidge	Endangered Species Advisory Committee
Dr Judy Caughley	Institute for Applied Ecology, Australian Capital Territory
Mr Peter Christie	Western Plains Zoo
Dr Tim Clancy	Queensland Department of Environment and Heritage[2]
Dr Tim Clark	Northern Rockies Conservation Co-operative/Yale University
Dr Dave Coates	Department of Conservation and Land Management, Western Australia
Dr Hal Cogger	Australian Museum, Sydney
Mr Peter Copley	Department of Environment and Natural Resources, South Australia
Ms Kate Cramond	Australian Nature Conservation Agency
Mr Jim Crennan	Australian Nature Conservation Agency
Mr Rick Davies	Black Hill Flora Centre, South Australia
Dr Lynn Day	Parks and Wildlife Commission of the Northern Territory
Ms Susanne Dennings	Malleefowl Preservation Group, Western Australia
Mr Terry Dennis	Department of Environment and Natural Resources, South Australia
Ms Elsa Dexter	Australian Nature Conservation Agency
Dr Chris Dickman	School of Biological Sciences and Institute of Wildlife Research, University of Sydney
Dr Kingsley Dixon	Kings Park and Botanic Garden, Western Australia
Mr Peter Dostine	Parks and Wildlife Commission of the Northern Territory
Ms Liz Dovey	Australian Nature Conservation Agency
Ms Anne Duncan	Australian Nature Conservation Agency
Mr Brendan Edgar	Australian Nature Conservation Agency
Ms Joanne Edney	NSW National Parks and Wildlife Service
Ms Felicity Faris	Threatened Species Network (Victoria)
Ms Leasie Felderhof	Queensland Department of Environment and Heritage[2]
Mr Jason Ferris	Inter.comm Scientific Event Co-ordination
Mr John Fisher	Department of Conservation and Natural Resources[1], Victoria
Mr John Gardner	Alcoa of Australia Ltd
Dr Stephen Garnett	Department of Environment and Natural Resources, South Australia
Ms Louise Gilfedder	Department of Environment and Land Management, Tasmania
Mr Stephen Harris	Department of Environment and Land Management, Tasmania
Dr Jim Hone	Applied Ecology Research Group, University of Canberra
Ms Christine Hopkins	Australasian Species Management Program
Dr Alan Horsup	Queensland Department of Environment and Heritage[2]
Dr Peter Jackson	Queensland Department of Primary Industries
Prof Peter Jarman	Department of Ecosystem Management, University of New England
Ms Ann Jelinek	Australian Nature Conservation Agency
Ms Katrina Jensz	Australian Nature Conservation Agency
Dr Chris Johnson	Zoology Department, James Cook University
Dr Ken Johnson	Parks and Wildlife Commission of the Northern Territory
Mr Dave Liddle	Parks and Wildlife Commission of the Northern Territory
Mr Graham Lightbody	Central Land Council
Ms Tanya Littlely	Conservation Council of South Australia
Mr Geoff Lundie-Jenkins	Queensland Department of Environment and Heritage[2]
Mr Ian Lunt	Botany Department, La Trobe University
Dr Charles Mackowski	State Forests of New South Wales
Dr Michael Mahoney	The University of Newcastle
Mr Bruce Male	Australian Nature Conservation Agency
Ms Fiona Mandelc	NSW National Parks and Wildlife Service
Ms Stephanie Maxwell	Consultant/World Wide Fund for Nature
Ms Nicki Mazur	Mawson Graduate Centre for Environmental Studies, South Australia

Mr Ian McClintock	National Farmers' Federation
Mr Keith McDonald	Queensland Department of Environment and Heritage[2]
Mr Peter Menkhorst	Department of Conservation and Natural Resources[1], Victoria
Ms Marg Moore	World Wide Fund for Nature
Assoc Prof Craig Moritz	Department of Zoology, University of Queensland
Mr Keith Morris	Department of Conservation and Land Management, Western Australia
Mr Tom Morris	Zapopan NL
Ms Sharon Nash	NSW National Parks and Wildlife Service
Dr Tim New	Endangered Species Advisory Committee
Mr Barry Nolan	Queensland Department of Environment and Heritage[2]
Ms Janet Owen	New Zealand Department of Conservation
Mr David Papps	Endangered Species Advisory Committee
Dr Rod Peakall	Australian National University
Dr Dave Pollard	Endangered Species Advisory Committee
Mr Jamie Pook	Australian Nature Conservation Agency
Mr David Rounsevell	Department of Environment and Land Management, Tasmania
Dr Andrew Sanger	Inland Fisheries Commission, Tasmania
Dr Alan Saunders	New Zealand Department of Conservation
Mr Michael Saxon	NSW National Parks and Wildlife Service
Ms Sarah Sharp	Australian Capital Territory Parks and Conservation Service
Dr Jeff Short	CSIRO Division of Wildlife and Ecology
Dr David Shorthouse	Australian Capital Territory Parks and Conservation Service
Mr Gary Slater	Healesville Sanctuary
Mr Andrew Smith	Royal Tasmanian Botanic Gardens
Mr Andrew Steed	NSW National Parks and Wildlife Service
Ms Sally Stephens	Australian Nature Conservation Agency
Mr Lloyd Van der Wallen	NSW National Parks and Wildlife Service
Ms Maria Vandergragt	Threatened Species Network (Queensland)
Dr Judy West	Endangered Species Advisory Committee
Dr John Woinarski	Parks and Wildlife Commission of the Northern Territory

[1]Now Department of Natural Resources and Environment
[2]Now Queensland Department of Environment

INDEX